STUDIES IN ORGANIC CHEMISTRY 18

BIO-ORGANIC HETEROCYCLES
Synthetic, Physical Organic and Pharmacological Aspects

PROCEEDINGS OF THE THIRD FECHEM CONFERENCE ON HETEROCYCLES
IN BIO-ORGANIC CHEMISTRY
BALATONFÜRED, HUNGARY, MAY 20–23, 1984

STUDIES IN ORGANIC CHEMISTRY

STUDIES IN ORGANIC CHEMISTRY 18

BIO-ORGANIC HETEROCYCLES
Synthetic, Physical Organic and Pharmacological Aspects

PROCEEDINGS OF THE THIRD FECHEM CONFERENCE ON
HETEROCYCLES IN BIO-ORGANIC CHEMISTRY
BALATONFÜRED, HUNGARY, MAY 20–23, 1984

Edited by

H. C. van der PLAS
LABORATORY OF ORGANIC CHEMISTRY, AGRICULTURAL UNIVERSITY,
WAGENINGEN, THE NETHERLANDS

L. ÖTVÖS
CENTRAL RESEARCH INSTITUTE FOR CHEMISTRY,
THE HUNGARIAN ACADEMY OF SCIENCES, BUDAPEST, HUNGARY

and

M. SIMONYI
CENTRAL RESEARCH INSTITUTE FOR CHEMISTRY,
THE HUNGARIAN ACADEMY OF SCIENCES, BUDAPEST, HUNGARY

ELSEVIER
AMSTERDAM–OXFORD–NEW YORK–TOKYO 1984

7226 - 1729

CHEMISTRY

Joint edition published by
Elsevier Science Publishers B. V., Amsterdam, The Netherlands
and

Akadémiai Kiadó, The Publishing House of the Hungarian Academy of Sciences, Budapest,
Hungary
The distribution of this book is being handled by the following publishers:
for the U.S.A. and Canada
Elsevier Science Publishing Company, Inc.
52, Vanderbilt Avenue
New York, New York 10017, U.S.A.
for the East European Countries, Democratic People's Republic of Korea, People's Republic of Mongolia,
Republic of Cuba and Socialist Republic of Vietnam
Kultura Hungarian Foreign Trading Company
P.O.B. 149, H-1389 Budapest 62, Hungary

for all remaining areas

Elsevier Science Publishers B.V.
1, Molenwerf
P.O. Box 211,
1000 AE Amsterdam, The Netherlands

Library of Congress Cataloging in Publication Data

FECHEM Conference on Heterocycles in Bio-organic
    Chemistry (3rd : 1984 : Balatonfüred, Hungary)
    Bio-organic heterocycles.
    (Studies in organic chemistry; 18)
    Includes bibliographies and indexes.
    1. Heterocyclic compounds—Congresses. 2. Bio-
organic chemistry—Congresses. I. Plas, H. C. van der.
II. Ötvös, L. (László), 1929-   . III. Simonyi, M. (Miklós),
1935-   . IV. Title. V. Series: Studies in organic
chemistry (Elsevier Science Publishers) ; 18.
[DNLM: 1. Heterocyclic Compounds—congresses.
W1 ST931 v. 18 / QD 399 F291 1984b]
QP550.F43   1984   574.19'2459   84-21259
ISBN 0-444-99585-4

Library of Congress Card Number:
ISBN 0-444-99585-4 (Vol. 18)
ISBN 0-444-41737-0 (Series)
© Akadémiai Kiadó, Budapest 1984

Printed in Hungary

PREFACE

When scientists in The Netherlands initiated a series
of FECHEM conferences on "Heterocycles in Bio-organic
Chemistry", they did not intend to put a heuristic emphasis
on interdisciplinarity. Since heterocycles represent an
enormous part of organic chemistry, the topic as stated above
defined immediately the indispensable role of synthetic work
within the scope of meetings. The amalgamation of heterocycles
with bio-organic chemistry, however, favoured those who are
usually also concerned with the fate of their compounds outside
of the preparative lab. The field of bio-organic chemistry
became limited as well by the above combination, excluding
peptides, terpenes, steroids (just to mention a few areas)
from the topic. Nevertheless, a substantial gain emerged from
this combination: however diverse the problems and methods of
the participants have been, their common platform was the
approach of a chemist, focusing attention on the structure and
properties of molecules or even on the role of certain chemical
bonds in the biological environment.

The third in these series of conferences - one every two
years - was organized by the Hungarian Chemical Society and
held at Balatonfüred during 20-23 May, 1984. This volume
presents the scientific material of the conference and we are
very grateful for the excellent co-operation of our contributors; the majority of the manuscripts was received during the
meeting.

The volume begins with the nine plenary lectures presented
at the Conference. Each of the plenary lectures is

V

complemented by a discussion and we thank Dr. G. Blaskó for collecting and organizing the discussion material. Of the forty-five poster presentations thirty-six short manuscripts have been received from twelve countries and included in the volume. One such paper is accompanied by a detailed treatise on a particular aspect of the original poster material. Editing has been slight for the most part. In some cases we had to contact contributors to clarify certain points.

We hope this book will convey the feeling we noted during the Conference: a deep desire by many to contribute to a cross-section representation of European work on heterocycles in bio-organic chemistry.

The Editors

CONTENTS

PLENARY LECTURES

CD AND STRUCTURE OF HETEROCYCLIC COMPOUNDS RELATED TO NATURAL PRODUCTS

SNATZKE G.

Lehrstuhl für Strukturchemie, Ruhruniversität Bochum
Postfach 10 21 48, D-4630 Bochum, FRG

Optically active heterocyclic compounds can belong to one of two classes: 1) those in which the heterocyclic ring(s) form(s) the chromophore, and 2) those where the heterocyclic ring(s) act(s) just as a chiral perturber. In each class we may then keep the "traditional" classification into molecules with inherently chiral chromophores (A), or such with inherently achiral, but chirally perturbed (by their molecular invironment) chromophores (B). E.g. glycals belong to class (1A), and so do diazepam derivatives, whereas simple furanosesquiterpenoids fall into class (1B). In aporphines the chromophore is inherently chiral, but does not contain the heterocyclic ring (class (2A)), and a phenyl glycoside belongs to class (2B).

There is no principle difference between class 1) and class 2); as long as we can identify the MOs of the chromophore, we can try to correlate absolute stereochemistry of the molecules with their chiroptical properties either a) by mere theoretical calculations, b) by empirical comparison, or c) by applying PMO theory. Examples from our laboratory could illustrate this. For an introduction into chiroptical methods and applications of PMO theory refer e.g. to literature [1].

EMPIRICAL CORRELATION

The empirical correlation has been applied e.g. to the determination of absolute configuration of the epoxydone group antibiotic G 7063-2 (1).

isolated from a _Streptomyces_ species [2]. It proved to be identical with a known compound, whose absolute configuration had not, however, been given. It contains a crossed conjugated chromophore, and for such it is known that the UV-spectrum is similar to that of the sum of the two individual "partial" spectra. We can thus refer to the "oxidoquinone"-chromophore, and the "conjugated amide"-chromophore. The latter absorbs only below 260 nm, the first should give, however, two n→π*-bands at longer wavelengths.

A reference compound with known absolute configuration containing the "oxidoquinone"-chromophore is (-)-terreic acid (2), whose CD-spectrum

1   2

above 300 nm indeed resembles closely that of 1: $\Delta\epsilon$ = +2.4 at 320 nm, $\Delta\epsilon$ = -2.1 at 366 nm; the corresponding values for 2 are +10.5 (327) and -6.6 (376). This proves then that the two substituted oxidoquinone chromophores have identical absolute stereochemistry. Below 300 nm the two CD-spectra differ, of course, appreciably. The difference between OH and NH_2 is spectroscopically irrelevant, although chemically these groups behave differently.

ISOCHROMANONES

Many isochromanones have been isolated from plants and microorganisms, and several of them are optically active. The absolute configuration of two of them, (+)-phyllodulcin (3) and (+)-agrimonolide (4) has been determined by chemical degradation [3], that of the others has been deduced from

4

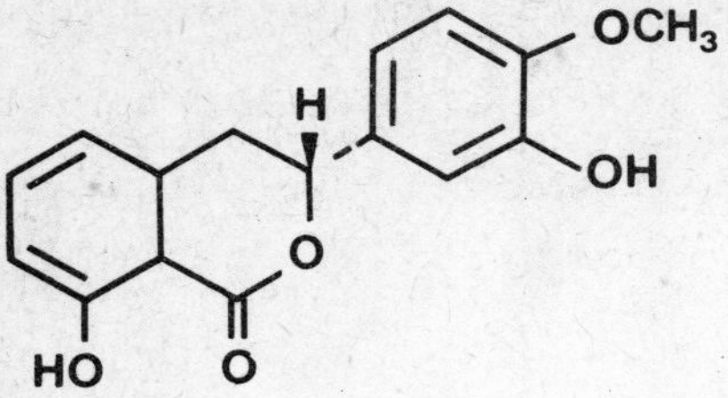

3

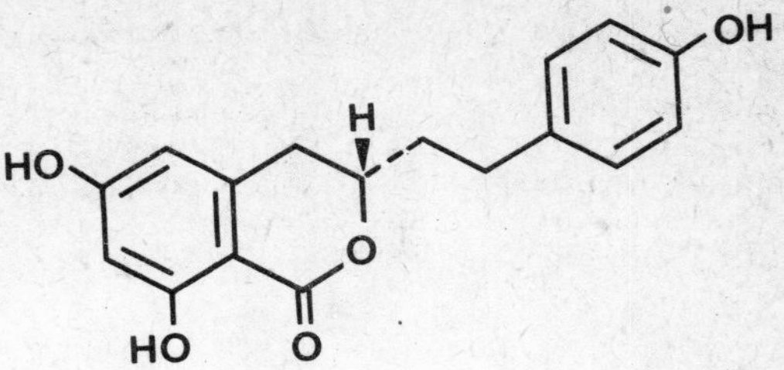

4

their chiroptical data by comparison with published CD- or ORD-data. In these compounds, the benzene ring of the isochromanone system may carry no, one, two, or even three OR-groups, and according to experience such strong pertur- bers – although themselves being achiral – may change appreciably the direction of the transition moments, and, therefore, also the signs of corresponding Cotton effects for substances with homochirally analogous molecules. Any direct comparison should thus be done only with a similar compound with identical substitution pattern of the aromatic ring in order to give compelling results.

This has been taken into consideration only for one case, however, in the past, namely for the determination of the absolute configuration of (–)- mellein (5), whose CD-spectrum was compared with that of (+)-3. The same CD-

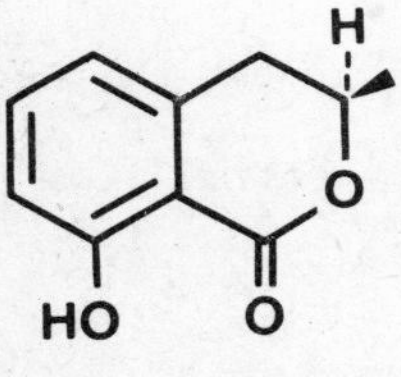

5

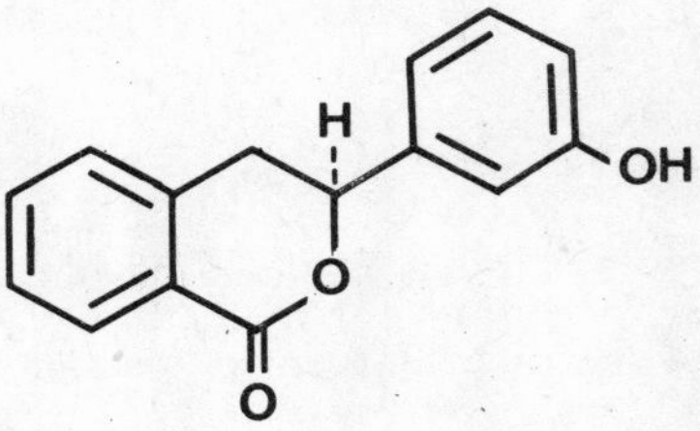

6

data served for the purpose of determining absolute configuration of the aglycon of dihydrohomalicin (6), which has no such HO-group on the mentioned benzene ring. By comparison with the data of the "monosubstituted" (-)-mellein (5) the absolute configuration was further derived for (-)-6-hydroxy mellein (7)

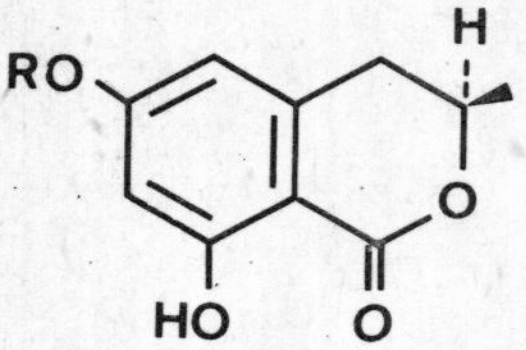

7 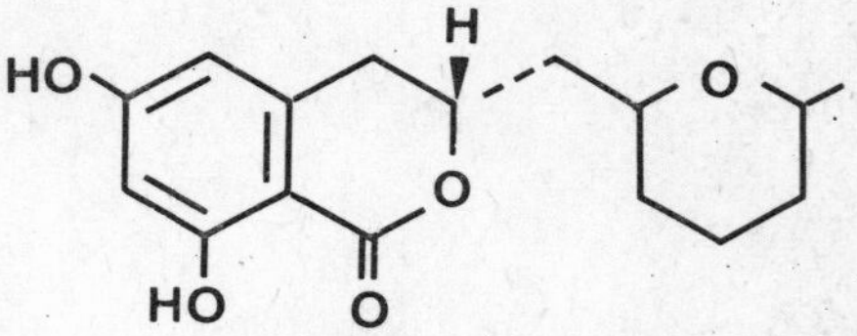   8

and its 6-methoxy analogue, as well as for (+)-asperentin (8), both containing two RO-groups, and for (-)-kigelin (9), with three such substituents. (-)-Fusa-rentin dimethyl ether (10), also carrying three such groups, was in similar way correlated with "disubstituted" (+)-asperentin (8).

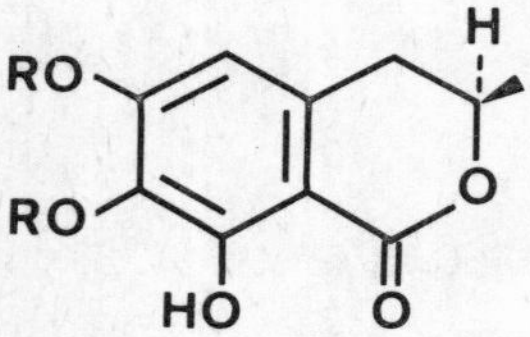

9 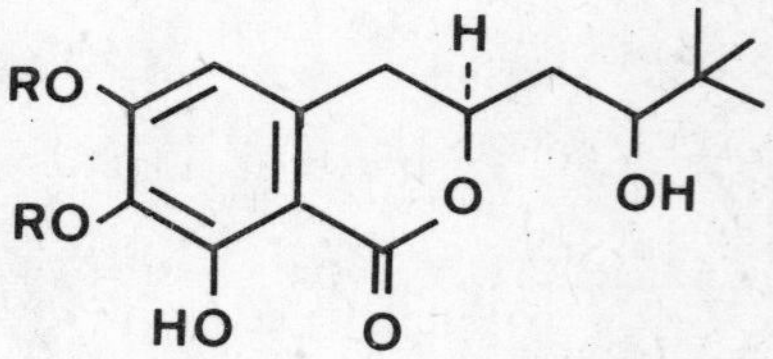   10

Although the CD-spectra of these compounds show several Cotton effects, and they are all shifted bathochromically by heavier substitution, com-parison with the correct "matching" model compound showed now at least for those cases, where data are available, that the original assignments (fortuitous-ly) were correct (comparison of 7 and 8 with 4, of 6 with 11).

In order to get a better insight into these CD-spectra we are syn-thesizing model compounds containing the steroid system as chiral part with an unequivocal configuration. At present three stereoisomers are available (11, 12,

6

11 = 2β H, 3α H
12 = 2β H, 3β H
13 = 2α H, 3α H

14

13), all without further substituents in the aromatic ring, but differing in the stereochemistry of linkage between the heteroring and ring A of the 5α-steroid.

Hitherto no band has been recognized in UV- or CD-spectra of such compounds containing this "benzoate" chromophore, which corresponds to the n→π* - transition. According to experience with conjugated esters we expect, however, such a band somewhere between 240 and 270 nm. A careful comparison of CD-spectra of 11 in polar and non-polar solvents revealed finally such a band around 255 nm (ethanol), which is negative; corresponding positive bands are obtained for the diastereomers 12 and 13. Besides this Cotton effect, a broader one with fine-structure is found at somewhat longer wavelengths (corresponding to the benzenoid α-band), and the long ago identified "conjugation" band gives CD around 230 nm; at still shorter wavelengths (206 nm) another strong Cotton effect can be observed.

The sign of this n→π* - band Cotton effect is the same as that for homochirally analogous tetralins (oxygen of isochromanone ring replaced by carbon). Direct comparison of CD-curves proved now that the published absolute configuration of (-)-dihydrohomalicin aglycon (6) is also correct. On the other hand, we could demonstrate that o,p-disubstitution of the aromatic ring within the same stereochemical series changes the signs of the Cotton effects of the

2*

$\pi \rightarrow \pi^*$ - bands, and only that of the $n \rightarrow \pi^*$ - band Cotton effect remains unaltered, as expected. This shows clearly the need for correct "substitution-matched" model compounds when one wants to apply chiroptical methods to the determination of absolute configuration of an aromatic optically active compound.

α-PYRONES

The characteristic UV-absorption of α-pyrones around 300 nm has already been used in the thirties for the identification of this moiety in the cardiotonic bufadienolides like e.g. bufalin (14). This absorption band is associated with the first $\pi \rightarrow \pi^*$ - transition of the heterocyclic chromophore, but we expect also the appearance of an $n \rightarrow \pi^*$ - band Cotton effect again, even when such a band has not yet been observed in the isotropic absorption spectrum. Its position should fall into the range from 250 to 280 nm, and indeed in all those CD-spectra of bufadienolides which we have measured [4] a shoulder or band around 260 nm of the same sign as that of the 300 nm Cotton effect could be seen. This is a typical chromophore of class (1B), and a sector rule should be valid, since the chromophore is planar and not anellated to another ring. X-ray diffraction data of several different bufadienolides are available, and they all show that the α-pyrone ring seems to have a rather fixed conformation on ring D. Since, thus, the empirical correlations between positions of perturbing atoms and signs of Cotton effects are known, we can try to understand them by application of PMO theory.

S_1 is (mainly) the $\pi_4 \pi_5^*$ - state which, in the point group C_s of the α-pyrone chromophore, belongs (like any other $\pi \pi^*$ - configuration, too) to A'-symmetry. S_2 is the $n \rightarrow \pi_5^*$ - state, which is of A"-symmetry. Although formally electrically dipole-allowed, the transition into S_2 is nevertheless expected to have only a negligible ε since the respective overlap integral is nearly zero. The magnetic transition moment, on the other hand, is large and it is directed along the O=C - axis. The missing electric transition moment along the same line can come from any $\pi \rightarrow \pi^*$ - transition; that one which is closest in energy and will thus contribute most is the $\pi_4 \rightarrow \pi_5^*$ - excitation. The chiral molecular surrounding of the chromophore reduces overall-symmetry to C_1 and mixes the transition into S_1 with that into S_2, which otherwise would be forbidden for symmetry reasons.

S_2 is the higher (in energy) of these two, thus the modified S_2 will result from the energetically unfavoured interaction. If we use a "test-charge" for finding the signs of the contributions of a substituent to the Cotton effect and choose e.g. attractive interaction with the electric transition moment, this means then that according to the "recipe" (cf. [1]) its interaction with the quadrupole associated with the magnetic transition moment must be taken as repulsive.

The π-MOs of α-pyrone are best approximated with the help of the LCBO method, and for the $S_0 \rightarrow S_1$ - transition we obtain then a quite strong electric transition moment $\vec{\mu}$, which is inclined by 60° to 70° against the direction of the O=C - bond. It is this $\vec{\mu}$ from which the electric transition dipole moment for the $S_0 \rightarrow S_2$ - transition has to be "stolen", and the corresponding MO-determined nodal surface, approximated as usual by a plane, is perpendicular to it. Together with the nodal plane of the n-orbital and the symmetry-determined (third) nodal plane (i.e. the plane of the chromophore) a somewhat "oblique" octant rule is thus built up for the $n \rightarrow \pi_5^*$ - band CD (Figure 1). Application of the aforementioned "recipe" gives us the signs, which are opposite to those of the well-known octant rule for saturated ketones (Figure 1). Putting the bufalin molecule into these sectors, using for that purpose the mentioned conformation of the side chain, one finds nearly all atoms in a "negative" sector (Figure 2), in full agreement with the measured negative Cotton effect.

A similar prediction of the CD-sign for the $S_0 \rightarrow S_1$ - transition is not possible, as this is associated with both an electric and a magnetic transition moment, thus, at least two terms had to be estimated. We just note the

9

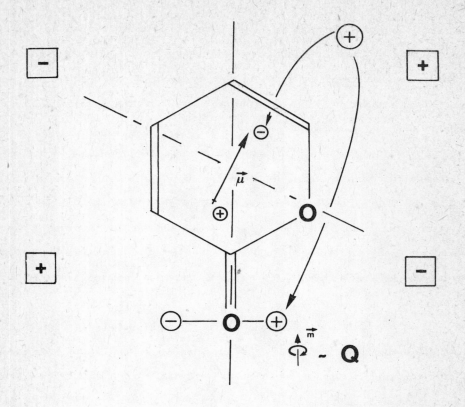

Figure 1.   Non-empirical sector rule for 260 nm band CD of α-pyrones.

The approximate direction of the electric transition moment for the 300 nm band, from which intensity is stolen is indicated by the arrow. A positive "test-charge" is assumed. If it interacts attractively with the transition dipole then it should interact repulsively with the quadrupole around the O=C group. The correlation between sign pattern of this quadrupole and the associated magnetic transition moment is known. For a perturbing group in a "right upper back octant" the angle between the two moments is acute, the corresponding contribution to the CD is thus positive.

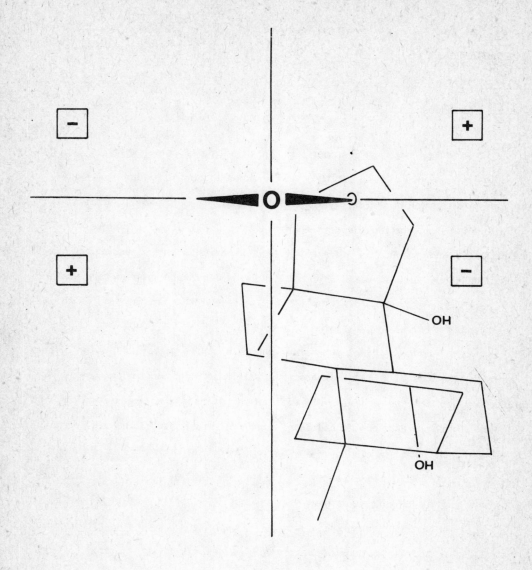

Figure 2. Projection of bufalin molecule (**14**) from O to C of O=C - group. Co-ordinates from X-ray data were used, the signs given are those for contributions to the n→π* band CD in back octants.

fact that the corresponding Cotton effect also has negative sign for the usual bufadienolides.

In-situ COMPLEXES OF DIOLS OF PYRANOSIDE STRUCTURE WITH TRANSITION METAL ACYLATES

The carboxylic group(s) in transition metal complexes like dimolybdenum tetraacetate $[Mo_2(O_2CCH_3)_4]$, its rhodium or ruthenium analogue can be replaced in solution not only by chiral acylates, but also by other suitable chiral bidentate moieties, leading thereby to optically active complexes showing Cotton effects within the different transition metal cluster absorptions [5]. Amongst these ligands are 1,2- and 1,3-diols, which react immediately at room temperature with the Mo_2-complex, only after prolonged heating with the Ru_2-complex, but not at all with Rh_2-acetate.

The molybdenum in-situ complexes show several Cotton effects between 700 and 250 nm, the most characteristic ones appearing around 400, 350, and 300 nm. As a result of investigation of many diastereomeric 2,3- and 3,4-diols with pyranoside structure by this method a rule was derived [6], that in each case the sign of the Cotton effect around 300 nm is identical with that of the torsional angle in the moiety (HO-)C-C(-OH) of the sugar (or any similar dihydroxy derivative). In most CD-spectra the 400 nm Cotton effect has the same sign; the 350 nm CD is opposite to it and is often detectable just as a CD-minimum between two CD-maxima. It is of the same magnitude as the 300 nm Cotton effect only if no axial RO- bond is present on one of the two carbon atoms flanking the glycol moiety. This same rule can also be applied to prim.-sec. and sec.-sec. aliphatic glycols if one fixes the glycol in that conformation where the torsional angle(s) (HO-)C-C(-C) is (are) 180°. This method is especially valuable for simple optically active lipids with 1-monosubstituted glycerol structure [7], for which optical rotations are very small and depend strongly on solvent and concentration.

12

ESTERS OF OPTICALLY ACTIVE ALCOHOLS WITH PYRIDINE DICARBOXYLIC ACIDS

Racemates of optically active sec. alcohols are frequently resolved via their hemiphthalates. The two CO_2R - moieties cannot be coplanar with the aromatic ring for steric reasons, thus the system forms an inherently chiral chromophore. We have recently shown [8] that the sign of the relatively strong Cotton effect around 245 nm ($n \rightarrow \pi^*$ - transition) can be correlated unequivocally with the stereochemistry of the alcohol, avoiding thus the necessity of preparing another "cottonogenic" derivative for the determination of absolute configuration.

Since the pyridine chromophore gives rise to an additional $n_N \rightarrow \pi^*$-band CD, we investigated whether a 2,3- or 3,4-pyridine dicarboxylic acid might be more useful for such a purpose [9]. This seems not to be the case, on the other hand, the same rule as for the (hemi-)phthalates holds also here: the sign of the CD-band around 235 to 240 nm is positive, if the generalized Fischer-projection is, as follows (L refers to the sterically largest group, M to one of medium size):

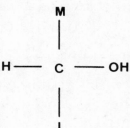

2- and 3-monoesters of the 2,3-dicarboxylic acid differ also in their CD-spectra: a second Cotton effect (of same sign) can only in the first case be distinctly seen around 270 nm.

Acknowledgement - My sincerest thanks are due to my many enthusiastic coworkers and to Mr. U. Wagner, who ran many of the CD-measurements. The financial support from Fonds der Chemischen Industrie, Deutsche Forschungsgemeinschaft, and HOECHST AG is gratefully acknowledged.

Leading literature references:

[1]   G.Snatzke, Chemie in unserer Zeit **15**,78(1981); **16**,160 (1982). - Angew.Chemie **91**,380 (1979).

[2]   G.C.S.Reddy, R.S.Sood, S.R.Nadkarni, J.Reden, B.N.Ganguli, H.W. Fehlhaber and G.Snatzke, Tetrahedron, in press.

[3]   All relevant references cited in S.Antus, G.Snatzke, and I.Steinke, Liebigs Ann.Chem. **1983**, 2247.

[4]   B.Green, F.Snatzke, and G.Snatzke, in preparation.

[5]   For a review cf. J.Frelek, A.Perkowska, G.Snatzke, M.Tima, U.Wagner, and H.P.Wolff, Spectros. Int.J. **2**,274 (1983).

[6]   J.Frelek, A.Lipták, G.Snatzke, and I.Vlahov, in preparation.

[7]   J.Frelek and G.Snatzke, Fresenius Z.anal.Chem. **316**,261 (1983).

[8]   M.Gross, J.Platzek, G.Snatzke, U.Vitinius, and R.Wiartalla, in preparation.

[9]   G.Snatzke and Ch.Tilios, in preparation.

DISCUSSION

PANDIT U.K. (Amsterdam, The Netherlands)

a, If a hydrogen bond from a nearby OH group is present in lactones, does this influence the intensity or the sign of the Cotton-effect?

b, We heard in the lecture that the sign of the appropriate Cotton-effect is influenced by the torsion angle in the sugar molecule discussed. On this basis, what can we say about the conformation?

c, When flexible molecules form complexes, could the former be forced to accomodate a different conformation in the complex?

SNATZKE G. (Bochum, German Federal Republic)

a, In general, hydrogen bonds may influence the magnitude of the Cotton-effects, but they do not change the sign.

b, We have investigated only pyranosides or related cyclo-hexanediols where the glycol torsion angle is approximately $+60°$, or $-60°$. At present, we are investigating furanosides and I hope to be able to answer this question soon. The torsion angle will definitely influence the magnitude of CD parameters, the value of which depends also on the properties of the complex. These are unknown and may vary from ligand to ligand.

c, It is a good thing that the complex forces the ligand into one and only one conformation, because just in such a case the following rule works: the sign of the Cotton-effect around 300 nm for aliphatic 1,2-diols is the same as that of the torsion angle HO-C-C-OH, provided that the torsion angle HO-C-C-C is approximately $180°$.

JACQUIER R. (Montpellier, France)

How wide is the scope of correlation between CD spectra and structure in the field of pyrone type compounds? E.g., what can be said if oxygen is replaced by sulphur?

SNATZKE G. (Bochum, German Federal Republic)
Alkyl substitutents will not change the picture because
they are very weak perturbers. If OR groups are present,
then their lone pairs have to be considered in the PMO
treatment. Replacement of oxygen by sulphur will, in
general, only shift all Cotton-effects to the red, but
will not change their signs.

van der PLAS   (H.C. Wageningen, The Netherlands)
a, What change can be observed in the CD spectrum if the
chromophore involves C=N or C=S double bonds instead of a
carbonyl group?
b, Could you compare the value of X-ray analysis and circular
dichroism in the determination of structure?

SNATZKE G. (Bochum, German Federal Republic)
a, The C=N chromophore gives a characteristic $n \to \Pi^*$ band
Cotton-effect around 250 nm. If it is built into a six-membered ring, the sign of the Cotton-effect is determined
unequivocally by the helicity of the ring. In thiocarbonyl
compounds, the bands are shifted appreciably to the red.
The sign of the $n \to \Pi^*$ band for carbonyls and thiocarbonyls
is the same in identical chiral environment.
b, Circular dichorism is a very good tool to determine
reliably absolute conformations of the chromophore or its
immediate vicinity, but there is always a second step
necessary, namely, the correlation of this absolute
conformation (the sign of a torsion   angle) with absolute
configuration ($\underline{R}$ or $\underline{S}$ at the centre of chirality) and there
mistakes could be made. If carefully applied, the coupled
oscillator theory (giving rise to a CD couplet) seems to
be the safest method available to determine in a
non-empirical way absolute configurations. The Bijvoet-
method is reliable only if rather different intensities
of the Friedel-reflections are calculated for the two
enantiomers.

SZÁNTAY Cs. (Budapest, Hungary)

Is it possible, for example, in the case of benzylisoquino-
lines that substitution (e.g. nitration) would result in a
change of sign of the Cotton-effects?

SNATZKE G. (Bochum, German Federal Republic)

The nitro group is actually a very strong perturber and,
at the same time, also a chromophore. Hence, the
introduction of a nitro group into any aromatic chromophore
will drastically change the shape of the CD curve.

ÖTVÖS L. (Budapest, Hungary)

I would have a more bio-organic and rather general question:
what is your opinion about the use of CD spectroscopy in
investigations on conformation and chiroptical properties
of ligands in the complex formed between small molecules
and biomolecules?

SNATZKE G. (Bochum, German Federal Republic)

CD is very useful for this purpose, especially if the
chromophore of a small molecule is inherently chiral but
the compound racemizes very quickly. An example is
diazepam each molecule of which is chiral, but the compound
itself is not optically active. In general, the small
molecule will be bound only in one conformation and this
will result in a strong CD which to the greatest extent
is due to the helicity of the small molecule itself and is
barely influenced by intermolecular chiral perturbance
by the protein.
The non-bound part of the substance will not give any CD
under usual conditions. The absolute conformation of the
bound molecule can be determined by comparison with the
CD of an optically active analogue, if the correlation
between its absolute conformation and CD is known.

CHIRAL HETEROCYCLIC SYNTHONS AS INTERMEDIATES FOR THE SYNTHESIS OF NATURAL PRODUCTS

HUSSON H.-P., GRIERSON D.S., ROYER J., BONIN M. and GUERRIER L.

Institut de Chimie des Substances Naturelles du C.N.R.S. 91190 Gif-sur-Yvette, France

The alkaloids constitute a large group of plant, and more recently, animal substances which display a wide diversity in structure. Despite this apparent multiformity of structural types, the piperidine ring appears as a common feature in the greater majority of these naturally occurring compounds (simple piperidines, spiropiperidines, decahydro-quinolines, azabicyclic systems, monoterpenoid indole alkaloids etc.).

The development of synthetic methodologies wherein preformed piperidine building units are used as "synthons" for the construction of more complex alkaloids structures is thus a challenging problem.

Over the last 20 years several synthetic approaches for the preparation of alkaloids involving the use of piperidine synthons have been developed which possess some general character. These strategies were for the most part based upon the chemistry of the enamine function of Δ^2 piperideines. By taking advantage of the enamine $\rightleftharpoons$ iminium ion tautomeric equilibrium of these tetrahydropyridines, substitution at the C-2 and C-3 positions is possible on reaction with nucleophiles and electrophiles, respectively.

It was our aim to develop a general approach to the synthesis of alkaloids based upon the chemistry of dihydropyridines. In principle, with such intermediates a wide variety of substitutions of the piperidine ring is possible. However, simple 1,2 (1,6)-dihydropyridines (and their corresponding 5,6-dihydropyridinium salts) or 1,4-dihydropyridines are inherently very unstable which limits severly the exploitation of their reactivity in synthesis. It was necessary, therefore, to overcome the problem associated with the instability of these molecules.

In concept, a logical way to achieve such a stabilization would be to convert them into stable equivalent forms which would either react in a manner identical to the parent system or permit its regeneration under appropriate reaction conditions. Toward this goal, we have studied the preparation and reactivity of two types of synthons: 1) 2-cyano Δ^3 pipe-rideines which are equivalent forms of 5,6-dihydropyridinium salts, and 2) 2-cyano-6-oxazolopiperidines which are prepared from chiral β-amino-alcohols and are equivalents of 1,4-dihydropyridines. Since these latter synthons are obtained in optically pure form their utilization for the enantiospecific synthesis of alkaloids becomes possible.

As enamines and iminium salts are often involved in the biogenesis of alkaloids, these synthons have served as key intermediates in the course of biomimetic syntheses.

SYNTHESIS AND CHEMISTRY OF 2-CYANO-Δ^3 PIPERIDEINES

Work in our laboratories[1] led to the discovery that the reaction of N-alkyl-3-piperideine N-oxides $\underline{1}$ with $(CF_3CO)_2O$ (modified Polonovski reaction) leads to the regiospecific formation of the corresponding 5,6-dihydropyridinium salts $\underline{2}$ in high yields (Scheme 1).

Reaction of these unstable intermediates with KCN in a two phase medium $(H_2O-CH_2Cl_2)$ buffered to pH 4.0 was found to produce the 1,2

addition adduct <u>3</u> in 50-75% yields (from 1). Formation of the alternate
and <u>a priori</u> thermodynamically more stable 1,4 cyanide addition adduct <u>4</u>
(isolated as its aminonitrile derivative <u>5</u>) was also possible by reaction
of <u>2</u> with Et$_2$AlCN in benzene [3].

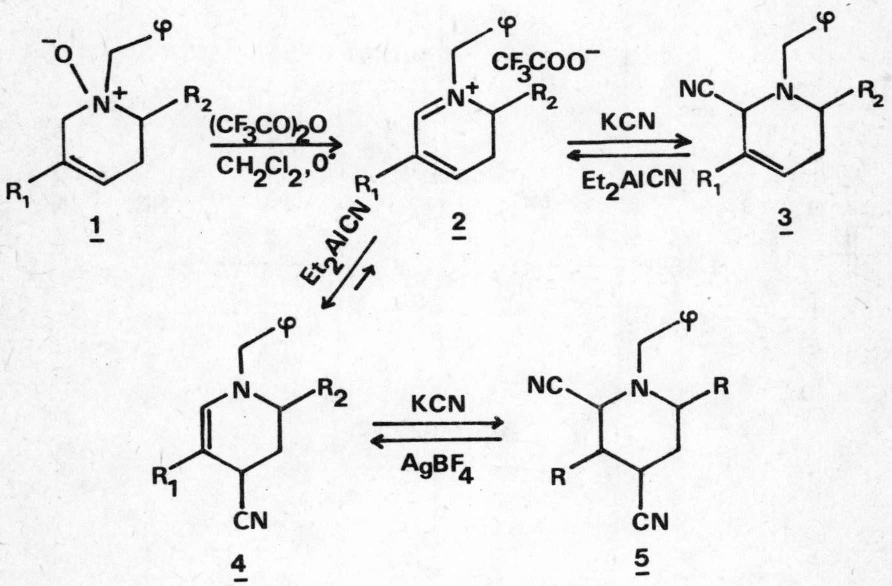

Scheme 1

POTENTIAL AND MASKED REACTIVITY
_____

　　　　The 2-cyano Δ^3 piperideines <u>3</u> were reacted with a series of
nucleophiles in order to examine their capacity to react as a "potential"
5,6-dihydropyridinium salt . The reaction of <u>3</u> with phenyl, phenylethynyl
and methylmagnesium bromides was observed to occur regiospecifically
affording the C-2 substituted products <u>6</u> in good to excellent yields
(scheme 2). In contrast,with β-dicarbonyl anions, indolmagnesium bromide
and C$_6$H$_5$Cu-BF$_3$ the C-4 addition products <u>7</u> were produced. Highly repro-

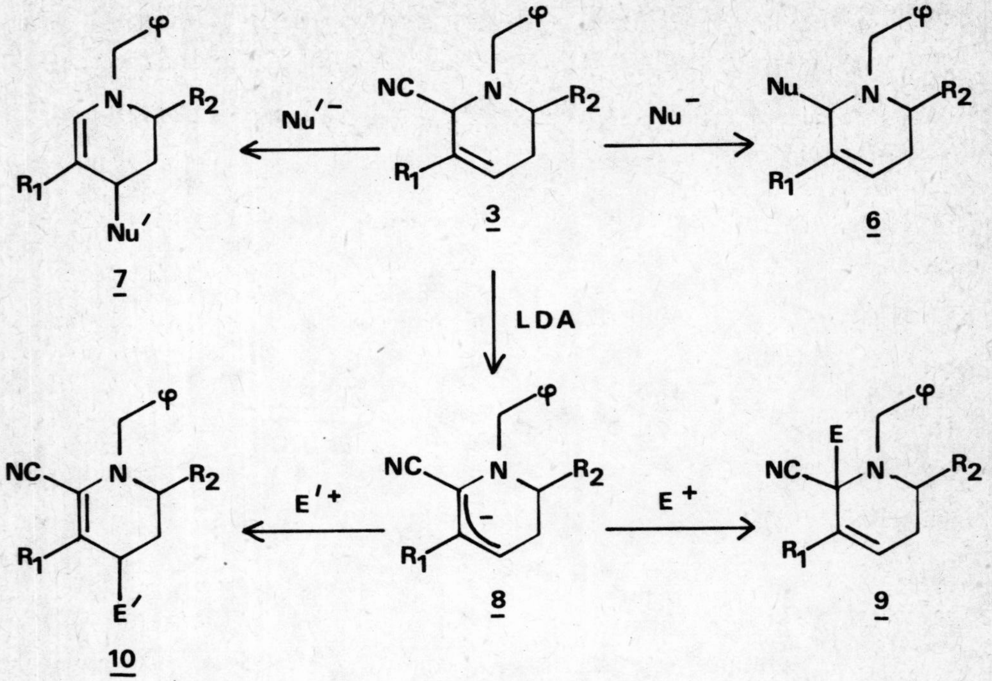

Scheme 2

ducible results were obtained when the reactions with 1,3-dicarbonyl anions
were conducted in the presence of stoichiometric amounts of $AgBF_4$ or with
catalytic quantities of $Pd(\varphi_3P)_4$, $Pt(\varphi_3P)_4$ and $ZnCl_2$.

Alternatively 3 was metalated with alkyllithium bases to
produce the ambient anion 8, thus providing an "umpolung" of the normal
electrophilicity at C-2 and C-4. Anion 8 reacted regiospecifically with a
series of alkyl and acyl halides and esters under kinetically controlled
conditions to give either the C-2 product 9 or the C-4 substituted product
10. Reaction of the anion 8 with benzaldehyde, butyrolactone and propion-
aldehyde occurred under equilibrating conditions. The temperature
dependent formation of C-2 addition products (-50° to -78°) of the former
two carbonyl reagents and their equilibration at higher temperature to the
thermodynamically more stable C-4 adducts 10 was also demonstrated.

Several representative syntheses of alkaloids where 2-cyano
Δ^3 piperideines served as versatile intermediates will be presented.

SYNTHESES OF (±) 20-EPIULEINE [2,3]

Examination of the uleine skeleton shows that the indole
component is connected directly to the C-2 position of the piperidine ring
and bonded to the C-4 position through a two carbon vinyl unit.

Taking these observations into account two strategies for the
construction of this alkaloid from 2-cyano Δ^3 piperidines become apparent
(Scheme 3).

In both cases the first step is the prior formation of the
bond between C-3 position of indole and C-2 position of the piperidine ring
via a Mannich type reaction. Utilisation of synthon 12 led, after
deprotection of the ketone function, to 14. Subsequent reaction of 14 with
EtMgBr and CuCl in THF afforded the epimeric ketones 15. Alternatively th

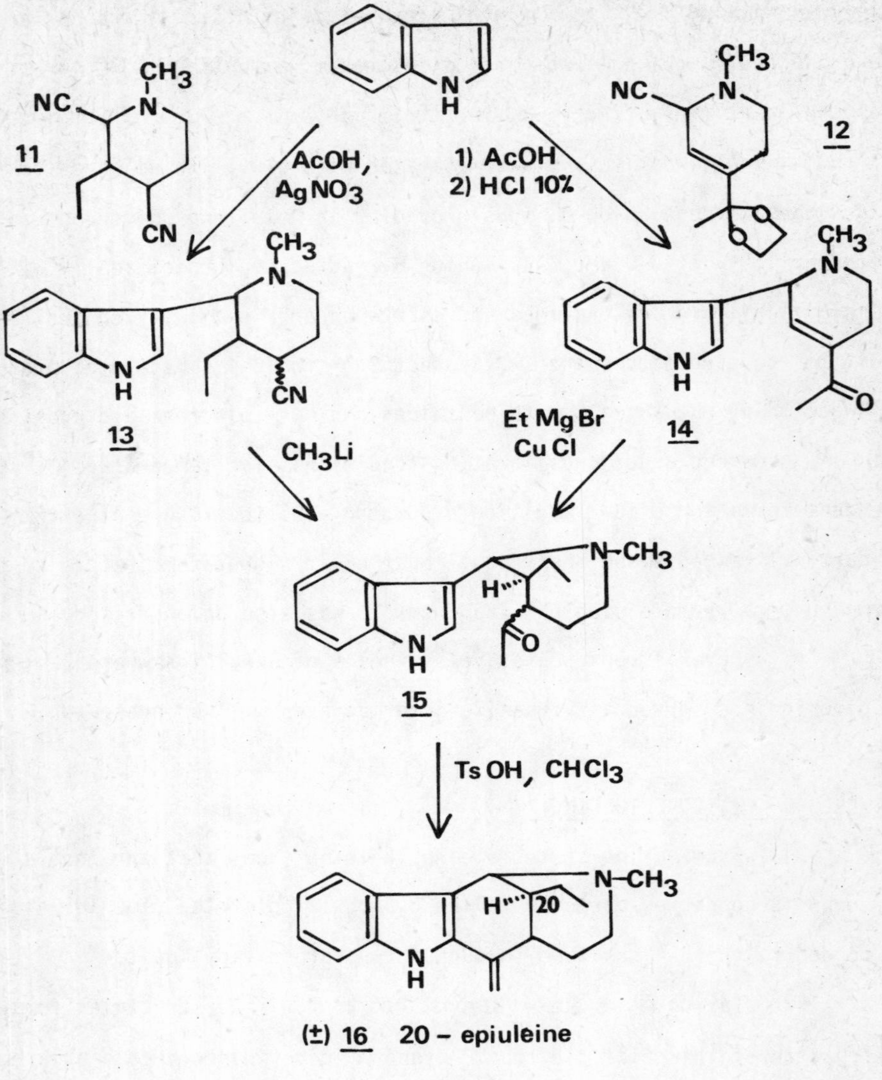

Scheme 3

4

dicyano synthon 11 readily available from 3-ethylpyridine gave 13 as a mixture of epimers on condensation with indole. Reaction of 13 with MeLi in ether led to formation of the corresponding imines which were hydrolysed to 15 by simple filtration through an alumina column. The ketones 15 (from both routes) were cyclized to 20-epiuleine 16 under acidic conditions.

SYNTHESIS OF CORYNANTHE ALKALOIDS [3,4]

Following the finding that 2-cyano Δ^3 piperideines react cleanly with β-dicarbonyl anions to give C-4 substituted enamines in high yield, a concise two steps strategy for the synthesis of the C-15 substituted indoloquinolizidines unit typical to many Corynanthe alkaloids was undertaken (Scheme 4).

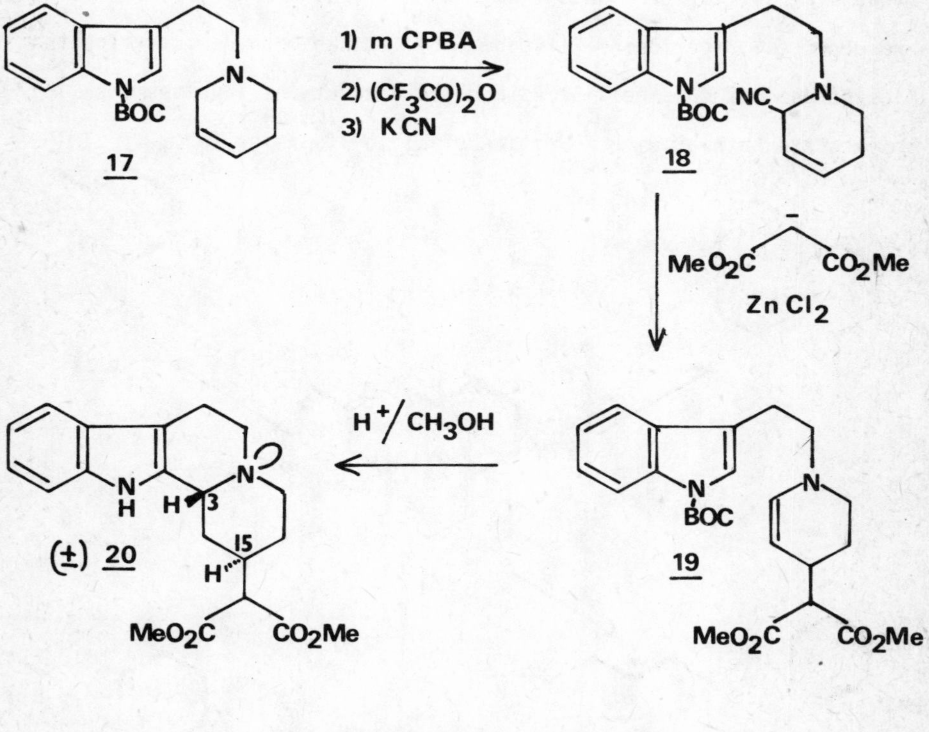

Scheme 4

Reaction of the $N_{(a)}$ t-butyloxycarbonyl (BOC) protected
aminonitrile 18 with sodium dimethylmalonate in the presence of $ZnCl_2$ as
catalyst [5] afforded 19 in nearly quantitative yield. Subsequent treatment
of enamine 19 with HCl/MeOH (60°, 8 hr) affected both the cleavage of the
BOC deactivating group, and cyclization giving the desired tetracyclic
indoloquinolizidine 20 in 50 % yield after purification by flash
chromatography on silica gel. This compound possesses the cis quinolizidine
(H-3, 15 trans) configuration typical to the Corynanthe alkaloid antirhine.

BIOMIMETIC SYNTHESIS OF (±) ADALINE AND BIOGENETICALLY RELATED ALKALOIDS [6,7,8]

The ladybug alkaloids adaline 22 and coccinelline 23 possess
structural similarities which suggest that they are derived from a common
precursor, Δ^2 piperideine 21 (scheme 5). It has been demonstrated that the
biosynthesis of coccinelline 23 and its derivatives involves the
polyacetate pathway with formation of 21 in its imine form.

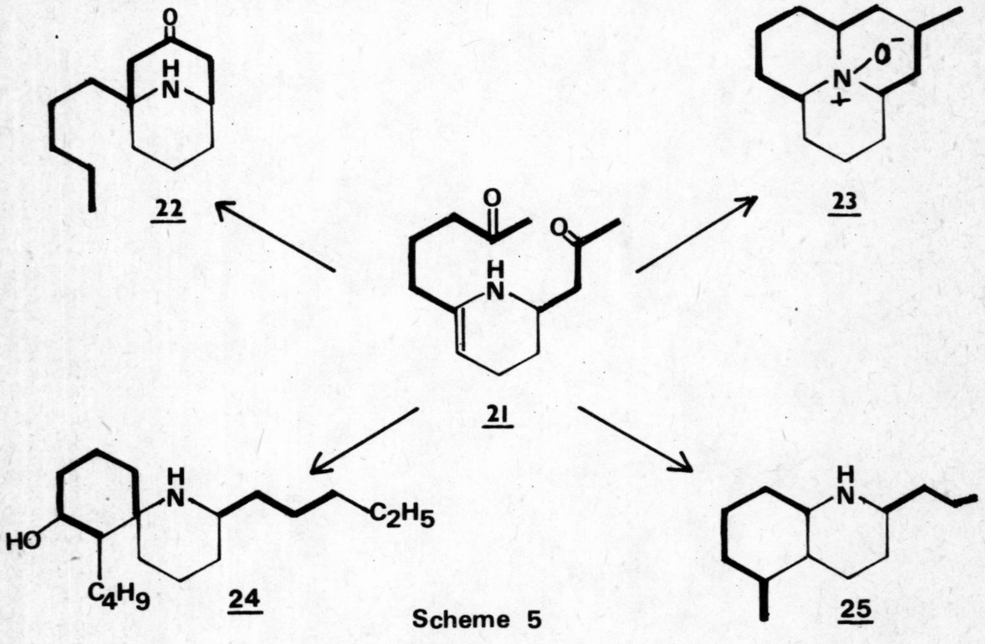

Scheme 5

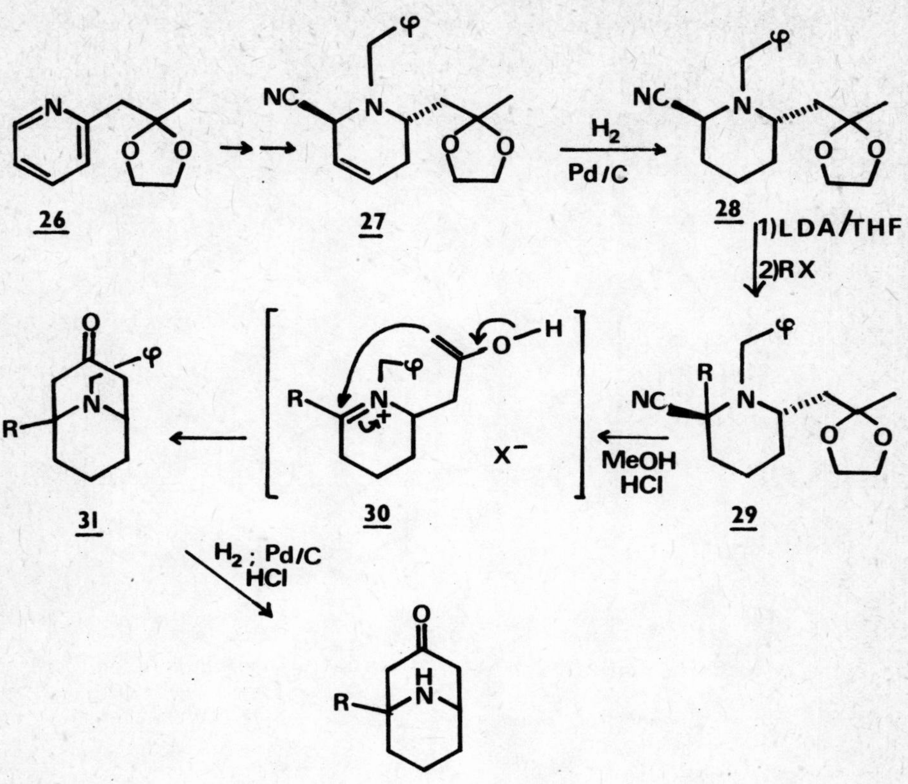

$(\pm)$ **22**   Adaline   R = C_5H_{11}

$(\pm)$ **32**   R = CH_3

Scheme 6

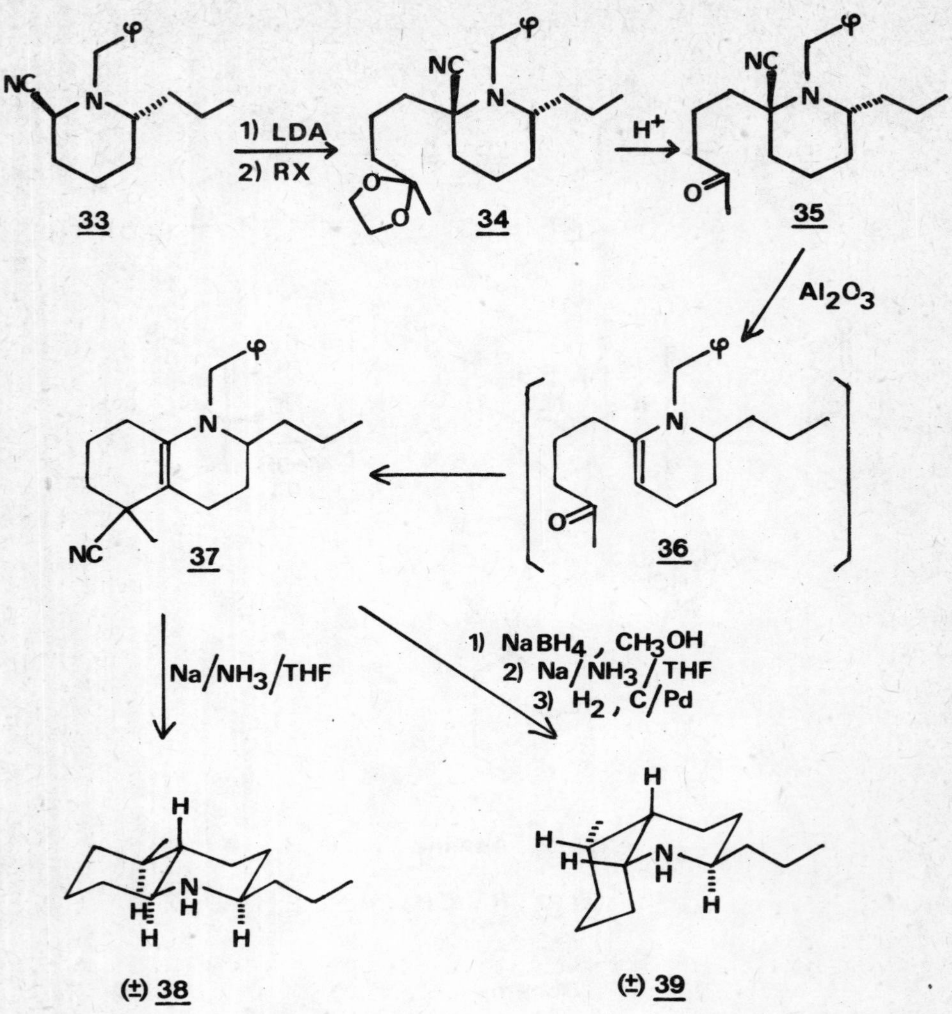

Scheme 7

It has also been proposed that the poison-dart frog toxins histrionicotoxin (depicted as its perhydroderivative 24) and pumiliotoxin-C 25 are derived from a Δ^2 piperideine of type 21. Thus, despite the diverse origin of these groups of compounds, a common element is involved in their respective biogenesis. It is this element which in turn suggests a common approach to their syntheses.

The key step in the synthetic route to adaline 22 and its analog 32 (scheme 6) was a biomimetic intramolecular reaction of the intermediate iminium-enol 30 derived from the amino nitrile equivalent 29 of enamine 21.

Similarly (Scheme 7), in a biomimetic approach to the pumiliotoxin C series a transformation of the aminonitrile 35 to the bicyclic intermediate 37 was accomplished by simple contact with alumina.

The cyclized intermediate 37 was then reduced stereo-specifically to trans decahydroquinoline 38 or stereoselectively to the cis compound 39 (gephyrotoxin conformation).

CHIRAL 1,4 DIHYDROPYRIDINE EQUIVALENTS [9]

In connection with our interest in the synthesis of 2,6-disubstituted piperidine alkaloids, we were prompted to consider the preparation of piperidine synthons based upon the 1,4-dihydropyridine system. It was evident that such synthons should be readily available and show non equivalent reactivities at the C-2 and C-6 positions (providing control over four carbon centers). Moreover, since biological activity of natural products is generally associated with only one enantiomer, the preparation of such synthons in chiral form would offer definite advantages.

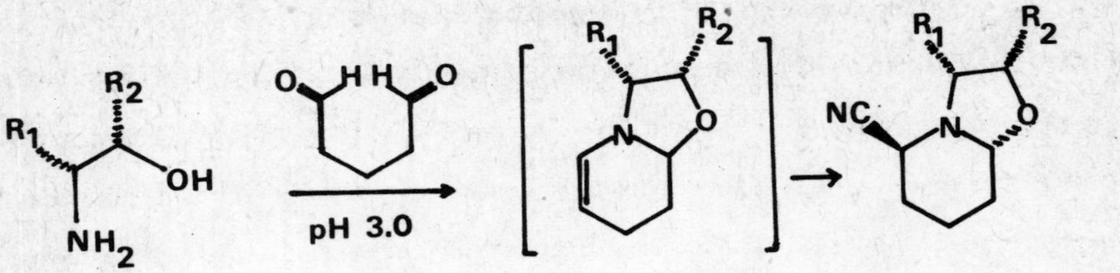
pH 3.0

40(+) $R_1 = CH_3$, $R_2 = C_6H_5$

41(−) $R_1 = C_6H_5$, $R_2 = H$

43 $R_1 = CH_3$, $R_2 = C_6H_5$

44 $R_1 = C_6H_5$, $R_2 = H$

Scheme 8

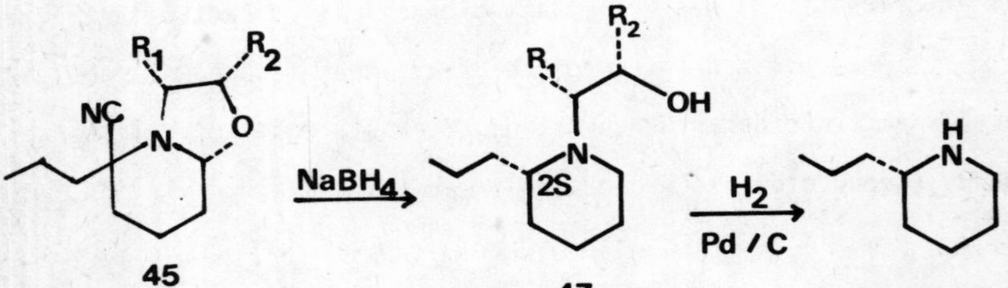
NaBH4
2S
H2
Pd / C

45 **46** **47** **48** **49** (+)Coniine

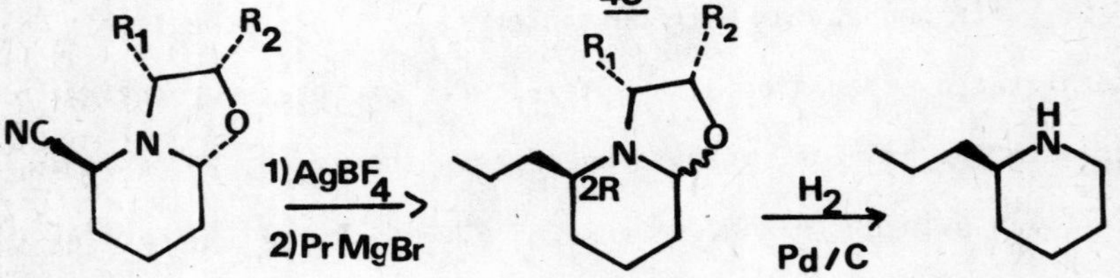
1) AgBF4
2) PrMgBr
2R
H2
Pd / C

43 **50** **51** (−)Coniine

Scheme 9

The Robinson-Schopf type condensation of glutaraldehyde with amino alcohols in the presence of KCN appeared a particularly attractive route for the preparation of synthons we were seeking. Thus, the reaction of (+) norephedrine 40 led in a "one pot" reaction to the formation of a single chiral crystalline 2-cyano-6-oxazolopiperidine 43. Similarly, the condensation of (-) phenylglycinol as a chiral component gave a single product 44 (scheme 8).

As required, chemo and stereoselective reaction at either the C-2 (α-amino nitrile) or C-6 (α-amino ether) centers of 43 and 44 could be achieved by an appropriate choice of reaction conditions. This is illustrated by the enantiospecific synthesis of both (+) and (-) enantiomers of coniine from 43 or 44 (scheme 9).

Alkylation of the anions of 43 and 44 with propyl bromide produced compounds 45 and 46 in nearly quantitative yields. Reaction of these products with $NaBH_4$ in EtOH gave alcohol 47 and 48 having 2S configuration (the 2R diastereomer < 10 %).

Under hydrogenolysis conditions the chiral auxiliary attached to the nitrogen of 48 was cleaved giving 2S (+) coniine 49. More drastic conditions (70 % H_2SO_4, 15 h) were used to cleave the chiral side chain of 47. Excellent chemical and optical yields (ee ≥ 98 %) were obtained.

The high stereoselectivity observed in the reaction of 45 and 46 with hydride ion implied a mechanism wherein there is prior formation of an iminium ion by elimination of the cyano group and subsequent approach of H^- under complete stereoelectronic control from the axial direction (upper face) to the more stable iminium conformer.

By the same mechanism, a propyl chain was introduced at C-2 of 43 in the opposite or R configuration on reaction with PrMgBr (after complexation of the cyano group with AgBF$_4$). Reductive opening of the oxazolidine 50 (NaBH$_4$) and cleavage of the chiral auxiliary gave (-) coniine 51 in high overall yield.

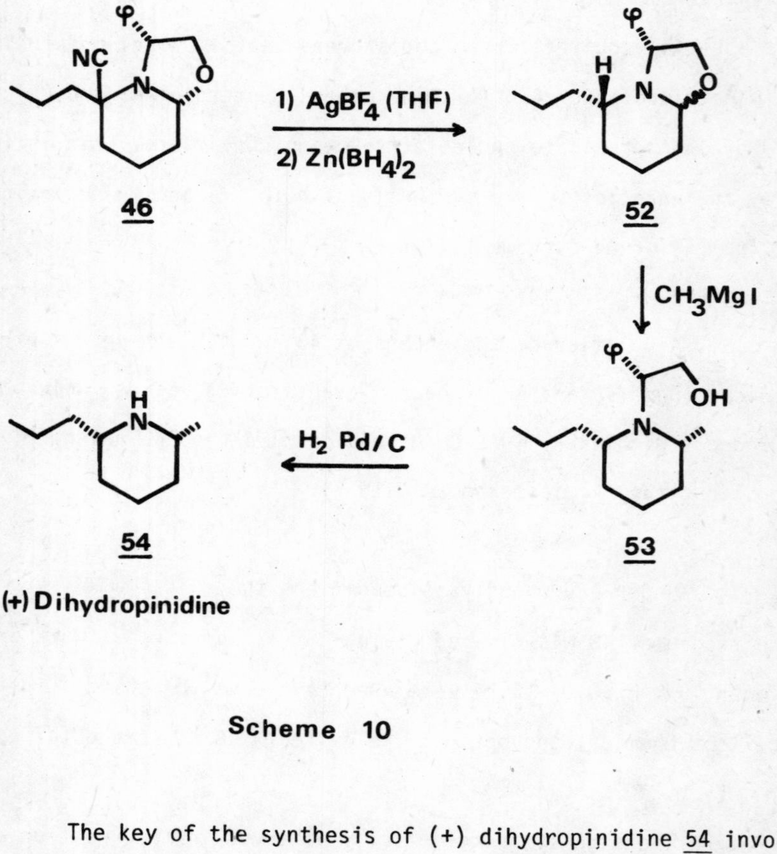

Scheme 10

The key of the synthesis of (+) dihydropinidine 54 involves the use of Zn(BH$_4$)$_2$ at low temperature for the selective removal of the cyano group of 46 (scheme 10). Reaction of methyl Grignard on the resultant 2-alkyl-6-oxazolopiperidine 52 gave the 2,6-cis-dialkylpiperidines 53 that after hydrogenolysis led to optically pure (+) dihydropinidine 54.

In a similar fashion, optically pure (-) dihydropinidine was prepared starting from alkylation of 44 with CH_3MgI followed by substitution with C_3H_7MgBr.

SYNTHESIS OF (-) MONOMORINE I [10]

Synthon 44 represents an ideal starting material for the chiral synthesis of the indolizidine system of monomorine, as hydrogenolysis produces a secondary nitrogen center capable of undergoing an intramolecular ring closure.

Alkylation of the anion of 44 with iodo-ketal 55 led to the formation of a single product 56 (scheme 11). The same reaction sequence as followed for dihydropinidine synthesis afforded (-) monomorine 59 whose absolute configuration is 3S, 5R, 9R according to our previous results.

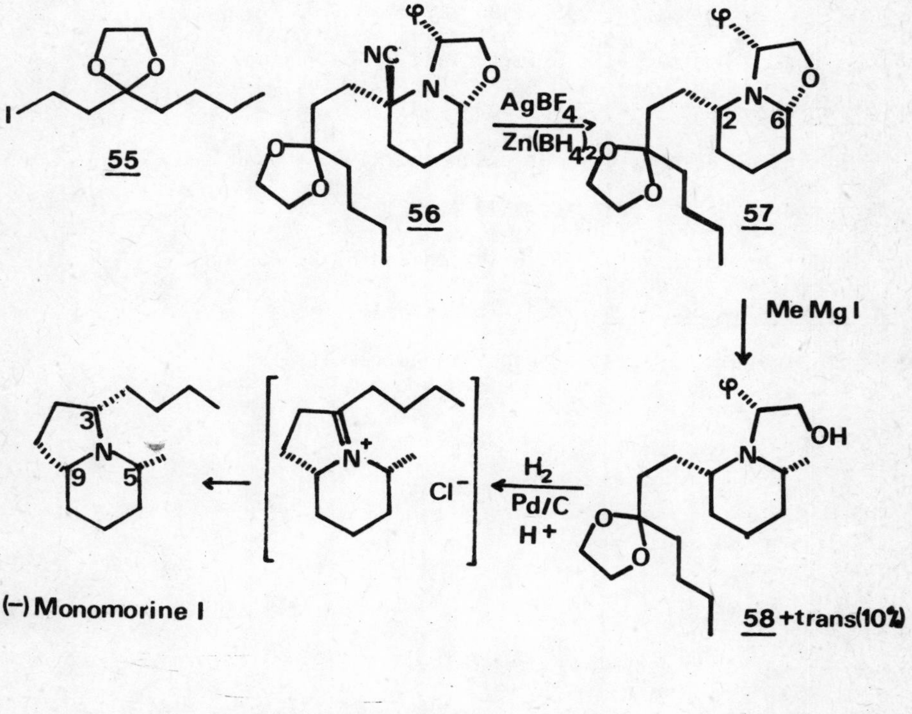

Scheme 11

33

REFERENCES

1 D.S. Grierson, M. Harris and H.-P. HUSSON
   J. Am. Chem. Soc., 102, 1064-1082 (1980).

2 M. Harris, R. Besselièvre, D.S. Grierson and H.-P. Husson
   Tetrahedron Lett., 22, 331-334 (1981).

3 D.S. Grierson, M. Harris and H.-P. Husson
   Tetrahedron, 39, 3683-3694 (1983).

4 D.S. Grierson, M. Vuilhorgne, G. Lemoine and H.-P. Husson,
   J. Org. Chem., 47, 4439-4452 (1982).

5 F. Guibé, D.S. Grierson and H.-P. Husson
   Tetrahedron Lett., 23, 5055-5058 (1982).

6 M. Bonin, R. Besselièvre, D.S. Grierson and H.-P. Husson
   Tetrahedron Lett., 24, 1493-1496 (1983).

7 D.H. Gnecco Medina, D.S. Grierson and H.-P. Husson
   Tetrahedron Lett., 24, 2099-2102 (1983).

8 M. Bonin, J. Romero, D.S. Grierson and H.-P. Husson
   Tetrahedron Lett., 23, 3369-3672 (1982).

9 L. Guerrier, J. Royer, D.S. Grierson and H.-P. Husson
   J. Am. Chem. Soc., 105, 7754-7755 (1983).

10 J. Royer and H.-P. Husson, unpublished results.

DISCUSSION

van der PLAS H.C. (**Wageningen**, The Netherlands)
   You suggested that in the conversion of 8 to 9 a one-
   electron transfer might be involved. What is your evidence?

HUSSON H.-Ph. (Gif-sur-Yvette, France)
   The only evidence we have of radical intermediates is the
   formation of different dimers of the 2-cyano-Δ^3-pyperideine.
   Moreover, alkylation of the anion with p-methoxy- and
   p-nitrobenzyl bromide gave different regiochemistry:
   p-methoxy led to C-2 substituted products, whereas p-nitro
   afforded C-4 substitution probably after migration from the
   C-2 position.

SZÁNTAY Cs. (Budapest, Hungary)
   The Polonovsky reaction results in iminium salts. Since
   there are possibilities to take off proton either from C-2,
   or from the benzylic position, do you know the real
   mechanism? Is a hydrogen shift probable? The reaction
   mechanism can perhaps be elucidated by deuterium labelling
   experiments. Has somebody carried out such kind of
   experiments?

HUSSON H.-Ph. (Gif-sur-Yvette, France)
   We have not studied the mechanism in detail, but elimination
   of the proton from C-2 is most favoured. If elimination
   occurs at the $CH_2\phi$ protons, one must assume a hydride
   transfer to explain the exo-endo isomerization of the
   iminium ion. As far as I know, such a hydride transfer
   has not yet been demonstrated.

BLASKÓ G. (Budapest, Hungary)
   What is the stereochemistry of the three chiral centers in
   the corynantheine type alkaloids synthesized?

HUSSON H.-Ph. (Gif-sur-Yvette, France)

We have achieved the synthesis of a series of indoloquinolizidines, in particular, the "inside series" wherein
the side chain is introduced at the wrong side of the
pyperideine ring (at position C-14). Today I discussed only
the antirhine type alkaloids and in this case we have
obtained the right stereochemistry (hydrogens at C-3 and
C-15 in <u>trans</u> position) with a <u>cis</u> quinolizidine configuration. The C-3, C-15 <u>cis</u> epimer was formed only in trace
amounts.

JACQUIER R. (Montpellier, France)

Can you comment on the mechanism of the removal of the
2-cyano group from the 2-cyano-Δ^3-pyperideines in acidic
medium? Have you tried other acid-catalyzed reactions?

HUSSON H.-Ph. (Gif-sur-Yvette, France)

Acidic medium is generally sufficient to remove the cyano
group. However, we cannot use protic acids for the
elimination of CN in the reaction of the β-dicarbonyl anion.
In this case, $AgBF_4$ or other Lewis acids are effective. We
have also used $Pd/P\phi_3/_4$ and $Pt/P\phi_3/_4$ in order to promote
π-allyl complex formation. In this case the reaction worked
also very well, but we were not able to demonstrate that
such a complex really did exist.

PANDIT U.K. (Amsterdam, The Netherlands)

Does the acid-catalyzed removal of the chiral oxazole
auxilliary go smoothly or are there side reactions?

HUSSON H.-Ph. (Gif-sur-Yvette, France)

The acid-catalyzed removal of the chiral auxilliary requires
drastic acidic conditions, however, no by-products were
observed. Therefore, an aziridinium intermediate can be
postulated.

REINHOUDT D.N. (Enschede, The Netherlands)

You have removed the CN group from the 4-position by using
Na/NH$_3$. In most cases this CN group is a tertiary cyanide.
Does your reaction also work for secondary cyanides?

HUSSON H.-Ph. (Gif-sur-Yvette, France)

In all cases the CN groups were tertiary. However,
secondary CN groups of α-aminonitriles are known to give
clean decyanation with sodium in liquid ammonia.

UNCOMMON AMINO ACIDS IN THE SYNTHESIS AND BIOSYNTHESIS OF NATURAL PRODUCTS

OTTENHEIJM H.C.J.

Laboratory of Organic Chemistry, University
of Nijmegen, Toernooiveld, 6525 ED Nijmegen,
The Netherlands

The chemistry of α-amino acids occurring in proteins is well established
and their biochemical importance is fully understood. Much less is known,
however, about the chemical behaviour of the variety of α-amino acids
featuring structures, which are uncommon to protein amino acids.

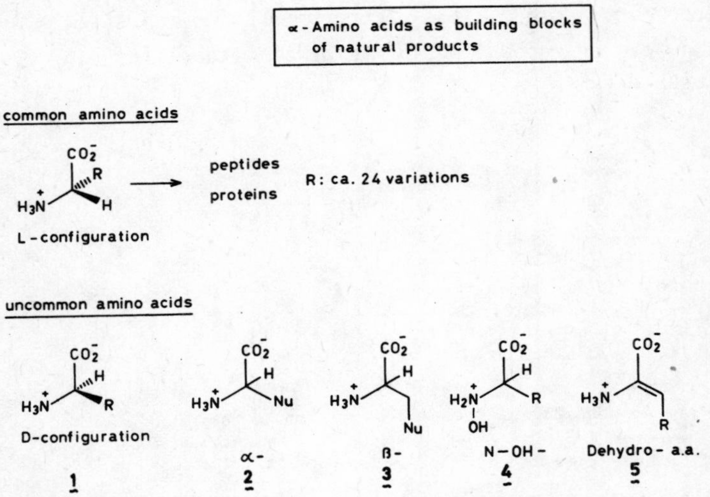

Fig 1

These 'uncommon' amino acids 1-5 have been shown to be characteristic
structural elements of several naturally occurring compounds. For example,
α-functionalized α-amino acids 2 are found in bicyclomycin and in fungal
metabolites of the epipolythiodioxopiperazine class, such as gliotoxin 6
and sporidesmin 7.

4*

uncommon amino acid moieties
of fungal metabolites

Fig 2

6 gliotoxin
2 α - a.a.

bicyclomycin
2 α - a.a.

7 sporidesmin B
2 α - a.a.

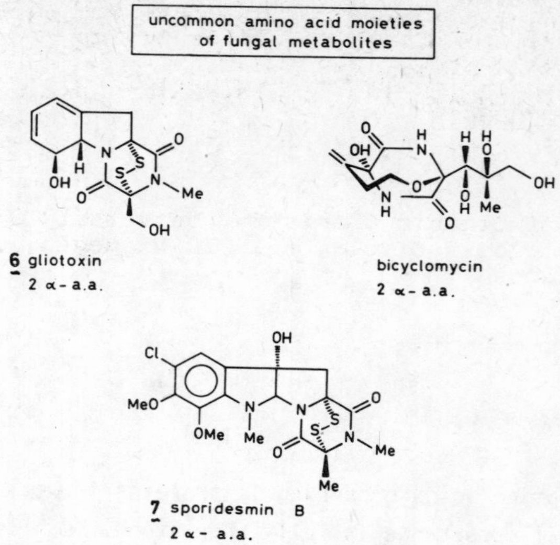

Numerous α,β-dehydroamino acids **5** have been identified in recent years as constituents of fungal metabolites[1]. In most of these metabolites D-amino acids **1** also occur. Another class of uncommon amino acids, *i.e.* N-hydroxy-amino acids **4**, can be recognized in microbial metabolites *e.g.* mycelian-amide and astechrome (Figure 3).

uncommon amino acid moieties
of fungal metabolites

geranyl

Fig 3

mycelianamide
N − OH a.a.
N − OH − dehydro a.a.

neo-echinulin B
2 dehydro a.a.

astechrome
N − OH a.a.

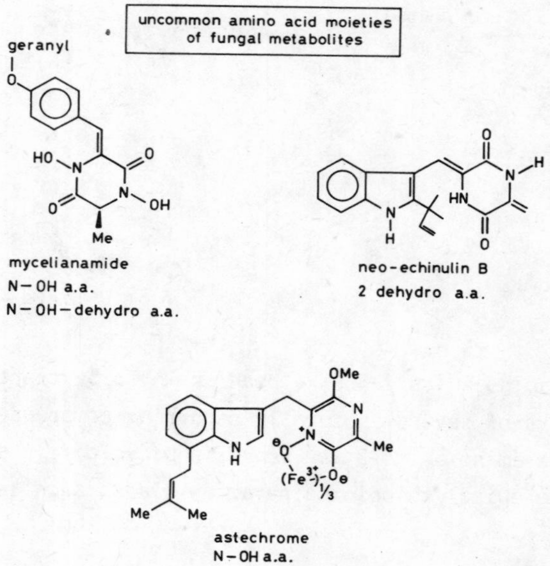

An intriguing question is whether there is a biogenetical or chemical relationship between L-amino acids and the uncommon amino acids 1-5. Let us
first address a possible biogenetical relationship.

One pathway for the formation of α-functionalized amino acids 2 from L-
amino acids is by direct oxidation, as has been discussed in the biosynthesis of the tripeptide part of the ergotalkaloids[2]. A second pathway
might involve a β-elimination reaction from serine, cysteine or threonine
to yield dehydroamino acids 5[1]. A third possible pathway is depicted in
Figure 4. It features oxidation of an amino acid to give an N-hydroxyamino
acid 4. Whereas the first two routes undoubtedly are of biogenetic relevance, the route presented in Figure 4 might also account for at least part
of the metabolism of amino acids in lower organisms.

Fig 4

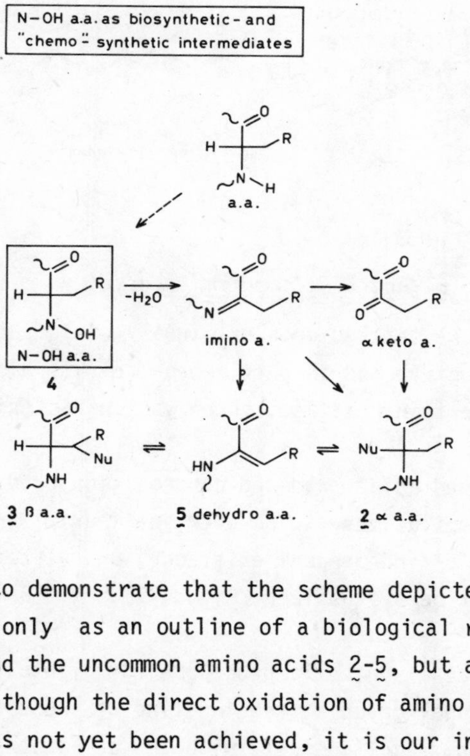

Herein we wish to demonstrate that the scheme depicted in Figure 4 deserves
attention not only as an outline of a biological relationship between
L-amino acids and the uncommon amino acids 2-5, but also as a chemosyn-
thetic chart. Although the direct oxidation of amino acids into N-hydroxyamino acids 4 has not yet been achieved, it is our intention to show that
N-hydroxyamino acids 4 are good synthons for the other uncommon amino acids
2, 3 and 5.

One of our other targets to be discussed is the synthesis of natural products featuring the monooxodithioacetal moiety 10 (Figure 5).

Fig 5

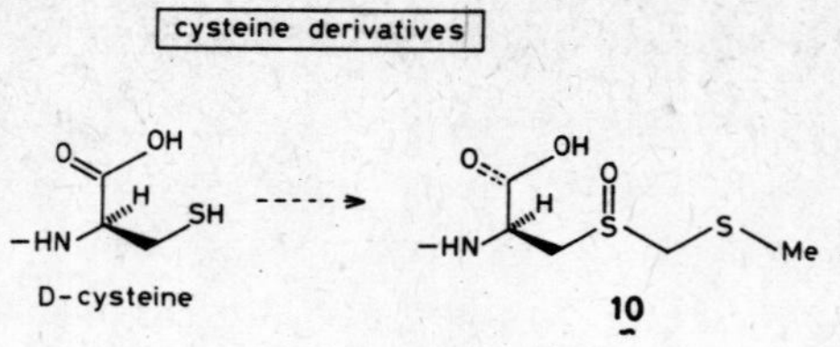

D-cysteine

10

So far only two representatives of this class of compounds are known, *i.e.* sparsomycin (8) and γ-glutamylmarasmin (9).

uncommon amino acid moieties
of fungal metabolites

Fig 6

8 sparsomycin
D- amino acid   ß- a.a.

9 Γ-glutamylmarasmine
ß-a.a.

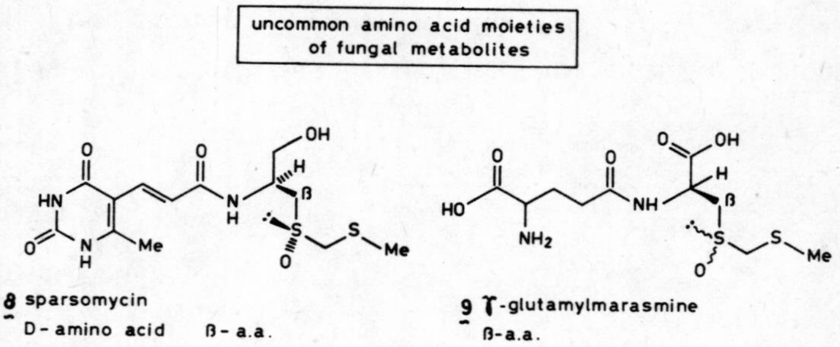

A *biogenetic approach to α-functionalized amino acids* 2

One of the targets that we settled upon originally was gliotoxin (6), in part because of its structure and in part because of its biological activity; it inhibits reverse transcriptase, an enzyme characteristic for RNA-viruses.

Gliotoxin can be viewed as an oxidized condensation product of two α-mercapto-α-amino acid derivatives 11. However, neither of these derivatives is evidently capable of independent existence; unacylated α-mercapto-α-amino acids could not yet be synthesized (Figure 7).

Accordingly, we felt that a synthetic procedure for gliotoxin analogs would have to create a functional group at the indoline C(2) position, convertible to a mercapto group, simultaneously with the acylation of the indoline nitrogen by an α-mercapto-α-amino acid equivalent. Our initial, successful approach[3] involved the addition of pyruvoyl chloride[4] to the imine bond of an indolenine and the intramolecular cyclization of an amide nitrogen with the pyruvoyl α-carbonyl group (see 12; Figure 8).

α- funct. amino acids

Fig 7

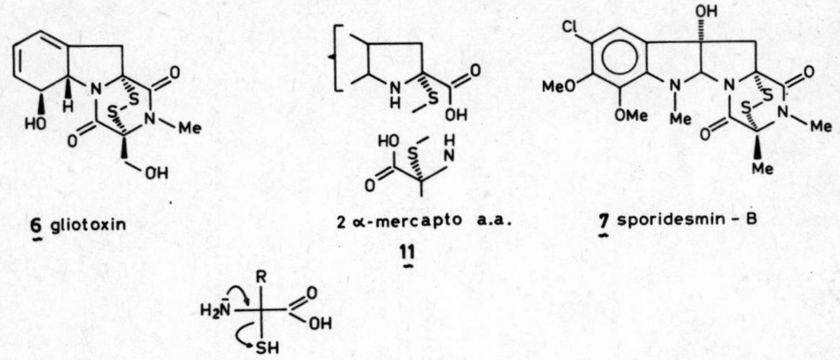

6 gliotoxin   2 α-mercapto a.a.   7 sporidesmin - B
                            11

Synthesis gliotoxin analogs

Fig 8

strategy:

12

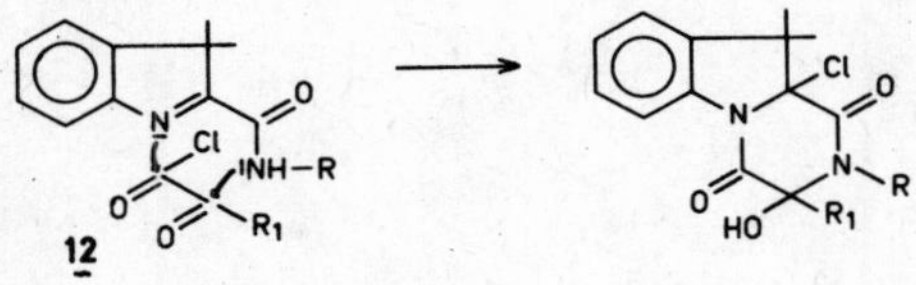

This approach - which is certainly not biogenetic - led us to speculate on
the biosynthesis of gliotoxin. Cyclo-L-Phe-L-Ser has been shown to be an
efficient precursor of gliotoxin (6), and further labeling studies have
demonstrated that the N-methyl group is derived from methionine, whereas
the sulfur atoms are delivered by cystine[5]. The most likely explanation
for the formation of the dihydroaromatic system has been provided by Neuss
et al[6], who invoked the intermediacy of a benzene oxide 13 (Figure 9).

We have proposed a mechanism for the introduction of the sulfur bridge[7].
This proposal features oxidation of the dioxopiperazine amide nitrogen to
form the hydroxamic acid 14 followed by dehydration to acylimine 15 (Figure 10).

Fig 9

Biosynthesis of gliotoxin

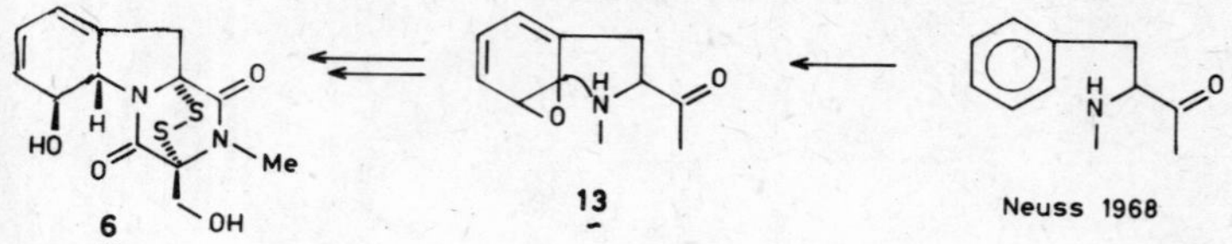

6

phenylalanine
serine
cystine
methionine

13

Neuss 1968

Fig 10

Biosynthesis of gliotoxin

an hypothesis

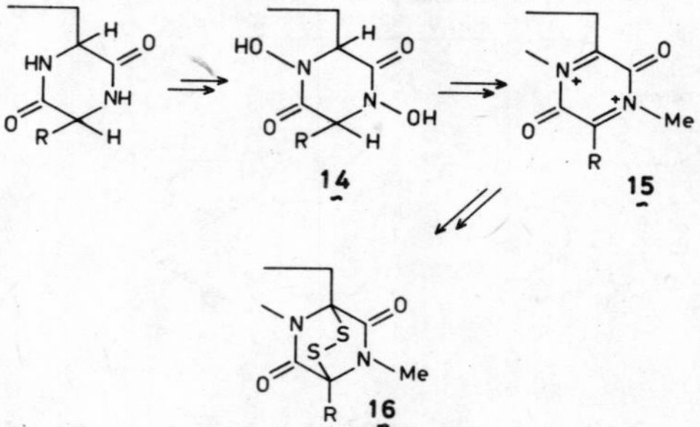

14

15

16

We reasoned that the proposed role of N-hydroxy-α-amino acid derivatives 4 in the biosynthetic conversion of α-amino acids into α-functionalized α-amino acids 2 might gain in probability if the latter could be obtained chemosynthetically by starting from N-hydroxy-α-amino acids 4. In Figure 11 some results toward this directive are depicted, which lent support to our hypothesis.

Selective reduction of the oxime C=N bond of 17 and subsequent N-acylation of 18 yielded N-hydroxy-α-amino acid derivatives 19. Treatment with base (t-BuOK in MeOH) afforded the α-methoxy-α-amino acid derivatives 2, a reaction for which we proposed the acylimine 20 as an intermediate[9]. If this acylimine is formed in the absence of a nucleophile (DBU in dioxane) it rearranges[10] into a dehydro-α-amino acid 5.

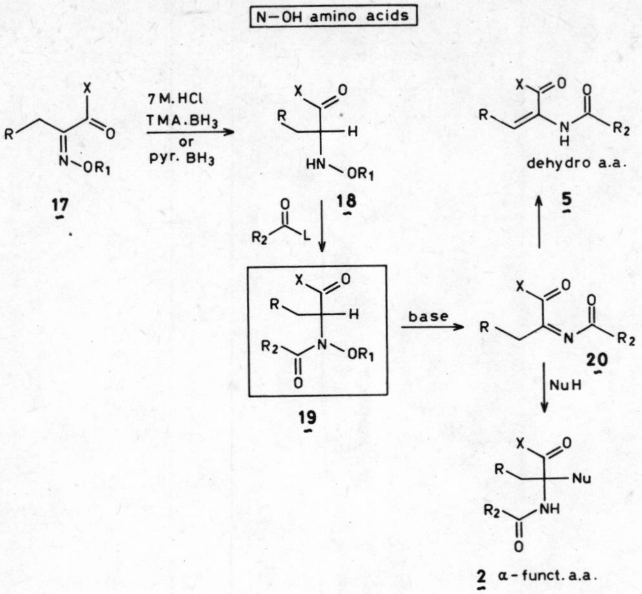

N—OH amino acids

Fig 11

Subsequently we were able to show[7] that this approach for α-functionaliza-
tion by transposition of an N-functionality could also be achieved in dioxopiperazines, e.g. 21, as outlined in Figure 12.

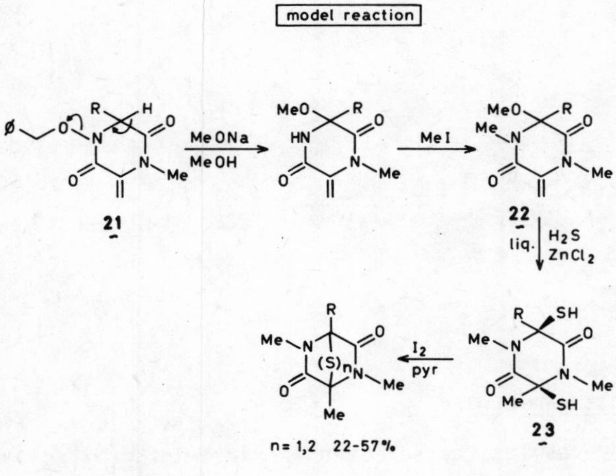

model reaction

Fig 12

The conversion 22→23 deserves an explanation. The Markownikoff-type H₂S addition to the exo-methylene bond of 22 was found to proceed in a diastereo-

selective fashion; only the *cis*-dithiol 23 could be detected. This zinc chloride catalyzed reaction could be explained in the following way (Figure 13):

regio- and stereospecific
H_2S – addition

Fig 13

α-addition via:

III X-ray

24 → 23

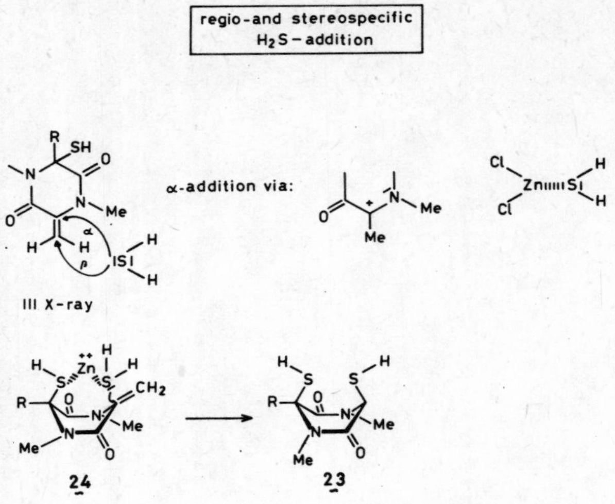

A zinc complex with the SH-group in 24 might direct the incoming SH-groups from the same face by complexation, yielding 23. A second role for the catalyst is to convert H_2S into a stronger acid; when 24 is exposed to H_2S in the absence of zinc chloride no reaction took place[3].

Presently, the conversion of di-N-hydroxydioxopiperazines[11] (25) into sulfur bridged dioxopiperazines 26 is being studied.

applications

Fig 14

biosynthesis
chemosynthesis

25   26

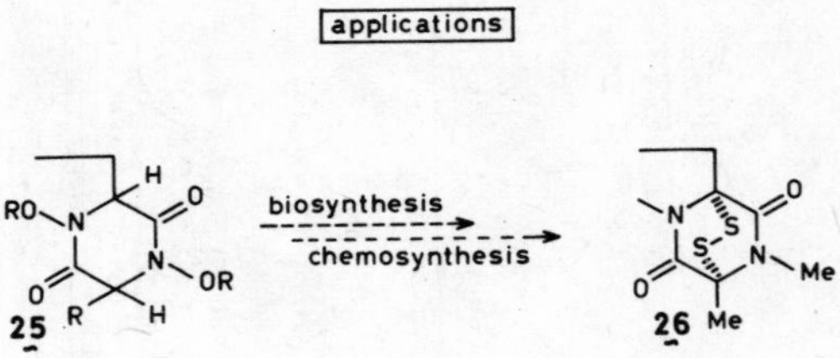

An approach to sporidesmin B (7)

Encouraged by these results, we selected sporidesmin B (7) as our next target and focussed on the conversion of N-hydroxytryptophan 27 into the corresponding α-functionalized amino acid.

Fig 15

N-OH Trp
27

Sporidesmin B
7

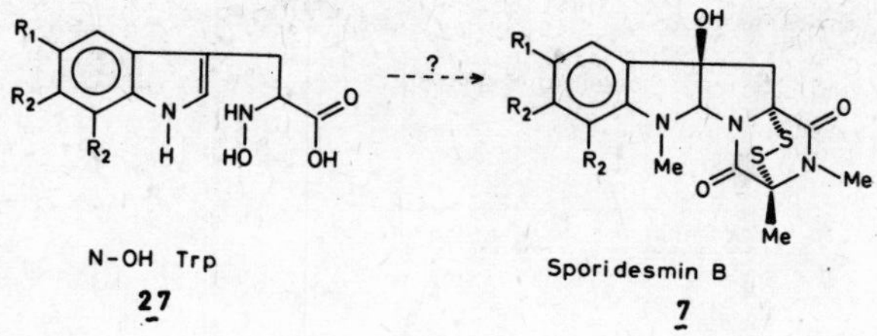

Two syntheses of 27 were developed[12]. One procedure started from indole-3-pyruvic acid which was converted into the corresponding O-benzyl oxime ethyl ester. Methylation of the indole nitrogen and reduction of the oxime C=N bond with $(CH_3)_3N-BH_3$ afforded the N-benzyloxy amino acid ester (cf 17→18). The second route[13] features the reaction of N-methylindole (28, R_1=H) with the transient nitroso olefin 30, prepared from 29. The adduct 31 cannot be isolated; base catalyzed rearomatization afforded 32. Reduction gave 33 in good yields.

Synthesis of N-OH Trp

method B

28   30   Na₂CO₃   31

29   32 R_2 70-90%

$(Me)_3N.BH_3$
EtOH/HCl

33   65-85%

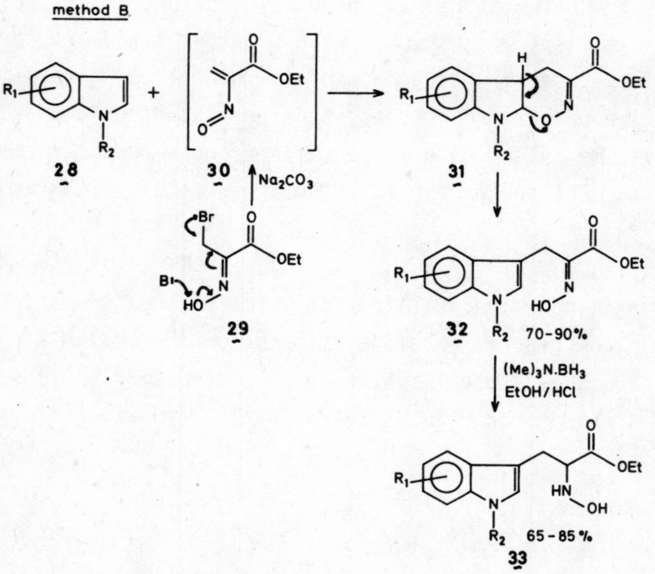

Pyruvoyl chloride[4] and 33 in CH_2Cl_2/ether reacted at room temperature to form 34; no O-acylation was observed. About 24 hrs after mixing the inter-

mediate was converted completely into 35. When stirred for several days
after the addition of methanol, the corresponding α-methoxy derivative is
formed, which was benzylated to give 36. Diastereomeric induction was observed; the indolyl group and methyl group are in a *cis*-relationship in
the major isomer (90%).

An approach to sporidesmin

Fig 17

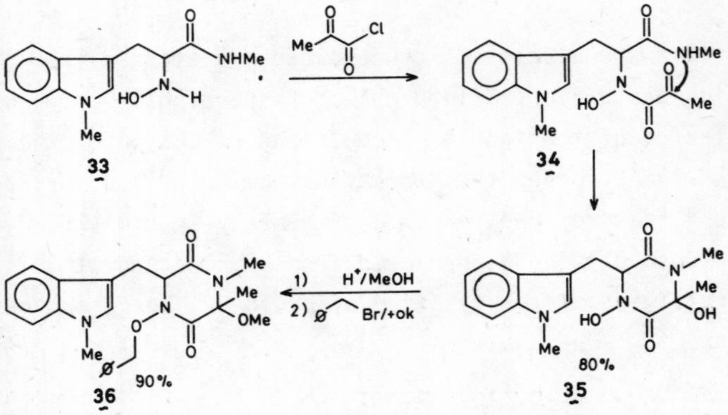

Transposition of the N-functionality in 36 to an α-functionality (→ 37)
was achieved by treatment with CH₃ONa/CH₃OH. The diastereomers were separated and each diastereomer was subjected to the following reaction. For
closing of the five membered ring singlet oxygen was used[14]. A reliable
procedure involves methylene blue as sensitizer and a reaction temperature
of -78 °C; reduction of the peroxo function led smoothly to 38 as outlined
in Figure 18.

Hopefully the remaining stages will directly follow the sequence outlined
in Figure 12 to allow conversion of 38 into the corresponding disulfide
bridged dioxopiperazine. A total synthesis of sporidesmin B (7) seems then
feasible by using the properly substituted indole derivative[15] instead of
28 (see Figure 16).

Fig 18

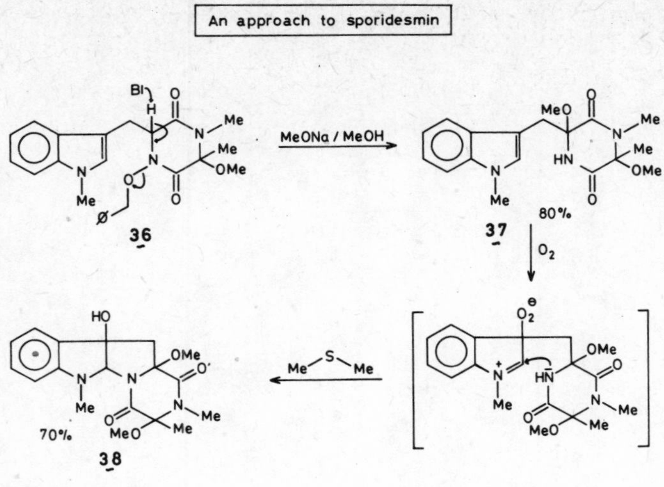

Reactions of the cycloadduct 39

In the synthesis of the N-hydroxy-tryptophan derivative 33 from indole (28) we have seen the primarily formed cycloadduct 31 as an intermediate that escaped isolation. Intrigued by the potential usefulness of this cyclo- adduct for other reaction schemes we have explored the chemistry of 39, formed by reaction of the corresponding indole derivative and the nitroso olefin 30. This cycloaddition proceeds smoothly even with the sterically demanding isopentenyl group at the C(2) position of indole; the adduct un- dergoes base catalysed ring opening followed by rearomatization to afford 40. Work is in progress to convert 40 into fungal metabolites of the neo- echinulin class[12] (Figure 19).

A new rearrangement was observed[16] in the cycloaddition of the nitroso ole- fin with indoles having a C(3)-alkylthio substituent. This reaction yielded 41, in which the thioalkyl group had migrated from C(3) to C(2). This re- arrangement occurs in good yields under mild conditions (room temperature, CH_2Cl_2, Na_2CO_3) and has been used recently by us[17] as a new approach to 2- (S-cysteinyl)tryptophan derivatives 43, which are characteristic structural elements of the toxic principles of members of the genus Amanita (Figure 20).

Reactions of the cycloadduct

Fig 19

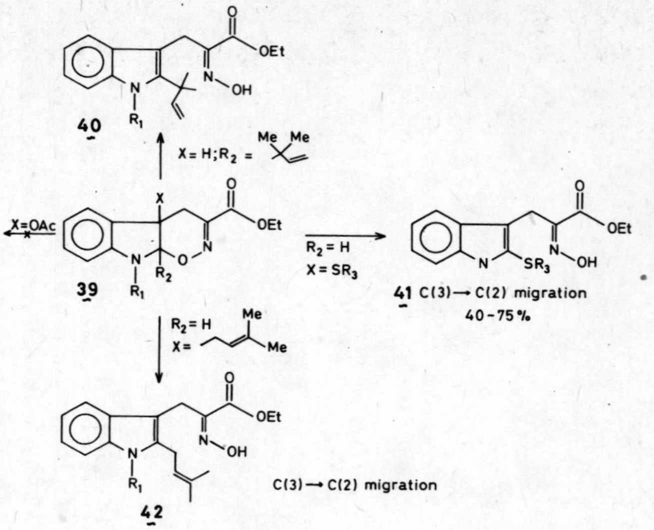

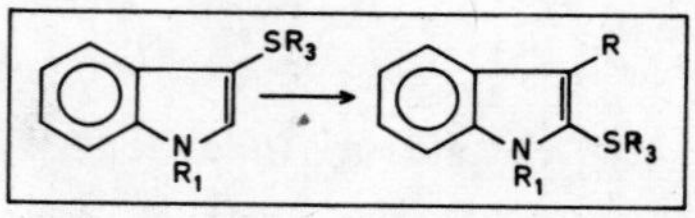

Fig 20

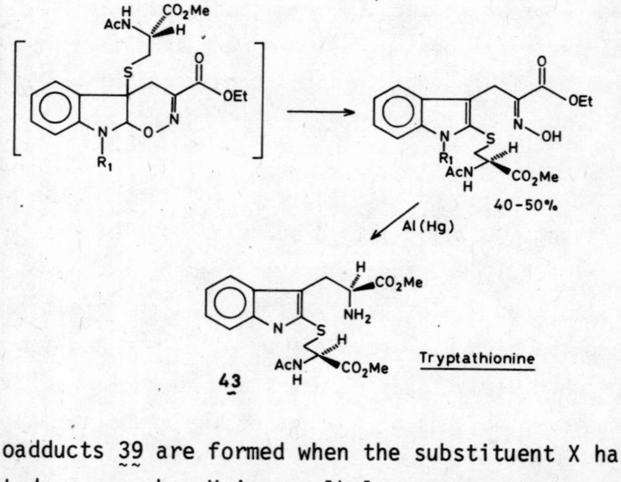

Stable cycloadducts 39 are formed when the substituent X has a low migratory aptitude, *e.g.* when X is an alkyl or acetoxy group. A borderline case arises when the alkyl group is an isopentenyl group; treatment with

trifluoroacetic acid causes an 1,2-allylshift to yield 42, a structural
feature of members of the fumitremorgen class.

The adduct 39 with X=OAc was also studied. It is a stable compound which
shows no tendency to rearrangement. Interestingly, the dihydro-1,2-oxazin
ring of 39 (X=OAc) is susceptible to nucleophilic attack: treatment with
hydrides and mercaptides afforded the indole derivative 44. We have used
this approach for a second synthesis of tryptathionine (43).

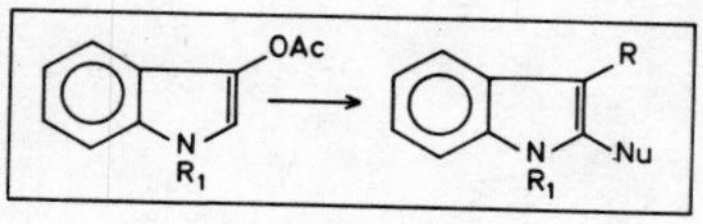

Fig 21

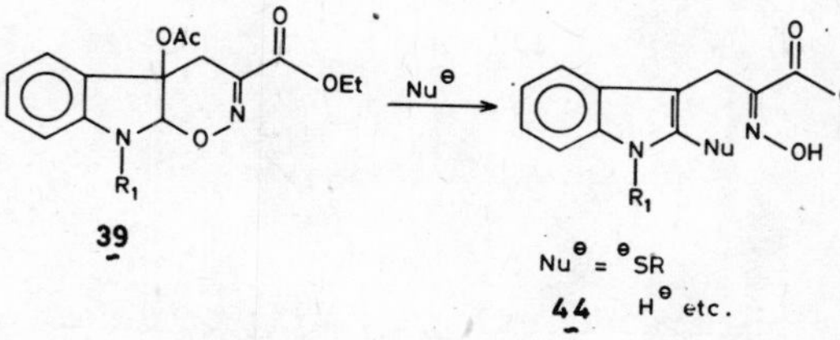

In conclusion the adduct 39 opens a new approach to indole alkaloids, both
in terms of strategy and methodology. We feel that N-hydroxyamino acids in
general are valuable synthons for a wide range of natural products.

Sparsomycin: synthesis and s.a.r. studies

Sparsomycin (8), a metabolite of *Streptomyces sparsogenes* or *Streptomyces
cuspidospores*, has attracted much attention, partly because of its unusual
structure (notice the -S(O)CH$_2$S-moiety rarely encountered in nature) and
partly because of its biological activity of which the antitumor activity
and the inhibition of protein biosynthesis are the most striking properties[18]. On the basis of spectroscopic and degradation studies[19] the presently accepted structure 8 was proposed; however, the chirality of the sulfoxide sulfur atom remained undetermined in these studies.

In an analysis of the synthetic problem (Figure 22) sparsomycin can be
considered as an amide derived from the acid component 45 and the amine
component 46. The latter can be viewed as a derivative of D-cysteine having

its $-CO_2H$ function reduced and its $-SH$ function alkylated and oxidized.

Sparsomycin – Retrosynthesis

Fig 22

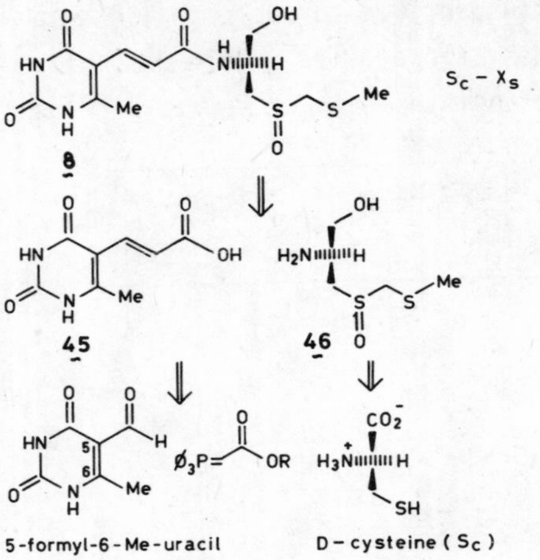

8

$S_c - X_s$

45   46

5-formyl-6-Me-uracil   D-cysteine (S_c)

Component 45 could be prepared by using a Wittig condensation[18]. More challenging was the synthesis of component 46, since the unsymmetrically substituted $-S(O)CH_2S-$ moiety is acid labile and may also undergo β-elimination
- thermal, or base-induced - to which sulfoxides are prone. An attractive
approach to a protected derivative of 46, i.e. 46A appeared to be nucleophilic ring opening of a cyclic sulfinate ester 47, a γ-sultine. This compound has a sulfur atom activated toward nucleophilic attack as well as a
protected alcohol function. This approach was a viable one: we have synthesized sultines of type 47 starting from D-cysteine and have studied their
ring opening reactions with nucleophiles.

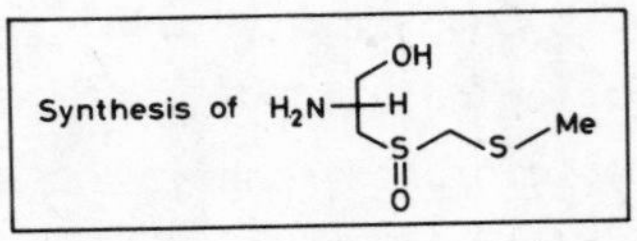

Synthesis of

Fig 23

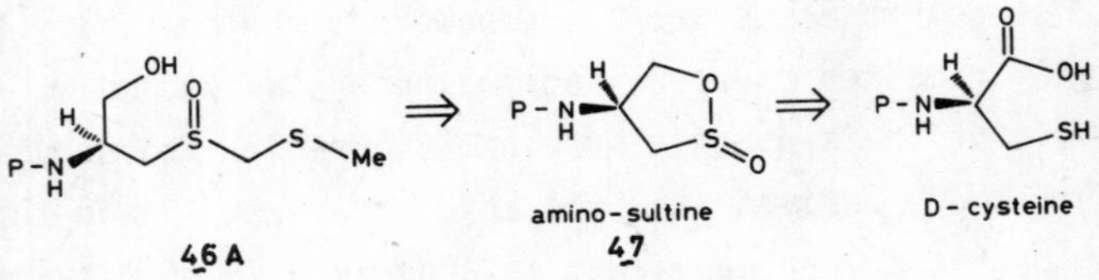

46 A   amino-sultine 47   D-cysteine

52

The N-protected D-cystinol derivative 49, prepared from 48 by $LiBH_4$ reduction followed by I_2 oxidation, was treated with three equivs. of NCS in AcOH to afford 47A and 47B (1:1 ratio) in 87% yield[20]. The structures - including the absolute configuration of the sulfur atoms - were assigned by CD spectroscopy and X-ray crystallographic analysis[20,21].

synthesis of amino sultines

Fig 24

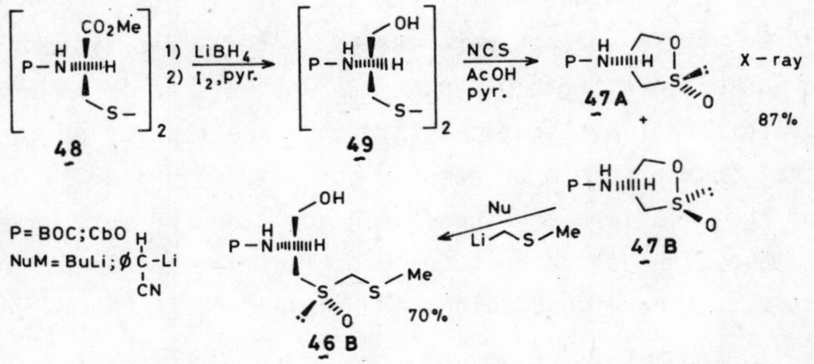

We were pleased to find that the sultines underwent nucleophilic ring opening smoothly and with inversion at sulfur; e.g. treatment of 47B with $MeSCH_2Li$ gave 46B in 70% yield. The amino alcohol 50B (Figure 25) was prepared quantitatively by treatment of 46B (P=t-BOC) with CF_3CO_2H at 0 °C and subsequent deprotonation with an ion-exchange resin. Coupling of 45 with 50B was achieved by means of DCC and hydroxybenztriazole to yield a compound that was identical with sparsomycin in all respects. From this we concluded that sparsomycin's sulfoxide sulfur atom must have the R-chirality.

Sparsomycin synthesis

Fig 25

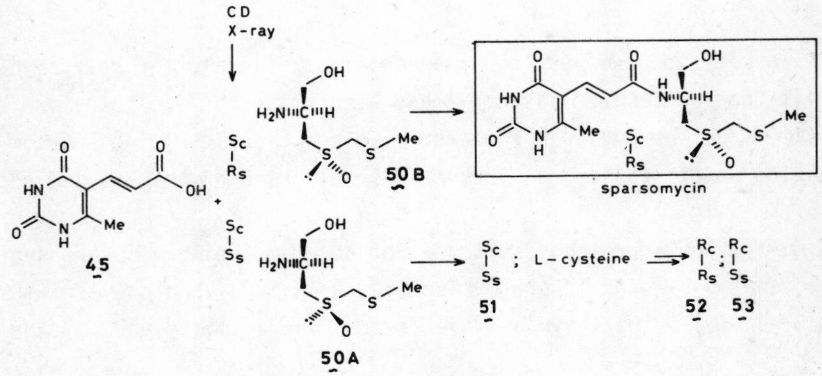

Sparsomycin's three stereomers 51-53 were prepared using the same approach. Finally, the synthesis of sparsomycin was optimized by making use of the finding[22] that sultines 47A and 47B undergo a clean epimerization at the sulfur atom when heated at 120-130 °C. Consequently, the 'wrong' isomer 47A was recycled by conversion into a 1:1 mixture of 47A and 47B, out of which 47B was readily isolated by flash column chromatography.

This approach and another one we had published previously[18] was flexible enough to allow the synthesis of 16 structural analogs of sparsomycin. Sparsomycin and these analogs were tested in cell-free systems for their ability to inhibit the protein synthesis[17] as well as in an *in vitro* clonogenic assay for their cytostatic activity[23]. The results of these assays indicate that sparsomycin's cytostatic activity is primarily due to an inhibition of the protein biosynthesis. In addition, we were able to draw conclusions about the structural features that are required for sparsomycin's activity. These conclusions are summarized in Figure 26.

s.a.r. studies

assays used × inhibition of protein biosynthesis in cell-free systems
× in vitro L 1210 clonogenic assay

Fig 26

54

++ : essential
+ : important
- : unimportant

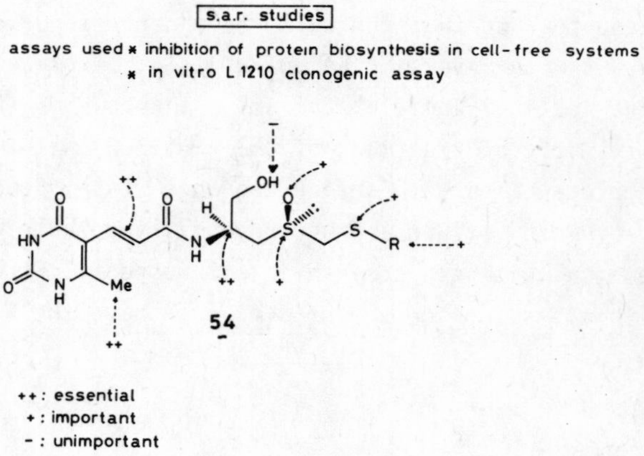

These results lend support to a mechanistic rationale for sparsomycin's activity on a molecular basis: it has been proposed[24] that the inhibitory activity might be due to a Pummerer-like reaction involving sparsomycin's sulfoxide moiety causing an irreversible blocking of the growing peptide chain.

Finally, our approach resulted - for the time being - in the synthesis of two analogs - benzylsparsomycine (54, R=CH$_2$C$_6$H$_5$) and octylsparsomycine (54, R=nC$_8$H$_{17}$) - that have a higher cytostatic activity and a lower toxicicity than sparsomycin. Work is in progress to obtain additional arguments for the introduction of the most promising analog of sparsomycin into the clinic.

In conclusion, our results underline the view that a flexible synthesis of a biologically active compound is a prerequisite for an analysis of the structural features that are essential for an optimal activity as well as for a study of the mode of action of that compound.

ACKNOWLEDGMENTS

I wish to thank my coworkers for their patience and dedication in collaborating in the work I have described. I am particularly indebted to Dr J.D.M. Herscheid, Dr R.M.J. Liskamp and to Mr R. Plate. We are grateful to Prof. R.J.F. Nivard for his support.

REFERENCES

1. The available data on α,β-dehydroamino acids and α-functionalized amino acids have been the subject of a survey by U. Schmidt, J. Häusler, E. Oehler, H. Poisel in 'Progress in the Chemistry of Natural Products', W. Herz, H. Grisebach, G.W. Kirby, Eds.; Springer Verlag: New York, 1979, Vol. 37, p. 251.

2. F.R. Quigley and H.G. Floss, J. Org. Chem. 1981, 46, 464.

3. H.C.J. Ottenheijm, J.D.M. Herscheid, G.P.C. Kerkhoff and T.F. Spande, J. Org. Chem. 1976, 41, 3433.

4. H.C.J. Ottenheijm and M.W. Tijhuis, Organic Syntheses 1983, 61, 1.

5. See references cited in 7.

6. N. Neuss, R. Nagarajan, B.B. Mallory and L.L. Huckstep, Tetrahedron Lett. 1968, 4467.

7. J.D.M. Herscheid, R.J.F. Nivard, M.W. Tijhuis and H.C.J. Ottenheijm, J. Org. Chem. 1980, 45, 1885.

8. M.W. Tijhuis, J.D.M. Herscheid and H.C.J. Ottenheijm, Synthesis 1980, 890.

9. J.D.M. Herscheid, R.J.F. Nivard, M.W. Tijhuis, H.P.H. Scholten and H.C.J. Ottenheijm, J. Org. Chem. 1980, 45, 1880.

10. J.D.M. Herscheid, H.P.H. Scholten, M.W. Tijhuis and H.C.J. Ottenheijm, Recl. Trav. Chim. Pays-Bas 1981, 100, 73.

5*

11. J.D.M. Herscheid, J.H. Colstee and H.C.J. Ottenheijm, J. Org. Chem. 1981, 46, 3346.

12. H.C.J. Ottenheijm, R. Plate, J.H. Noordik and J.D.M. Herscheid, J. Org. Chem. 1982, 47, 2147.

13. The versatile cycloaddition of a nitroso olefin to a carbon-carbon double bond has been described for the first time by T.L. Gilchrist, D.A. Longham, T.G. Roberts, J. Chem. Soc., Chem. Commun. 1979, 1089.

14. M. Nakagawa, S. Kato, S. Kataoka, S. Kodato, H. Watanabe, H. Okajima, T. Hino and B. Witkop, Chem. Pharm. Bull. (Japan) 1981, 29, 1013.

15. J.W. Blunt, A.F. Erasmuson, R.J. Ferrier and M.H.G. Munro, Aust. J. Chem. 1979, 32, 1013.

16. R. Plate, H.C.J. Ottenheijm and R.J.F. Nivard, J. Org. Chem. 1984, 49, 540.

17. Manuscript in preparation.

18. H.C.J. Ottenheijm, R.M.J. Liskamp, S.P.J.M. van Nispen, H.A. Boots and M.W. Tijhuis, J. Org. Chem. 1981, 46, 3273 and references cited therein.

19. P.F. Wiley and MacKellar, J. Org. Chem. 1976, 41, 1858.

20. R.M.J. Liskamp, H.J.M. Zeegers and H.C.J. Ottenheijm, J. Org. Chem. 1981, 103, 1720.

21. H.C.J. Ottenheijm, R.M.J. Liskamp, P. Helquist, J.W. Lauher and M.S. Shekhani, J. Am. Chem. Soc. 1981, 103, 1720.

22. R.M.J. Liskamp, H.J. Blom, R.J.F. Nivard and H.C.J. Ottenheijm, J. Org. Chem. 1983, 48, 2733.

23. R.M.J. Liskamp, J.H. Colstee, H.C.J. Ottenheijm, P. Lelieveld and W. Akkerman, J. Med. Chem. 1984, 27, 301.

24. G.A. Flynn and R.J. Ash, Biochem. Biophys. Res. Commun. 1983, 114, 1.

DISCUSSION

PANDIT U.K. (Amsterdam, The Netherlands)

Is the "cycloaddition" of indole to nitroso olefin a concerted or a multistep process? The 2-3 double bond of indole is an enamine bond and cannot be regarded as a simple olefin.

OTTENHEIJM H.C.J. (Nijmegen, The Netherlands)

I have no convincing evidence that would permit a choice between a concerted and a multistep process. A dipolar intermediate is indeed commonly accepted for this type of cycloaddition. However, there is no doubt about the nature of the cycloadduct, which we have isolated in a few cases.

REINHOUDT D.N. (Enschede, The Netherlands)

You have shown a very interesting application of the cycloaddition reaction of nitroso alkenes. Have you studied the analogous reaction of indoles with nitro alkenes? Reactions of this type have been studied, e.g. with enamines.

OTTENHEIJM H.C.J. (Nijmegen, The Netherlands)

The reaction you propose is an interesting one; we have not yet studied it. It is part of our research program.

van der PLAS H.C. (Wageningen, The Netherlands)

a, You showed the interesting stereoselective addition of H_2S and explain it by assuming that the incoming second molecule of H_2S coordinates with the Zn^{2+} used in your reaction. Have you also found that other dicationic species like Mg^{2+} is able to direct the addition in a specific way?

b, I saw in one of your slides the highly stereoselective reduction of the $HO-N=CH-CO_2Et$ group by aluminium. Is that reaction really stereoselective? I expect the formation of a diastereomeric mixture.

OTTENHEIJM H.C.J. (Nijmegen, The Netherlands)
   a, We have studied other catalysts, beside $ZnCl_2$. From the
   Lewis acids studied, only $ZnCl_2$ gave good results. For
   instance, $MgCl_2$ failed. CF_3COOH gave the desired adduct
   also, however, in low yield (50%).
   b, You are right, my slide was unwillingly misleading. What
   I meant to say was that during the oxime reduction a
   mixture of diastereomers is formed (69%). The
   S(TRP)-R(Cys) isomer was isolated by HPLC.

HUSSON H.-Ph. (Gif-sur-Yvette, France)
   Concerning the mechanism of action of sparsomycin, you
   postulated a Pummerer rearrangement with the transfer of
   an acyl group from a proteinic substrate. Do you know some
   precedents for such facile reactions?

OTTENHEIJM H.C.J. (Nijmegen, The Netherlands)
   I do not know of any precedent. However, we have studied
   this possibility with a model compound and found that
   a Pummerer-type reaction can take place under mild reaction
   conditions. This work is still in progress.

PREOBRAZHENSKAYA M.N. (Moscow, U.S.S.R.)
   Inhibitors of protein synthesis can be cytotoxic, but in
   most cases the inhibition does not produce specific
   antitumour activity. There must be a certain process which
   leads to antitumour specificity for sparsomycin. May it be
   a selective transport to cancer cells?

OTTENHEIJM H.C.J. (Nijmegen, The Netherlands)
   Your remark is well taken. The antitumour activity of
   sparsomycin is most likely due to its inhibitory action
   on the protein synthesis. So it is cytotoxic. However, it
   seems to have at least some selectivity for transformed
   cells. This selectivity may well be the result of an
   abnormal membrane permeability of transformed cells.

ÖTVÖS L. (Budapest, Hungary)

You have mentioned a ring opening reaction where an oxime
is formed. Is this reaction stereospecific? Have you
studied the configuration of the oxime?
I think the reaction discussed must be stereospecific.

OTTENHEIJM H.C.J. (Nijmegen, The Netherlands)

This reaction has not been studied in that details.

Bio-Organic Heterocycles
van der Plas H.C., Ötvös L., and Simonyi M. eds

AMANITA MUSCARIA IN MEDICINAL CHEMISTRY
ANALOGUES OF GABA AND GLUTAMIC ACID DERIVED
FROM MUSCIMOL AND IBOTENIC ACID

KROGSGAARD-LARSEN P.

Department of Chemistry BC, Royal Danish School
of Pharmacy, 2, Universitetsparken, DK-2100
Copenhagen Ø, Denmark

INTRODUCTION

Certain amino acids play a key role in the function of the
mammalian central nervous system (CNS). Whilst 4-aminobutyric
acid (GABA) is the major inhibitory neurotransmitter in the
brain, glutamic acid (GLU), and possibly also aspartic acid,
appear to be the major excitatory neurotransmitters (Roberts
et al. 1976, Krogsgaard-Larsen et al. 1979b, Di Chiara and
Gessa 1981, Mandel and DeFeudis 1983). GABA is known to be
involved in the central regulation of a variety of physiological functions including cardiovascular mechanisms (DeFeudis
1981) and the sensation of pain (Hill et al. 1981, Kendall et
al. 1982, Grognet et al. 1983, Kjaer and Nielsen 1983) and anxiety (Hoehn-Saric 1983). Furthermore, accumulating evidence
suggests that the symptoms associated with epilepsy, Huntington's chorea, and spasticity are caused by impaired function,
and possibly degeneration, of GABA neurones (Roberts et al.
1976, Krogsgaard-Larsen et al. 1979b, Morselli et al. 1981,
Meldrum 1982, Mondrup and Pedersen 1983). As a result of
these observations there is a pharmacological interest in a-
gents with GABA stimulating effects, including GABA agonists.

Based on extensive animal studies it is assumed that the alterations of neuronal pathways and the neuronal degenerations
observed in patients suffering from epilepsy or Huntington's
chorea to a certain extent are caused by hyperactivation of
central GLU receptors located on the degenerating neurones
(Coyle et al. 1981). Consequently, these receptors have been

61

brought into focus as potential pharmacological sites of attack. Antagonists at these receptors may have therapeutic interest, and the availability of such compounds and specific GLU agonists is a necessary condition for satisfactory receptor characterization (Watkins 1981, Krogsgaard-Larsen and Honoré 1983). Furthermore, the observation that the neuronal release of GLU appears to be regulated by GABA receptors (Meier et al. 1983), possibly located on GLU nerve terminals, adds further interest to GABA receptors as sites for pharmacological intervention.

In the present review certain aspects of the design and dedelopment of specific heterocyclic GABA and GLU agonists are described. Furthermore, the clinical studies on the GABA agonist 4,5,6,7-tetrahydroisoxazolo[5,4-c]pyridin-3-ol (THIP) are briefly reviewed.

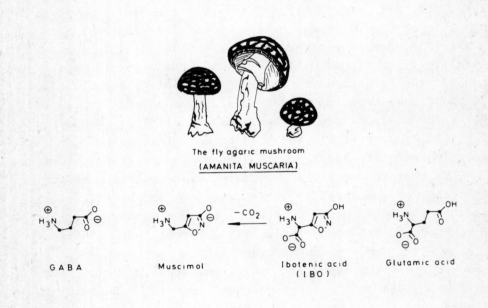

The fly agaric mushroom
(AMANITA MUSCARIA)

GABA   Muscimol   Ibotenic acid   Glutamic acid
                                     (IBO)

Fig. 1. The structures of the Amanita muscaria constituents ibotenic acid and muscimol. The structural similarities between muscimol and GABA and between ibotenic acid and glutamic acid are indicated.

62

THE CENTRALLY ACTIVE HETEROCYCLIC CONSTITUENTS OF
AMANITA MUSCARIA

The mushroom Amanita muscaria Fr. (Fig. 1), the fly agaric,
has psychotropic effects, which manifest themselves after consumption of the fresh or dried mushrooms (Waser 1979). The use
of the fly agaric as an inebriant among certain Siberian tribes
may represent a degeneration of traditions, which played part
in ancient Indo-European religious ceremonies (Wasson 1979).
The plant Soma, which was deified and consumed by Indo-European
priests, probably is identical with Amanita muscaria Fr. (Wasson 1968).

Two decades ago the isolation and structure determination of
the centrally active constituents of Amanita muscaria Fr. were
reported (Takemoto et al. 1964, Eugster et al. 1965, Bowden and
Drysdale 1965). The compounds were shown to be zwitterionic
3-isoxazolol derivatives, a new type of naturally occurring
compounds. Ibotenic acid, (RS)-α-amino-3-hydroxy-5-isoxazole-
acetic acid (IBO), which is an analogue of GLU, is synthesized
by the mushroom, whereas muscimol (5-aminomethyl-3-isoxazolol),
which is not present in the fresh mushroom, is formed in the
dried plant material by decarboxylation of IBO (Fig. 1), a process which is not catalyzed by enzymes in the mushroom (Eugster
1969). The relative importance of IBO and muscimol for the
central effects in man after consumption of the mushroom is unknown (Waser 1979, Wasson 1979). Muscimol is known to penetrate the blood-brain barrier (BBB) (Moroni et al. 1982), and
it is possible that muscimol, formed by decarboxylation of IBO
in vivo, is primarily responsible for the central effects of
the mushroom. These effects are known to pass into the urine
of consumers of the mushroom (Wasson 1979). Another person may
drink this urine to enjoy the same effect, and the urine of
three or four successive drinkers may thus be consumed without
noticeable loss of inebriating effect (Wasson 1979). Unchanged
IBO is likely to be the active constituent of this urine, and
IBO may thus be acting as a "pro-drug" for muscimol.

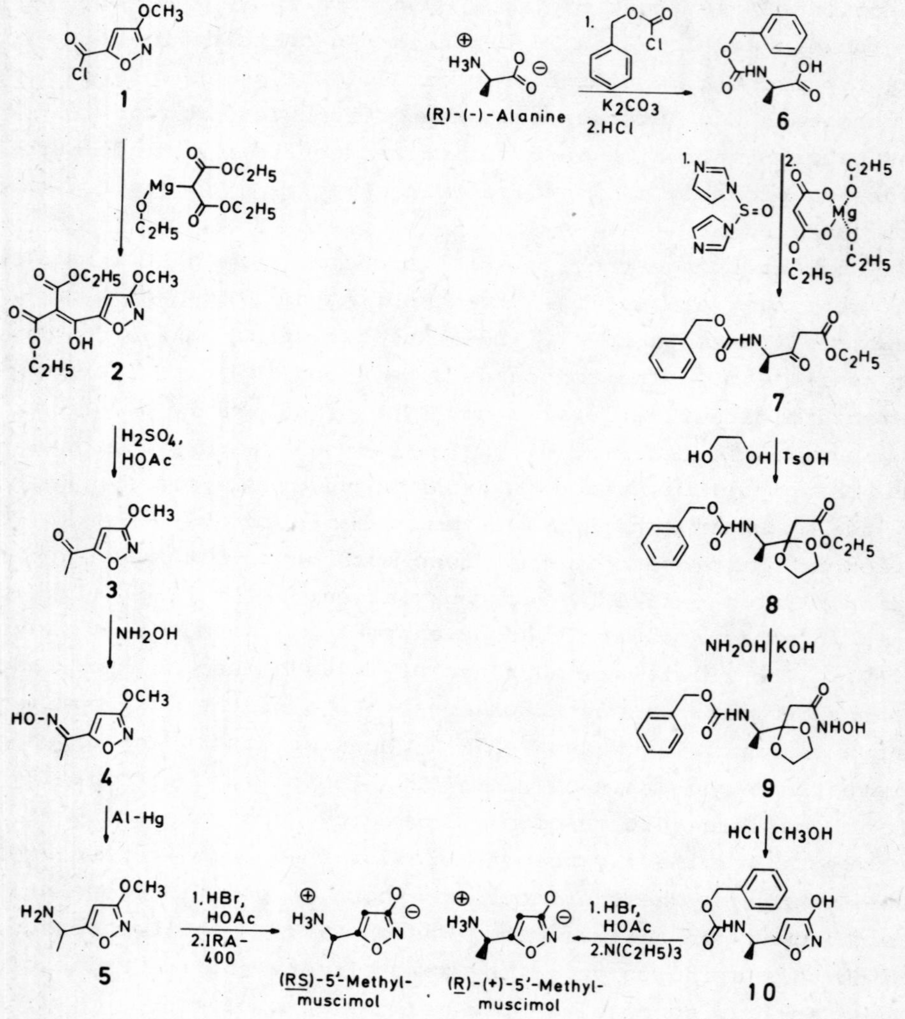

Fig. 2. Scheme for the syntheses of (RS)- and (R)-(+)-5-(1-aminoethyl)-3-isoxazolol (5'-methylmuscimol).

As illustrated in Fig. 1 muscimol is a heterocyclic analogue of
GABA. Muscimol is a powerful agonist at the postsynaptic GABA
receptors sensitive to the antagonist bicuculline methochloride
(BMC) (Curtis et al. 1971, Krogsgaard-Larsen et al. 1979a).
This very effective receptor interaction of muscimol can also
be detected in vitro, muscimol being an inhibitor of the receptor binding of GABA at concentrations in the low nanomolar
range (Fig. 5).

Muscimol is, however, not a specific GABA agonist. It also
interacts with the GABA transport (uptake) systems, though with
much lower affinity than GABA (Krogsgaard-Larsen et al. 1979a,
Schousboe et al. 1979), and muscimol is a substrate for the
GABA-metabolizing enzyme GABA-T (Fowler et al. 1983). Muscimol
actually is transported by the neuronal GABA uptake system
(Johnston et al. 1978), and metabolites formed within the GABA
nerve terminals may contribute to the toxicity of muscimol in
man and animals.

However, the potency of muscimol as a GABA agonist and its
ability to penetrate the BBB have made muscimol an important
lead structure for the design of specific GABA agonists with
favourable pharmacokinetic properties.

Syntheses of muscimol analogues

During the past decade a variety of muscimol analogues have
been synthesized and tested for GABA-ergic activities (Krogsgaard-Larsen 1981,1983, Krogsgaard-Larsen and Falch 1981,
Krogsgaard-Larsen et al. 1981a,1983a,b).

In Fig. 2 the syntheses of (RS)- and (R)-(+)-5-(1-amino-
ethyl)-3-isoxazolol (5'-methylmuscimol) are outlined (Krogsgaard-Larsen and Christensen 1974, Krogsgaard-Larsen et al.
1978). The synthesis of the latter compound illustrates a
widely used synthetic route to 3-isoxazolols based on β-oxo-
esters (see also Fig. 9). The appropriate β-oxoester 7 of
known absolute configuration was prepared from (R)-(-)-alanine.
Compound 7 was converted in two steps into the ketalized hydroxamic acid 9, which was cyclized to give 10 and subsequently
deprotected without detectable racemization to give (R)-(+)-5'-

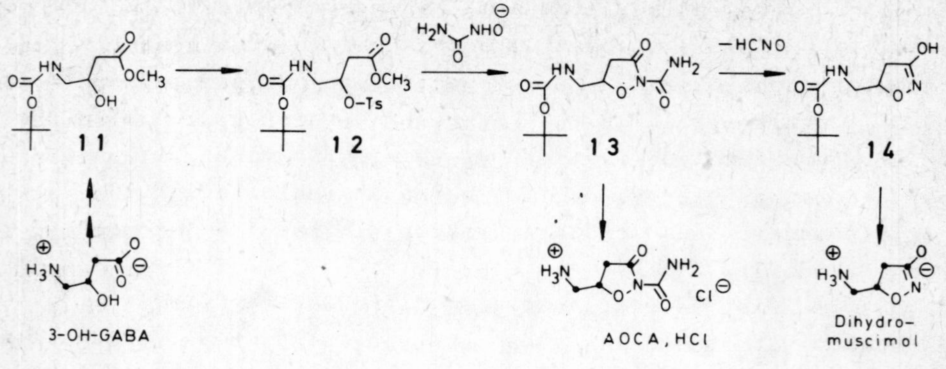

Fig. 3. Scheme for the synthesis of (RS)-dihydromuscimol.

methylmuscimol. The (S)-(-)-isomer of this muscimol analogue was synthesized following an analogous procedure (Krogsgaard-Larsen et al. 1978).

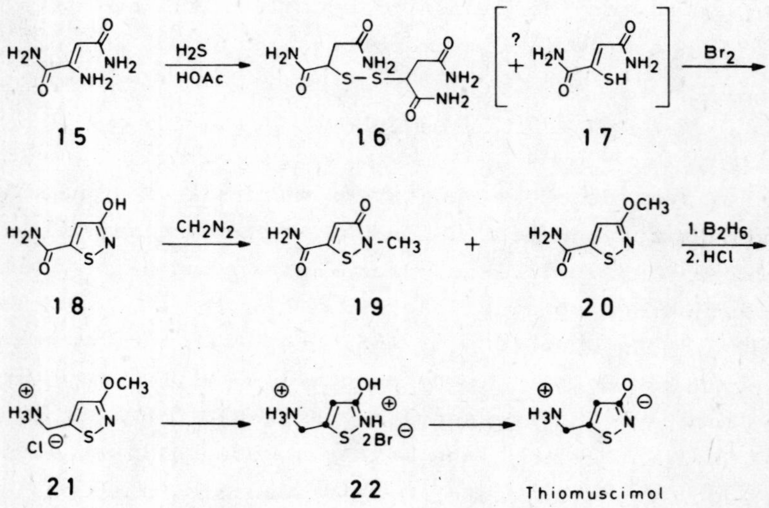

Fig. 4. Scheme for the synthesis of thiomuscimol.

Fig. 3 outlines the synthesis of (RS)-dihydromuscimol (Krogsgaard-Larsen et al. 1978). 3-Hydroxy-4-aminobutyric acid (3-OH-GABA) was stepwise protected to give 12. In the key step of the sequence 12 was converted into 13 via an elimination-addition-cyclization reaction. In accordance with this proposed mechanism, treatment of boc-protected trans-4-amino-crotonic acid methyl ester or the (R)-(-)-form of 12 with hydroxyurea under the same conditions gave (RS)-13 (Brehm et al. 1983). Deprotection of 14, prepared by treatment of 13 with an appropriate amine to trap the cyanate ion, to give (RS)-dihydromuscimol hydrochloride was accomplished under anhydrous acid conditions.

The synthesis of thiomuscimol (Lykkeberg and Krogsgaard-Larsen 1976) is illustrated in Fig. 4. Treatment of aminofumaramide 15 with hydrogen sulphide gave a complex reaction mixture containing among various other products 16 and probably 17. Treatment of this mixture with bromine gave in relatively low yields 18, which was converted into a separable mixture of 19 and 20. Compound 20 was transformed into thiomuscimol in three steps.

Structure-activity relationship for muscimol analogues

The structure-activity studies summarized in Fig. 5 indicate that strict structural constraints are imposed on agonists at BMC-sensitive receptors. Whilst (RS)-dihydromuscimol and thiomuscimol are approximately equipotent with muscimol, isomuscimol is much weaker, and 23 is inactive (Krogsgaard-Larsen et al. 1979a). These remarkable differences in activity reflect that the degrees of delocalization of the negative charges of isomuscimol and 23 are much higher than those of muscimol or dihydromuscimol, the structures of which are similar to that of GABA (Krogsgaard-Larsen et al. 1979a).

Whilst (S)-(-)-5'-methylmuscimol is about thirty times more potent than the (R)-(+)-isomer in inhibiting GABA receptor binding these optical antipodes are approximately equipotent as BMC-sensitive GABA agonists (Krogsgaard-Larsen et al. 1983a). 4-Methylmuscimol is a very weak GABA agonist, and 24 is totally

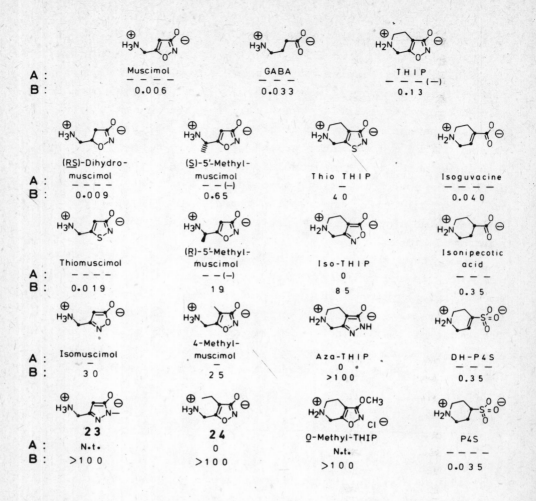

	Muscimol	**GABA**	**THIP**
A:	– – – –	– – –	– – – (–)
B:	0.006	0.033	0.13

	(RS)-Dihydro-muscimol	(S)-5'-Methyl-muscimol	Thio THIP	Isoguvacine
A:	– – – –	– – (–)	–	
B:	0.009	0.65	40	0.040

	Thiomuscimol	(R)-5'-Methyl-muscimol	Iso-THIP	Isonipecotic acid
A:	– – – –	– – (–)	0	– – –
B:	0.019	19	85	0.35

	Isomuscimol	4-Methyl-muscimol	Aza-THIP	DH-P4S
A:	–	–	0	
B:	30	25	>100	0.35

	23	**24**	O-Methyl-THIP	P4S
A:	N.t.	0	N.t.	– – – –
B:	>100	>100	>100	0.035

Fig. 5. Structures and effects on BMC-sensitive GABA receptors of GABA, muscimol, THIP, and a number of related compounds. A: The GABA agonist activity relative to that of GABA (---) as determined microelectrophoretically (Curtis et al. 1971). B: Inhibition of GABA receptor binding in vitro (IC_{50}, μM) using rat brain membranes (Falch and Krogsgaard-Larsen 1982).

inactive (Krogsgaard-Larsen et al. 1975).

The potency and specificity of THIP as a GABA agonist (Krogsgaard-Larsen 1977, Krogsgaard-Larsen et al. 1977,1979a,

1983a) indicate that this rigid compound essentially reflects
the active conformations of muscimol and GABA at the BMC-sensitive GABA receptors. Even minor alterations of the structure
of THIP result in compounds with much lower activity, thio-THIP
being a weak GABA agonist (Krogsgaard-Larsen et al. 1983b) and
iso-THIP, aza-THIP and O-methyl-THIP being virtually devoid of
receptor affinity (Krogsgaard-Larsen and Falch 1981).

Replacement of the 3-isoxazolol anion of THIP by carboxylate
or sulphonate groups did, on the other hand, lead to a series
of potent and specific BMC-sensitive GABA agonists (Krogsgaard-
Larsen et al. 1977,1979a,1983a). Introductions of a double
bond into the molecules of isonipecotic acid and piperidine-4-
sulphonic acid (P4S) have opposite effects (Fig. 5). Isoguvacine is an order of magnitude more potent than isonipecotic
acid as an inhibitor of GABA receptor binding, whereas the unsaturated analogue of P4S, 1,2,3,6-tetrahydropyridine-4-sulpho-
nic acid (DH-P4S), is proportionally weaker than P4S. Electro-
physiological studies have disclosed the same relative order of
potency of these compounds as BMC-sensitive GABA agonists.

Pharmacological and clinical studies on THIP

Like muscimol, THIP penetrates the BBB very easily in animals
(Krogsgaard-Larsen 1981, Moroni et al. 1982) and in man
(Schultz et al. 1981, Christensen et al. 1982). THIP is well
tolerated by various animal species (Christensen et al. 1982),
it is active in man after oral administration, and it is ex-

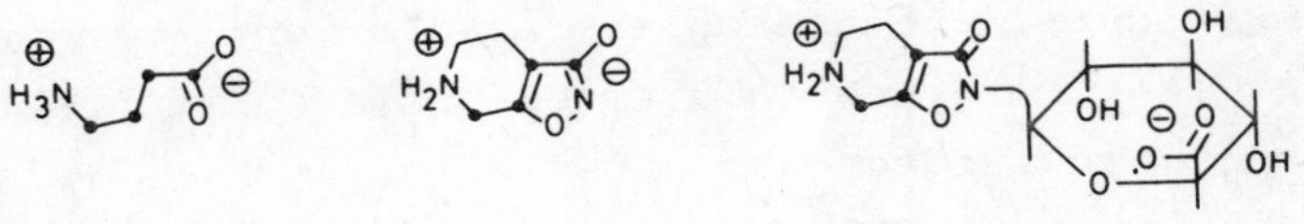

GABA   THIP   THIP-N-Glucuronide

Fig. 6. The structures of GABA, THIP, and the proposed metabolite of THIP, THIP-N-glucuronide.

6 Plas

creted unchanged and, to some extent, in a conjugated form in the urine of animals and man (Schultz et al. 1981). The conjugate of THIP has not been isolated in a pure form yet, but it has chromatographic properties identical with those of synthetic THIP-N-glucuronide (Fig. 6).

THIP has been the subject of extensive pharmacological studies in animals, which have disclosed anticonvulsant, non-opioid analgesic, and anorexic effects, and THIP is active in animal models used for studies of ethanol withdrawal symptoms and spasticity (for reviews see Christensen et al. 1982, Krogsgaard-Larsen 1983).

THIP is a clinically active non-opioid analgesic (Kjaer and Nielsen 1983) and an antispastic agent (Mondrup and Pedersen 1983), and THIP has relatively weak anxiolytic properties in man (Hoehn-Saric 1983). THIP is, however, only a very weak antiepileptic agent (Petersen et al. 1983) and it does not affect the symptoms in choreic patients significantly (Foster et al. 1983), probably reflecting as yet unclarified interactions between the central GABA, GLU, and dopamine systems.

The antispastic effects of THIP and its anxiolytic and potent non-opioid analgesic effects would seem to have therapeutic interest. In any case, the clinical studies on THIP have shown that GABA agonists with appropriate pharmacokinetic properties may have interest in certain clinical situations.

IBOTENIC ACID AS A LEAD FOR THE DEVELOPMENT OF SPECIFIC GLUTAMIC ACID AGONISTS

Based on the relative sensitivity of various naturally occurring and synthetic GLU analogues to different antagonists the central excitatory amino acid receptors are, at present, most conveniently subdivided into three classes (Fig. 7) (Watkins 1981, Davies et al. 1982, Krogsgaard-Larsen and Honoré 1983):

(1) Quisqualic acid (QUIS) receptors, at which (S)-glutamic acid diethyl ester (GDEE) acts as a relatively selective antagonist,

(2) N-methyl-D-aspartic acid (NMDA) receptors, which can be blocked by a number of compounds including 2-amino-5-phos-

phonovaleric acid (2APV) and α-aminoadipic acid (α-AA), and
(3) kainic acid (KA) receptors, which can be studied <u>in vitro</u>
using radioactive KA as a ligand (Simon et al. 1976).

Whilst the physiological relevance of the NMDA and, in particular, the KA receptors is unclear, the QUIS receptors are
assumed to represent the physiological postsynaptic GLU receptors, and, consequently, there is an interest in specific agonists and antagonists at these GDEE-sensitive receptors. QUIS
itself is, however, far from being an ideal tool for studies of
these receptors, primarily because QUIS also binds tightly to
the KA receptors (Simon et al. 1976).

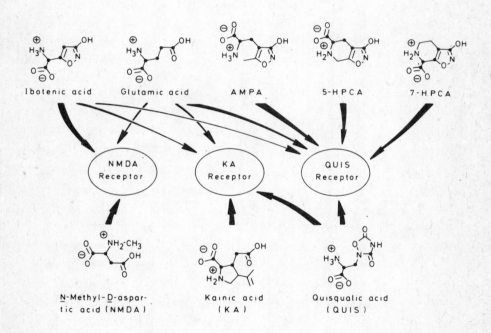

<u>Fig. 7.</u> An illustration of the interaction of glutamic acid,
ibotenic acid, and a variety of naturally occurring and synthetic analogues with different types of excitatory amino acid receptors.

	GLU	Conformation	
		I Ibotenic acid	II
Agonist activity	+ +	+ + +	
Antagonism GDEE:	Yes	Weak	
2APV or a-AA:	Weak	Yes	

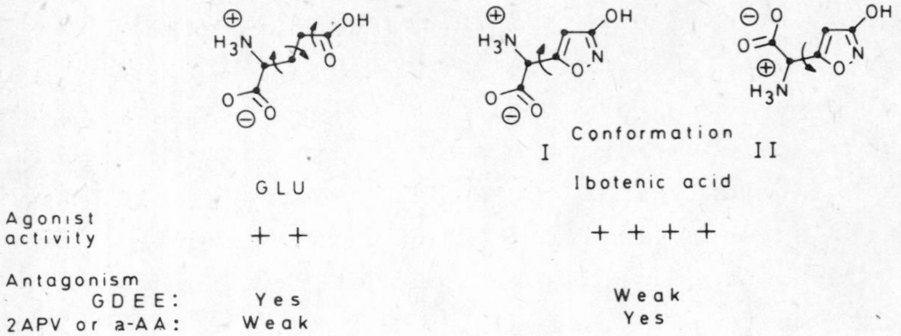

Fig. 8. An illustration of the different degrees of conformational mobility of glutamic acid and ibotenic acid. The relative potencies of glutamic and ibotenic acids as neuronal excitants and the sensitivity of these effects to different antagonists were determined microelectrophoretically (Curtis et al. 1979).

Ibotenic acid (IBO) interacts with all types of excitatory receptors (Figs. 7 and 8) (Curtis et al. 1979, Krogsgaard-Larsen et al. 1981b). CNDO/2 molecular orbital calculations have shown that a conformation of the fully charged molecule of IBO similar to conformation II (Fig. 8) is energetically favourable in the conservative state (Borthwick and Steward 1976). Allowance for an aqueous medium does, however, predict a change in the conservative-molecule minimum-energy conformation from II into I (Fig. 8) (Borthwick and Steward 1976). These conformations may represent active conformations of IBO at different excitatory receptors, and in an attempt to distinguish these actions of IBO and with the aim of optimizing its effect on the QUIS receptors a number of analogues of IBO have been synthesized and tested.

In the bicyclic IBO analogue (RS)-3-hydroxy-4,5,6,7-tetrahydroisoxazolo[5,4-c]pyridine-7-carboxylic acid (7-HPCA) (Figs. 7 and 9) the structural equivalent to IBO has been locked in a conformation similar to conformation I (Fig. 8). In Fig. 9 the

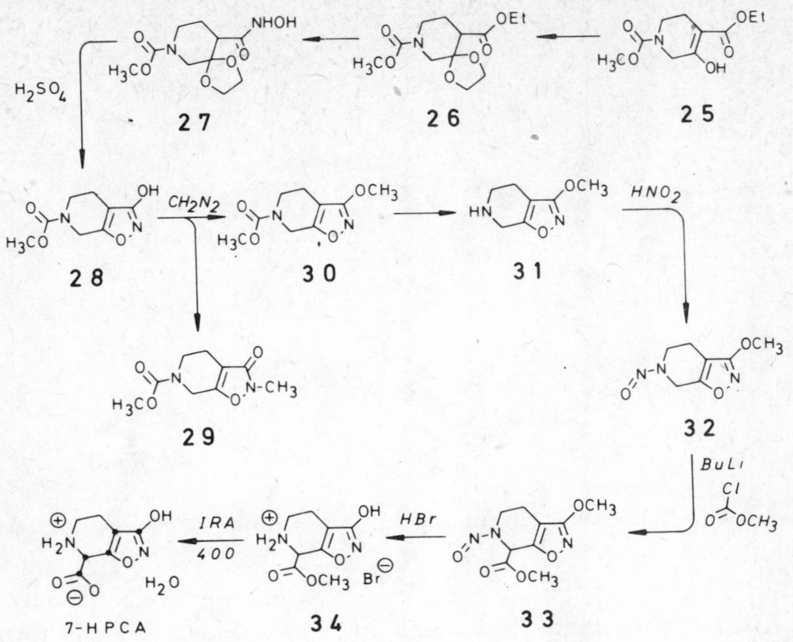

H2SO4

27 NHOH

26 OEt

25 OEt OH

28 OH CH2N2

30 OCH3

31 OCH3 HNO2

29 N-CH3

32 OCH3

BuLi Cl O OCH3

33 OCH3

34 OH HBr

7-HPCA IRA 400 H2O

Fig. 9. Scheme for the synthesis of 7-HPCA, a conformationally immobilized analogue of IBO.

last steps of the reaction sequence for the preparation of 7-HPCA are outlined (Krogsgaard-Larsen et al. 1984b). Treatment of 31, obtained by alkaline hydrolysis of 30, with nitrous acid transformed it into the N-nitroso derivative 32. As indicated, a methoxycarbonyl group was introduced regiospecifically into position 7 of 32 to give 33, which was stepwise deprotected to give 7-HPCA hydrate. In order to avoid decarboxylation of 7-HPCA, the last reaction step was an ester hydrolysis of 34 on a strongly basic ion exchange resin without significant loss of the carboxylate group.

Fig. 10 illustrates the use of 3-methoxy-4,5-dimethylisoxazole (35) as a starting material for the syntheses of the IBO analogues (RS)-3-hydroxy-4,5,6,7-tetrahydroisoxazolo[5,4-c]-pyridine-5-carboxylic acid (5-HPCA), (RS)-α-amino-3-hydroxy-5-

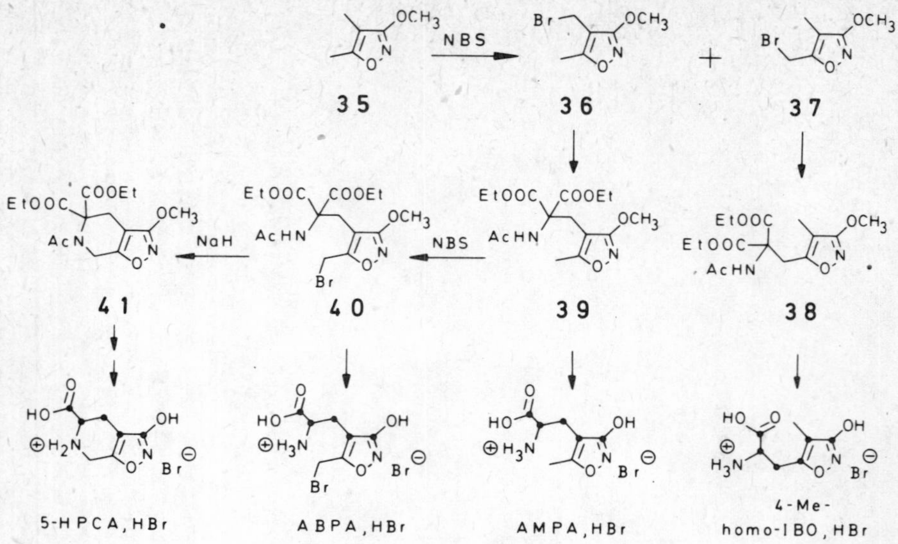

Fig. 10. Scheme for the synthesis of some analogues of IBO.

bromomethyl-4-isoxazolepropionic acid (ABPA), (RS)-α-amino-3-
hydroxy-5-methyl-4-isoxazolepropionic acid (AMPA), and (RS)-α-
amino-3-hydroxy-4-methyl-5-isoxazolepropionic acid (4-Me-homo-
IBO) (Hansen and Krogsgaard-Larsen 1980, Honoré and Lauridsen
1980, Krogsgaard-Larsen et al. 1984a,b). A crude mixture of
36 and 37, obtained by NBS bromination of 35, was converted
into a separable mixture of 39 and 38, respectively. Cyclization of the brominated compound 40, prepared from 39, under
appropriate conditions gave 41. The intermediates 38-41 were
deprotected under acid conditions to give the hydrobromides of
4-Me-homo-IBO, AMPA, ABPA, and 5-HPCA, respectively.

Structure-activity relationship for ibotenic acid analogues

A comparison of the activity of GLU and a number of IBO analogues as agonists at the GDEE-sensitive QUIS receptors and as
inhibitors of the binding of radioactive GLU, KA, and AMPA is
illustrated in Fig. 11. AMPA obviously is the most potent ex-

COMPOUND	AGONIST ACTIVITY ON CAT SPINAL NEURONES GDEE SENSITIVE (Rel. Potency)	INHIBITION OF THE BINDING OF RADIOACTIVE		
		GLU	KAINIC ACID	AMPA
			(IC_{50}, μM)	
GLU	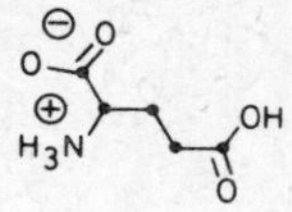 + +	0.5	0.4	1
(S)- AMPA	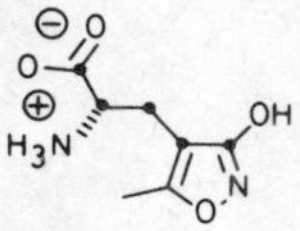 + + + + + +	>100	>100	0.6
(R)- AMPA	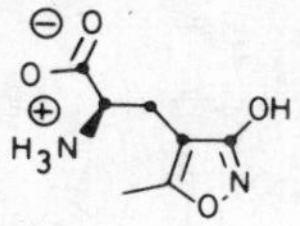 + + +	>100	>100	5
Homo- IBO	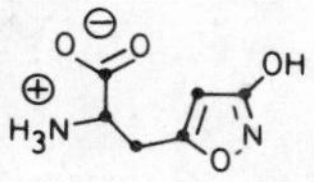 + (+)	7.5	>100	16
4-Me- homo- IBO	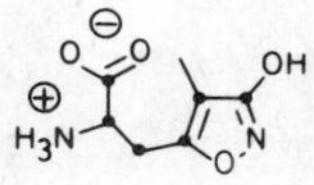 + + +	45	>100	2

Fig. 11. Structure, biological, and in vitro activity of GLU and a number of analogues of IBO. The agonist activities at GDEE-sensitive QUIS receptors on cat spinal neurones were mea-sured microelectrophoretically (Krogsgaard-Larsen et al. 1980, 1982,1984b). The binding studies using radioactive GLU (Honoré et al. 1981), kainic acid (Simon et al. 1976, Krogsgaard-Larsen et al. 1980), and AMPA (Honoré et al. 1982) were performed as described elsewhere.

citant, and neither this compound nor any of the other ana-logues interact significantly with the NMDA receptors in vivo or with the KA receptor sites in vitro (Krogsgaard-Larsen et al. 1980,1982,1984b, Honoré et al. 1981, Hansen et al. 1983).

There is no simple correlation between the in vivo effects of the compounds and their ability to inhibit the binding of radioactive GLU and KA. Whilst this is in agreement with the

KA receptor sites being distinctly different from the receptors associated with the neuronal excitation by GLU (Watkins 1981, Krogsgaard-Larsen and Honoré 1983), there is no obvious explanation of the conspicuous discrepancy between the potencies of the compounds in vivo and as inhibitors of GLU binding. It is likely that the excitatory actions of the compounds are mediated by a subpopulation of GLU receptors of relatively low density (Krogsgaard-Larsen and Honoré 1983). The density of the ^{3}H-AMPA binding sites is only ten per cent of that of the ^{3}H-GLU sites (Honoré et al. 1982). This observation and the correlation between the potencies of the compounds as GDEE-sensitive neuronal excitants and as inhibitors of ^{3}H-AMPA binding (Fig. 11) is consistent with the view that the ^{3}H-AMPA binding sites represent the physiological postsynaptic GLU receptors (Krogsgaard-Larsen and Honoré 1983).

Structural requirements of the QUIS receptors assumed to represent the physiological postsynaptic glutamic acid receptors

As mentioned above there is some evidence supporting the view that the QUIS receptors (Fig. 7) primarily represent the physiologically relevant postsynaptic GLU receptors (Watkins 1981, Davies et al. 1982, Krogsgaard-Larsen and Honoré 1983). Whilst IBO is a selective agonist at the NMDA receptors (Curtis et al. 1979), all of the IBO analogues illustrated in Figs. 10-12 interact selectively with the QUIS receptors (Krogsgaard-Larsen et al. 1980,1982,1984a,b, Hansen et al. 1983). Consequently, structure-activity studies on these IBO analogues with more or less rigid structures have shed light on the structural requirements for activation of the physiological GLU receptors.

It is reasonable to assume that 7-HPCA and, thus, conformation I of IBO (Figs. 8 and 12) reflects the conformation of IBO, and GLU, during their interaction with the physiological GLU receptors. Although the structures of AMPA, 5-HPCA, and 7-HPCA are different (Fig. 7), a comparison of the fully charged molecules of these compounds, which occur at physiological pH, reduces this difference considerably (Fig. 12). Ac-

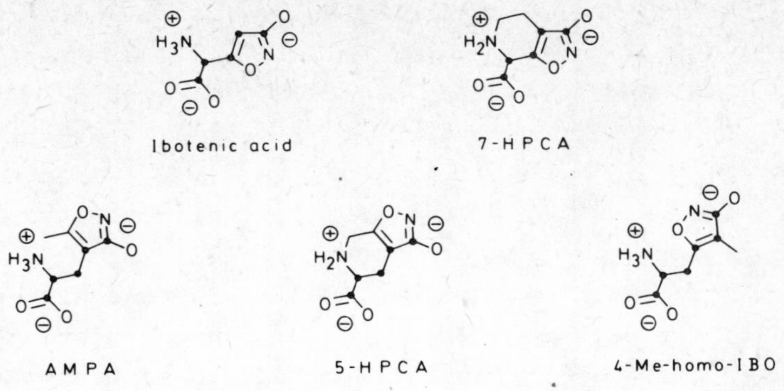

Fig. 12. Structures of the fully charged molecules of IBO and of a number of IBO analogues.

cordingly, 7-HPCA (Krogsgaard-Larsen et al. 1984b), AMPA, and 5-HPCA (Krogsgaard-Larsen et al. 1980,1982,1984a) are very similar in potency and selectivity as GDEE-sensitive neuronal excitants. In the light of the somewhat different structure of 4-Me-homo-IBO (Fig. 12) it is understandable that this compound is weaker than AMPA (Fig. 11), 7-HPCA, and 5-HPCA as a GDEE-sensitive GLU agonist.

Based on the described information about the structural requirements for activation of the GDEE-sensitive GLU receptors, the design and development of antagonists at these receptors are in progress.

ACKNOWLEDGEMENTS

This work was supported by The Danish Medical Research Council. The collaboration with professor D.R. Curtis, Canberra, Australia and with Drs. T. Honoré, E.Ø. Nielsen, and J. Lauridsen and the secretarial assistance of Mrs. B. Hare are gratefully acknowledged.

REFERENCES

Borthwick, P.W. and Steward, E.G. (1976) Ibotenic acid: further observations on its conformational modes. J. Mol. Struct. 33, 141-144.

Bowden, K. and Drysdale, A.C. (1965) A novel constituent of Amanita muscaria. Tetrahedron Lett. 727-728.

Brehm, L., Jacobsen, P., Johansen, J.S. and. Krogsgaard-Larsen, P. (1983) Heterocyclic GABA agonists. Synthesis and crystal structure of (RS)-5-(N-t-butyloxycarbonylaminomethyl)-3-oxo-isoxazolidine-2-carboxamide, a derivative of dihydromuscimol. J. Chem. Soc. Perkin Trans. I. 1459-1464.

Christensen, A.V., Svendsen, O. and Krogsgaard-Larsen, P. (1982) Pharmacodynamic effects and possible therapeutic uses of THIP, a specific GABA agonist. Pharm. Weekbl. Sci. Ed. 4, 145-153.

Coyle, J.T., Bird, S.J., Evans, R.H., Gulley, R.L., Nadler, J.V., Nicklas, W.J. and Olney, J.W. (1981) Excitatory amino acid neurotoxins: selectivity, specificity, and mechanisms of action. Neurosci. Res. Prog. Bull. 19, 333-427.

Curtis, D.R., Duggan, A.W., Felix, D. and Johnston, G.A.R. (1971) Bicuculline, an antagonist of GABA and synaptic inhibition in the spinal cord of the cat. Brain Res. 32, 69-96.

Curtis, D.R., Lodge, D. and McLennan, H. (1979) The excitation and depression of spinal neurones by ibotenic acid. J. Physiol. (London) 291, 19-28.

Davies, J., Evans, R.H., Jones, A.W., Smith, D.A.S. and Watkins, J.C. (1982) Differential activation and blockade of excitatory amino acid receptors in the mammalian and amphibian central nervous systems. Comp. Biochem. Physiol. 72C, 211-224.

DeFeudis, F.V. (1981) GABA and "neuro-cardiovascular" mechanisms. Neurochem. Int. 3, 113-122.

Di Chiara, G. and Gessa, G.L. (1981) Glutamate as a Neurotransmitter, Raven Press, New York.

Eugster, C.H. (1969) Chemie der Wirkstoffe aus dem Fliegenpilz. Fortschr. Chem. Org. Naturst. 27, 261-321.

Eugster, C.H., Müller, G.F.R. and Good, R. (1965) Wirkstoffe aus Amanita muscaria: Ibotensaeure und Muscazon. Tetrahedron Lett. 1813-1815.

Falch, E. and Krogsgaard-Larsen, P. (1982) The binding of the GABA agonist [^{3}H]THIP to rat brain synaptic membranes. J. Neurochem. 38, 1123-1129.

Fowler, L.J., Lovell, D.H. and John, R.A. (1983) The reaction of muscimol with 4-aminobutyrate aminotransferase. J. Neurochem. 41, 1751-1754.

Foster, N.L., Chase, T.N., Denaro, A., Hare, T.A. and Tamminga, C.A. (1983) THIP treatment of Huntington's disease. Neurology 33, 637-639.

Grognet, A., Hertz, F. and DeFeudis, F.V. (1983) Comparison of the analgesic actions of THIP and morphine. Gen. Pharmacol. 14, 585-589.

Hansen, J.J. and Krogsgaard-Larsen, P. (1980) Isoxazole amino acids as glutamic acid agonists. Synthesis of some analogues and homologues of ibotenic acid. J. Chem. Soc. Perkin Trans. I. 1826-1833.

Hansen, J.J., Lauridsen, J., Nielsen, E. and Krogsgaard-Larsen, P. (1983) Enzymatic resolution and binding to rat brain membranes of the glutamic acid agonist, α-amino-3-hydroxy-5-methyl-4-isoxazolepropionic acid. J. Med. Chem. 26, 901-903.

Hill, R.C., Maurer, R., Buescher, H.H. and Roemer, D. (1981) Analgesic properties of the GABA-mimetic THIP. Eur. J. Pharmacol. 69, 221-224.

Hoehn-Saric, R. (1983) Effects of THIP on chronic anxiety. Psychopharmacology 80, 338-341.

Honoré, T. and Lauridsen, J. (1980) Structural analogues of ibotenic acid. Syntheses of 4-methylhomoibotenic acid and AMPA, including the crystal structure of AMPA, monohydrate. Acta Chem. Scand. B34, 235-240.

Honoré, T., Lauridsen, J. and Krogsgaard-Larseń, P. (1981) Ibotenic acid analogues as inhibitors of [^{3}H]glutamic acid binding to cerebellar membranes. J. Neurochem. 36, 1302-1304.

Honoré, T., Lauridsen, J. and Krogsgaard-Larsen, P. (1982). The binding of [^{3}H]AMPA, a structural analogue of glutamic acid, to rat brain membranes. J. Neurochem. 38, 173-178.

Johnston, G.A.R., Kennedy, S.M.E. and Lodge, D. (1978) Muscimol uptake, release and binding in rat brain slices. J. Neurochem. 31, 1519-1523.

Kendall, D.A., Browner, M. and Enna, S.J. (1982) Comparison of the antinociceptive effect of GABA agonists: evidence for a cholinergic involvement. J. Pharmacol. Exp. Ther. 220, 482-487.

Kjaer, M. and Nielsen, H. (1983) The analgesic effect of the GABA-agonist THIP in patients with chronic pain of malignant origin. A phase-1-2 study. Br. J. Clin. Pharmacol. 16, 477-485.

Krogsgaard-Larsen, P. (1977) Muscimol analogues. II. Synthesis of some bicyclic 3-isoxazolol zwitterions. Acta Chem. Scand. B31, 584-588.

Krogsgaard-Larsen, P. (1981) 4-Aminobutyric acid agonists, antagonists, and uptake inhibitors. Design and therapeutic aspects. J. Med. Chem. 24, 1377-1383.

Krogsgaard-Larsen, P. (1983) GABA agonists: Structural, pharmacological and clinical aspects. In: Metabolic Relationship between Glutamine, Glutamate and GABA in the CNS, eds. Hertz, L., Kvamme, E., McGeer, E. and Schousboe, A., pp. 537-557. Alan R. Liss, Inc., New York.

Krogsgaard-Larsen, P. and Christensen, S.B. (1974) Organic hydroxylamine derivatives. XII. Structural analogues of 4-aminobutyric acid (GABA) of the isoxazole enol-betaine type. Synthesis of some 3-hydroxy-5-(1-aminoalkyl)isoxazole zwitterions. Acta Chem. Scand. B28, 636-640.

Krogsgaard-Larsen P., Falch, E. (1981) GABA agonists. Development and interactions with the GABA receptor complex. Mol. Cell. Biochem. 38, 129-146.

Krogsgaard-Larsen, P. and Honoré, T. (1983) Glutamate receptors and new glutamate agonists. Trends Pharmacol. Sci. 4, 31-33.

Krogsgaard-Larsen, P., Johnston, G.A.R., Curtis, D.R., Game, C.J.A. and McCulloch, R.M. (1975) Structure and biological activity of a series of conformationally restricted analogues of GABA. J. Neurochem. 25, 803-809.

Krogsgaard-Larsen, P., Johnston, G.A.R., Lodge, D. and Curtis, D.R. (1977) A new class of GABA agonist. Nature 268, 53-55.

Krogsgaard-Larsen, P., Larsen, A.L.N. and Thyssen, K. (1978)
GABA receptor agonists. Synthesis of muscimol analogues including (R)- and (S)-5-(1-aminoethyl)-3-isoxazolol and (RS)-
5-aminomethyl-2-isoxazolin-3-ol. Acta Chem. Scand. B32,
469-477.

Krogsgaard-Larsen, P., Hjeds, H., Curtis, D.R., Lodge, D. and
Johnston, G.A.R. (1979a) Dihydromuscimol, thiomuscimol, and
related heterocyclic compounds as GABA analogues.
J. Neurochem. 32, 1717-1724.

Krogsgaard-Larsen, P., Scheel-Krüger, J. and Kofod, H. eds.
(1979b) GABA-Neurotransmitters. Pharmacochemical, Biochemical
and Pharmacological Aspects, Munksgaard, Copenhagen.

Krogsgaard-Larsen, P., Honoré, T., Hansen, J.J., Curtis, D.R.
and Lodge, D. (1980) New class of glutamate agonist structurally related to ibotenic acid. Nature 284, 64-66.

Krogsgaard-Larsen, P., Brehm, L. and Schaumburg, K. (1981a)
Muscimol, a psychoactive constituent of Amanita muscaria, as
a medicinal chemical model structure. Acta Chem. Scand. B35,
311-324.

Krogsgaard-Larsen, P., Honoré, T., Hansen, J.J., Curtis, D.R.
and Lodge, D. (1981b) Structure-activity studies on ibotenic
acid and related muscimol analogues. In: Glutamate as a Neurotransmitter, eds. Di Chiara, G. and Gessa, G.L., pp.
285-294. Raven Press, New York.

Krogsgaard-Larsen, P., Hansen, J.J., Lauridsen, J., Peet, M.J.,
Leah, J.D. and Curtis, D.R. (1982) Glutamic acid agonists.
Stereochemical and conformational studies of DL-α-amino-3-
hydroxy-5-methyl-4-isoxazolepropionic acid (AMPA) and related
compounds. Neurosci. Lett. 31, 313-317.

Krogsgaard-Larsen, P., Falch, E., Peet, M.J., Leah, J.D. and
Curtis, D.R. (1983a) Molecular pharmacology of the GABA receptors and GABA agonists. In: CNS Receptors - From Molecular
Pharmacology to Behaviour, eds. Mandel, P. and DeFeudis, F.V.
pp. 1-13. Raven Press, New York.

Krogsgaard-Larsen, P., Mikkelsen, H., Jacobsen, P., Falch, E., Curtis, D.R., Peet, M.J. and Leah, J.D. (1983b) 4,5,6,7-Tetrahydroisothiazolo[5,4-c]pyridin-3-ol and related analogues of THIP. Synthesis and biological activity. J. Med. Chem. 26, 895-900.

Krogsgaard-Larsen, P., Johansen, J.S., Lauridsen, J., Brehm, L. and Curtis, D.R. (1984a) Synthesis and biological and in vitro activity of some ibotenic acid analogues. J. Med. Chem. (in press).

Krogsgaard-Larsen, P., Nielsen, E.Ø. and Curtis, D.R. (1984b) Ibotenic acid analogues. Synthesis and biological and in vitro activity of conformationally restricted agonists at central excitatory amino acid receptors. J. Med. Chem. 27 (in press).

Lykkeberg, J. and Krogsgaard-Larsen, P. (1976) Structural analogues of GABA. Synthesis of 5-aminomethyl-3-isothiazolol (thiomuscimol). Acta Chem. Scand. B30, 781-785.

Mandel, P. and DeFeudis, F.V. eds. (1983) CNS Receptors - From Molecular Pharmacology to Behavior, Raven Press, New York.

Meier, E., Drejer, J. and Schousboe, A. (1983) Inhibition by GABA of evoked glutamate release from cultured cerebellar granule cells. In: Glutamine, Glutamate, and GABA in the Central Nervous System, eds. Hertz, L., Kvamme, E., McGeer, E.G. and Schousboe, A., pp. 509-516. Alan R. Liss, Inc., New York.

Meldrum, B. (1982) Pharmacology of GABA. Clin. Neuropharmacol. 5, 293-316.

Mondrup, K. and Pedersen, E. (1983) The acute effect of the GABA-agonist, THIP, on proprioceptive and flexor reflexes in spastic patients. Acta Neurol. Scand. 67, 48-54.

Moroni, F., Forchetti, M.C., Krogsgaard-Larsen, P. and Guidotti, A. (1982) Relative disposition of the GABA agonists THIP and muscimol in the brain of the rat. J. Pharm. Pharmacol. 34, 676-678.

Morselli, P.L., Löscher, W., Lloyd, K.G., Meldrum, B. and Reynolds, E.H. eds. (1981) Neurotransmitters, Seizures, and Epilepsy, Raven Press, New York.

Petersen, H.R., Jensen, I. and Dam, M. (1983) THIP: a single-blind control trial in patients with epilepsy. Acta Neurol. Scand. 67, 114-117.

Roberts, E., Chase, T.N. and Tower, D.B., eds. (1976) GABA in Nervous System Function, Raven Press, New York.

Schousboe, A., Thorbek, P., Hertz, L. and Krogsgaard-Larsen, P. (1979) Effects of GABA analogues of restricted conformation on GABA transport in astrocytes and brain cortex slices and on GABA receptor binding. J. Neurochem. 33, 181-189.

Schultz, B., Aaes-Jørgensen, T., Bøgesø, K.P. and Jørgensen, A. (1981) Preliminary studies on the absorption, distribution, metabolism, and excretion of THIP in animal and man using ^{14}C-labelled compound. Acta Pharmacol. Toxicol. 49, 116-124.

Simon, J.R., Contrera, J.F. and Kuhar, M.J. (1976) Binding of [^{3}H]kainic acid, an analogue of L-glutamate, to brain membranes. J. Neurochem. 26, 141-147.

Takemoto, T., Nakajima, T. and Sakuma, R. (1964) Isolation of a flycidal constituent "ibotenic acid" from Amanita muscaria and A. pantherina. J. Pharm. Soc. Jpn. 84, 1233-1234.

Waser, R.G. (1979) The pharmacology of Amanita muscaria. In: Ethnopharmacologic Search for Psychoactive Drugs, eds. Efron, D.H., Holmstedt, B. and Kline, N.S., pp. 419-439. Raven Press, New York.

Wasson, R.G. (1968) Soma. Divine Mushroom of Immortality, Harcourt Brace Jovanovich, New York.

Wasson, R.G. (1979) Fly agaric and man. In: Ethnopharmacologic Search for Psychoactive Drugs, eds. Efron, D.H., Holmstedt, B. and Kline, N.S., pp. 405-414. Raven Press, New York.

Watkins, J.C. (1981) Pharmacology of excitatory amino acid receptors. In: Glutamate: Transmitter in the Central Nervous System, eds. Roberts, P.J., Storm-Mathisen, J. and Johnston, G.A.R., pp. 1-24. John Wiley & Sons, Chichester & New York.

DISCUSSION

van der PLAS H.C. (Wageningen, The Netherlands)

In all of your slides you have depicted the 3-oxygenated
isoxazoles as 3-hydroxyisoxazoles. Does the 4-isoxazoline-
3-one tautomeric form not exist?

KROGSGAARD-LARSEN P. (Copenhagen, Denmark)

The tautomerism of this class of compound has been quite
extensively studied. In all cases the 3-hydroxyisoxazole
form has been proved to be totally dominating. In the
charged form of this system, however, the negative charge
is more or less evenly distributed on the oxygen and
nitrogen atoms.

MAKSAY G. (Budapest, Hungary)

Is there any correlation between the pK_A values of GABA
analogues and their potency as GABA agonists?

KROGSGAARD-LARSEN P. (Copenhagen, Denmark)

The pK_A values of GABA agonists can actually be varied   .
within broad ranges. Thus, very potent GABA agonist activity can be detected for GABA analogues with pK_A-I values
between 1 and 6 and with pK_A-II values between 8 and 11.
We have not tried to establish a detailed correlation
along these lines, but we do know that if you block either
of the protolytic groups of a GABA agonist, you will loose
activity almost completely. At least, this is the observation in our laboratories from a number of studies.

SIMONYI M. (Budapest, Hungary)

You mentioned muscimol and isomuscimol representing
different degrees of delocalization of the negative charge
and these compounds possess different potencies as GABA
agonists. Is there some relation between the degree of
delocalization of the charges of GABA agonists and their
activity?

KROGSGAARD-LARSEN P. (Copenhagen, Denmark)
Yes. An increasing degree of delocalization of the
negative as well as the positive charges of GABA analogues
is tantamount to a decreased GABA agonist activity. We
have studied in particular the effects of delocalizing the
negative charge. The fact that isomuscimol and iso-THIP
are very much weaker than muscimol and THIP as GABA
agonists reflects that the negative charges of the iso-
compounds are highly delocalized. Whilst isomuscimol is a
very weak GABA agonist, iso-THIP actually has a weak
antagonist profile.

MAKSAY G. (Budapest, Hungary)
Could you give an explanation for the toxicity of
muscimol? How can THIP, an accepted GABA agonist
antagonize the myoclonic jerks elicited by muscimol?

KROGSGAARD-LARSEN P. (Copenhagen, Denmark)
Muscimol is a substrate for the neuronal GABA uptake
system, and it is also a substrate for the intracellularly
located enzyme GABA transaminase. It is likely that
metabolites of muscimol formed within presynaptic terminals are mainly responsible for the toxicity of muscimol.
The antagonism by THIP is not yet clear.

HOLCZINGER L. (Budapest, Hungary)
I have three questions.
a, How specific is THIP as a GABA agonist?
b, Have you made dose - response curves to characterize the
activity of your GABA agonists?
c, How far is the GABA system involved in schizophrenia?

KROGSGAARD-LARSEN P. (Copenhagen, Denmark)
a, THIP interacts with the GABA system and does not interact
with any other central neurotransmitter system. Within
the GABA system, THIP acts exclusively at the GABA
receptors; it has no affinity for any other GABA synaptic
mechanism.

b, You can make dose - response curves on the basis of binding studies but not on the basis of _in_ _vivo_ micro electro-
   phoretic experiments.
c, GABA malfunctions may play a role in schizophrenia;
   reduced GABA levels have been demonstrated in certain
   areas of the schizophrenic brain.

ÖTVÖS L. (Budapest, Hungary)
   I would be interested in GABA analogues with two substituents at the same carbon atom. Have such compounds been
   synthesized?

KROGSGAARD-LARSEN P. (Copenhagen, Denmark)
   I think that one such compound has been prepared and shown
   to be inactive, but I am not completely sure.

PREOBRAZHENSKAYA M.N. (Moscow, USSR)
   Do tumour-active 3-chloro-2-isoxazoline-substituted
   glycine analogues interact with the central glutamic acid
   system?

KROGSGAARD-LARSEN P. (Copenhagen, Denmark)
   Yes; at least one such compound having the glycine residue
   in the 5-position of the ring has been shown to activate
   the glutamic acid receptors.

TELEGDI J. (Budapest, Hungary)
   Which type of amino acid acylase enzyme did you use for
   the resolution of the specific glutamic acid agonist
   AMPA ?

KROGSGAARD-LARSEN P. (Copenhagen, Denmark)
   The enzyme we applied was an immobilized form of amino
   acid acylase. It deacetylated N-acetylmethionine with
   certain degree of selectivity.

SEARCH FOR ANTITUMOUR AND ANTIVIRAL COMPOUNDS AMONG ANALOGUES OF NUCLEOSIDES

PREOBRAZHENSKAYA M.N.

All-Union Cancer Research Center of the USSR Academy
of Medical Sciences, Moscow 115478, USSR

The present report deals with a series of investigations carried out in the laboratory: a) synthesis and elucidation of structure-activity relationship among 5-substituted α- and β-2'-deoxyuridines; b) synthesis and investigation of acyclic C-nucleoside analogues; c) investigation of pyrazolo(3,4-d)pyrimidines and their nucleosides.

In recent years 5-substituted 2'-deoxyuridines (5-XdUrd) have exhibited a wide variety of biological activities which is largely due to their capability to inhibit enzymes of thymidine (dThd) metabolism. These enzymes are important targets for chemotherapeutic agents. A series of 5-XdUrd were found active against DNA viruses (herpetic and others). Antiherpetic compounds can be also considered as anticancerogens, because herpes viruses have been associated with certain types of human malignancies. An analogue of dThd exhibits its antimetabolite properties only after transformation into 5'-phosphate (5-XdUMP) catalyzed by dThd kinase. The level of dThd kinase in herpes virus infected cells is higher and the enzyme is less specific than in host cells. This difference provides a biochemical basis for development of specific

antiherpetic compounds selectively activated in virus-infected cells. Targets for antiviral action of 5-XdUrd may be different. Several 5-XdUMP are inhibitors of dTMP synthetase;
5-XdUDP can affect nucleoside diphosphate reductase, some
5-XdUTP inhibit DNA-polymerase or incorporate into DNA. The
main biochemical mechanism that inactivates these analogues
of dThd is phosphorolysis by dThd or Urd phosphorylase [1].
Biochemical pathways of dThd and 5-XdUrd and the possible
sites of inhibition are shown in Fig. 1.

Until recently only ß-anomers of dThd analogues were
considered as potential antimetabolites, while α-anomers
did not attract the interest of investigators. We examined
a series of α- and ß-5-XdUrd with different substituents
in the 5th position [2-5]. As parent compounds for preparation of 5-XdUrd we used uracils with 5-fluoroalkyl, 5-fluoro-
alkoxymethyl, 5-alkyl, 5-hydroxyalkyl, 5-alkoxymethyl,
5-Me_3Si and 5-Et_3Si substituents [4,6-10]. Modified uracils
were transformed into 2,4-bis-O-trimethylsilyl derivatives,
which were glycosylated by 2-deoxy-3,5-di-O-p-toluyl-α-D-
ribofuranosyl chloride in the presence of $SnCl_4$ in dichloro-
ethane or acetonitrile; α- and ß-anomers were separated by
absorption chromatography and then deacylated (Fig. 2). Anomeric homogeneity of deprotected nucleosides was controlled
with reverse-phase HPLC.

Biological assays of the above anomeric 5-XdUrd revealed that several α-anomers with branched 5-substituents inhibit HSV-1 replication in vitro, while their ß-counter-
parts have less or no activity. A pair of anomeric

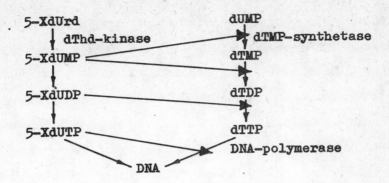

Fig. 1 Biochemical pathways of dThd and 5-XdUrd and the possible sites of inhibition

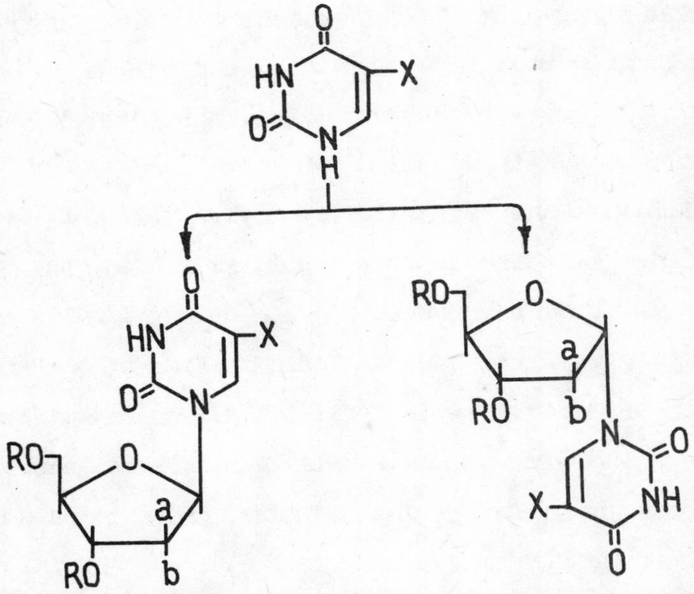

X: $CH_2CH_2CF_3$, CH_2CHFCF_3, $CH_2OCH_2CF_3$, $CH_2OCH_2CF_2CF_2H$,

$CH_2OCH_2CH_2CH_2OH$, $CH(CH_3)_2$[*●], $C(CH_3)_3$, $C(OH)(CH_3)_2$[*],

$C(OH)(CF_3)_2$[*], $CF(CF_3)_2$[●], $CH(CH_3)C_2H_5$, $CH_2CH(CH_3)_2$[*],

$CH(OH)CH(CH_3)_2$, $CH(OC_2H_5)CH(CH_3)_2$, $Si(CH_3)_3$[*], $Si(C_2H_5)_3$

Fig.2. Anomeric 5-XdUrd derivatives . The sign[●] shows
        antiherpetic β-anomer (I), the sign[*] marks anti-
        herpetic α-anomer (II).

89

5-iC$_3$F$_7$dUrd, in which α - but not ß-anomer is inactive, represents an exception. The most active antiherpetic compound is α-5-Me$_3$SidUrd [11]. α-Me$_3$CdUrd and α-5-Et$_3$SidUrd are inactive, as well as anomeric 5-XdUrd with linear alkyl substituents. The bulk of 5-substituent and its remoteness from uracil ring seems to be the essential features for antiviral activity of α-5-XdUrd. Indeed, the C-Si bond (1,8 Å) is longer than the C-C bond (1,5 Å), the steric effect of Me$_3$Si- group is therefore similar to that of CH(CH$_3$)$_2$, CH$_2$CH(CH$_3$)$_2$, and C(OH)(CF$_3$)$_2$ groups. It can be supposed that α-Me$_3$CdUrd and α-Et$_3$SidUrd are not the substrates for viral enzymes (e.g. for dThd kinase), since the branched C-atom in α-Me$_3$CdUrd is too close to the uracil ring whereas in α-Et$_3$SidUrd the substituent is too bulky.

ß-5-iC$_3$F$_7$dUrd has cytotoxic and antiviral activity in vitro, but in vivo this compound does not inhibit the growth of transplanted tumours of mice, though it retains antiviral properties in vivo . In this case the biological activity is supposedly based on reactivity of C-F bonds and irreversible inhibition of some enzymes, as it was demonstrated for ß-CF$_3$dUrd, the phosphate of which is an irreversible inhibitor of dTMP synthetase [12].

Since only one anomer is active, this prompted us to elaborate stereospecific methods of synthesis of these deoxynucleosides. ß-5-C$_3$F$_7$ dUrd was prepared starting from 1-ß-D-ribofuranosyl-5-perfluoroisopropyluracil (IV), which was obtained by ribosylation of uracil (III) by trimethylsilyl method. When (IV) was treated with SOCl$_2$, a mixture of diastereomeric 2',3'-O-sulphinates (V) was obtained, which yielded with NaOAc in DMF 2,2'-anhydronucleoside (VI). The latter was treated with HCl in DMF to give 2'-deoxy-2'-chlororiboside (VII), which was reduced to I (X = iC_3F_7) by Bu_3SnH (Fig. 3) [9].

For the stereospecific synthesis of α-5-XdUrd, 1-α-D-arabinofuranosyluracils (VIII) were chosen to start with because glycosylation of uracils with the use of 2,3,5-tri-O-acylarabinofuranosyl derivatives leads exclusively to α-anomers. The transformation of 5-substituted uracil α-arabinonucleosides via their 2,2'-anhydroderivatives (IX) was similar to that described for the β-series and led to corresponding α-deoxynucleosides (Fig. 4.) [13].

The p.m.r. spectra at 360 MHz of anomeric 5-XdUrd were investigated. Splitting pattern of H-1' proton signals is typical for anomeric 2'-deoxynucleosides: quartet for α-anomer with $J_{1'2'_a} \simeq 7$ Hz and $J_{1'2'_b} \simeq 2$ Hz and pseudotriplet for β-anomer with $J_{1'2'_a} \simeq J_{1'2'_b} \simeq 7$ Hz. The additional characteristic feature of p.m.r. spectra of α-anomers are signals of $H-2'_a$ and H-4' shifted downfield by 0,5 and 0,3 ppm, respectively, due to deshielding by aglycon, whereas $H-2'_b$ proton resonates upfield as compared with β-anomers. These characteristics can be helpful in distinguishing between α- and β-anomers in 2'-deoxynucleosides especially when splitting patterns of anomeric protons do not follow the "triplet-quartet rule".

In CD spectra of 5-XdUrd, Cotton effect of B_{2u} transition ($\sim$ 270 nm) is positive for β- and negative for α-anomers.

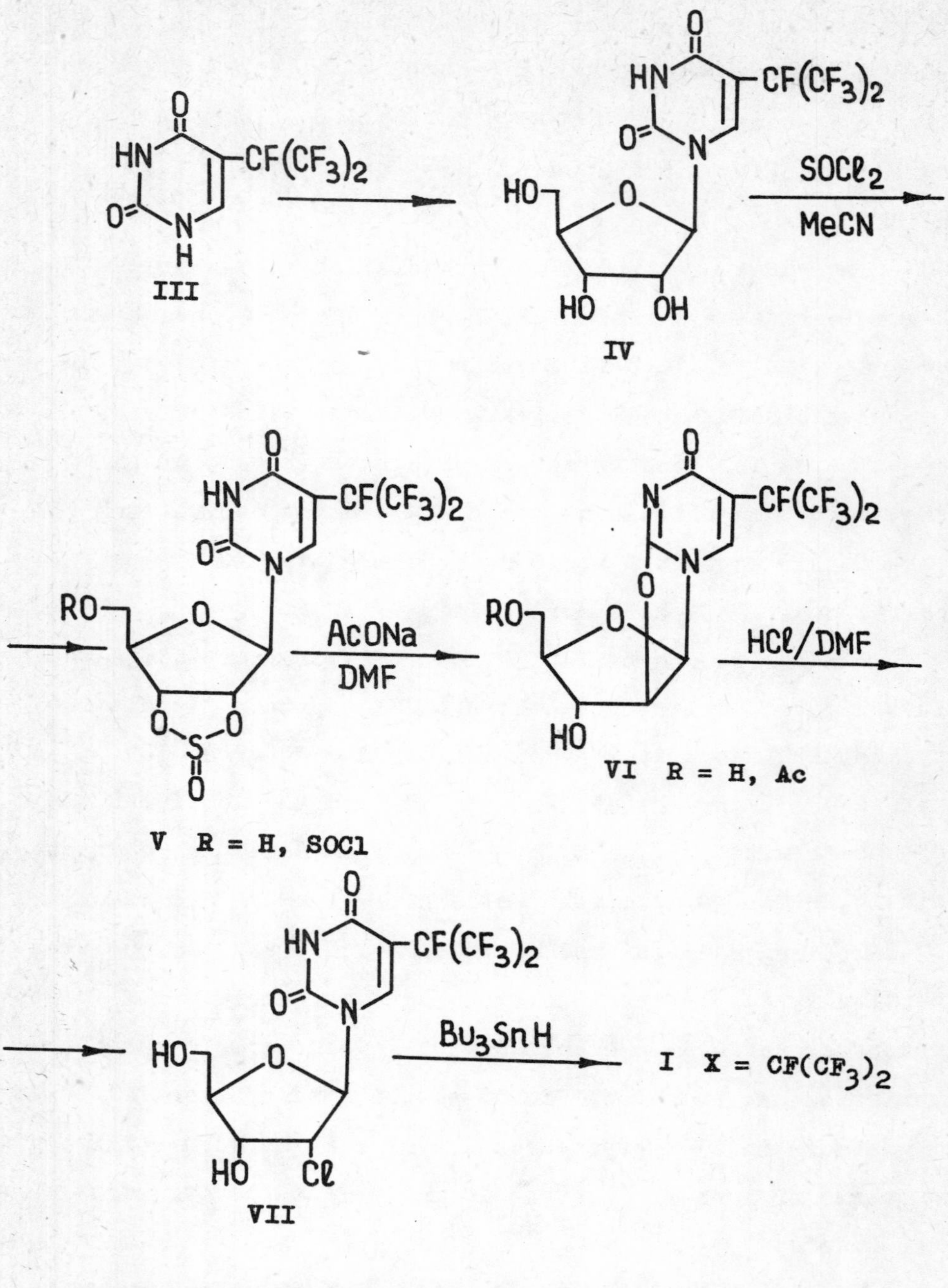

Fig.3. Synthesis of β-5-CF(CF$_3$)$_2$dUrd

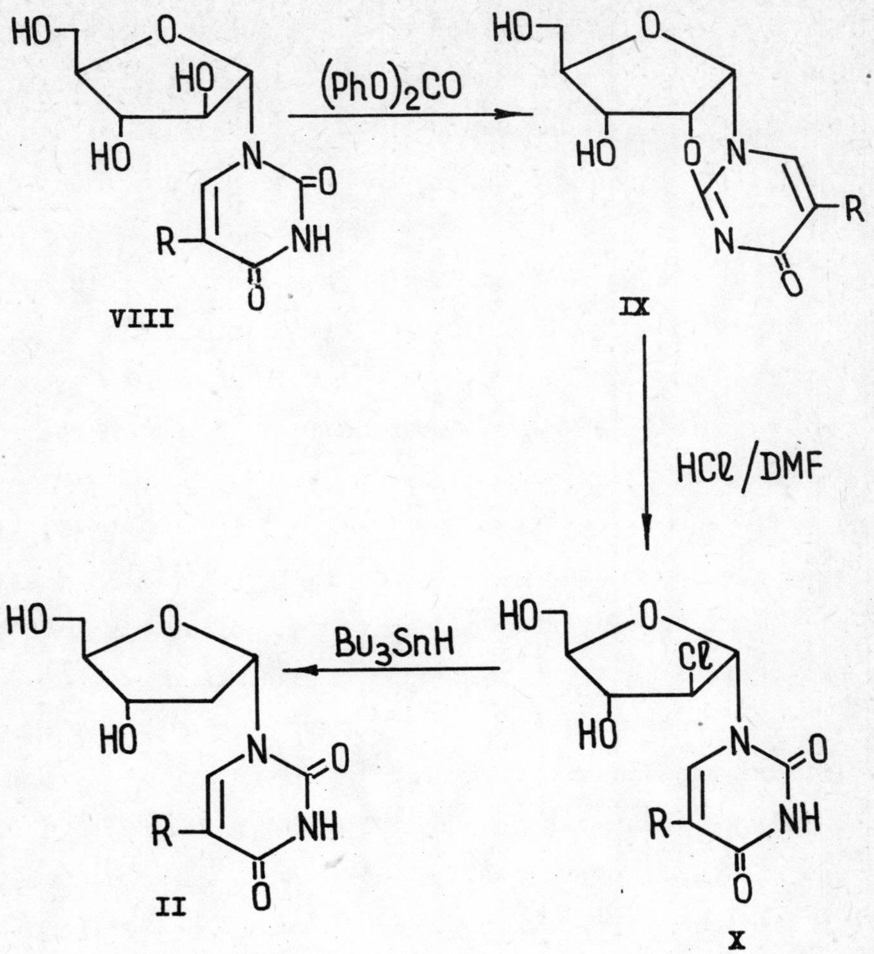

Fig.4.   Synthesis of α-5-XdUrd

R = H, Me or Me₃Si

CD spectra temperature dependence and calculation of syn-anti equilibrium in water demonstrate that predominant population of anti-conformer exists both for α- and ß-anomers (94-95%). In methanol solutions syn-population increases up to 10-20% [14].

α-Me$_3$SidUrd, which demonstrates the greatest potency and selectivity, is hardly toxic for chicken fibroblasts (ID$_{50}$ 2000 μg/ml), LD$_{50}$ in mice by i.p. administration is about 2 g/kg. It inhibits HSV-I replication in vitro at concentration range 30-250 μg/ml. α-5-Me$_3$SidUrd has proven to be highly efficient in the topical therapy of various HSV-1 infections, i.e. cutaneous HSV-1 infection of rabbits and hairless mice, HSV-1 keratitis in rabbits and guinea pigs. It is also efficacious in the intravenous therapy of herpetic encephalitis of cotton rats: 35% of the animals survived when α-Me$_3$SidUrd treatment was started 1 hour after virus inoculation and was continued daily until day 4 postinfection in the dose of 400 mg/kg (all untreated control animals died). Topical treatment with the compound has also shown a significant effect on herpes genitalis (HSV-2) in guinea pigs, which is comparable with the effect of acycloguanosine (XI).

α-5-Me$_3$SidUrd as well as antiherpetic α-5-C(OH)(CF$_3$)$_2$ dUrd are not phosphorylated by mice dThd kinase. All α-5-XdUrd are not splitted by Urd or dThd phosphorylases isolated from human or mice tissues. ß-Anomers with bulky substituents (for example ß-5-iC$_3$F$_7$dUrd) are not substrates for phosphorylases either. It suggests that α-5-Me$_3$SidUrd is a substrate for specific herpetic enzymes; the compound is more toxic for virus-infected than for normal cells and dThd protects the virus from the effect of α-5-Me$_3$SidUrd. Though effective antiviral concentration of the compound in vitro is higher than for β-5-(E)-(2-bromovinyl)dUrd or acycloguanosine, our α-deoxynucleoside has a pronounced effect in vivo owing to high metabolic stability.

Very often 5-substituted dCyd derivatives inhibit HSV-1 replication at similar or slightly higher concentrations than their dUrd counterparts because they not only inhibit dCyd metabolyzing enzymes but after deamination form active 5-XdUrd. α- and β-Me$_3$SidCyd were obtained by deoxyribosylation of 5-Me$_3$Si-cytosine. They were proved to have no antiviral activity. It can be supposed that the bulky 5-substituent prevents deamination of α-5-Me$_3$SidCyd as it has been found for α-5-EtdUrd [15].

1-β-D-Ribofuranosyl- and 1-α-D-arabinofuranosyl-5-
-Me$_3$Si-uracils were not active against HSV-1 as well.

Analogues of nucleosides containing sugar-like acyclic substituent are of great interest. The most active antiviral antimetabolites of this series are 9-[(2-hydroxyethoxy)methyl]guanine (acycloguanosine, ACG) (XI), 9-[(1,3-dihydroxy-
-2-propoxy)methyl]guanosine (2'-nor-2'-deoxyguanosine) (XII),
9-(2,3-dihydroxypropyl-1)adenine and some others [16-18].
ACG is phosphorylated by viral dThd kinase and after transformation into its 5'-triphosphate inhibits viral DNA polymerase sparing normal cells.

In our laboratory, condensation of Me$_3$Si-derivatives of 5-fluorouracil or 5-Me$_3$Si-uracil (XIII) with 1,3-di-O-benzyl-

-2-acetoxymethylglycerol (XIV) by the method [19] led after de-O-benzylation to acyclic analogues of 5-FdUrd and 5-Me$_3$SidUrd (XV) (Fig. 5).

We reported a simple and convenient procedure for the synthesis of hitherto unknown series of acycloanalogues of pyrimidine C-nucleosides of anticipated biological potentialities [20]. Some of pyrimidine C-nucleosides, for example ψ-isoCyd (XVI) exhibit valuable biological properties. By interaction of 5-hydroxymethyluracil (XVII) with 1,2-ethanediol or 1,3-propanediol in the presence of H$_2$SO$_4$ 5-[(2-hydroxyethoxy)methyl]- (XVIII) and 5-[(3-hydroxypropoxy)methyl]-uracil (XIX) were isolated (Fig. 6). A way to prepare ψ-isoCyd acycloanalogues is based on the usage of ψ-Urd analogues. A recently reported method for this transformation was attempted [21]. Under the action of (MeO)$_2$CHNMe$_2$ the compounds (XVIII) or (XIX) were 1,3-N-dimethylated to (XX) or (XXI), respectively which by the action of guanidine gave 5-[(2-hydroxyethoxy)methyl]- (XXII) or 5-[(3-hydroxypropoxy)methyl]-isocytosine (XXIII). By trimethylsilylation and consequent methylation of Me$_3$Si-derivatives of (XVIII or XIX) with CH$_3$I acyclic analogues of 1-N-methyl-ψ-uridine (XXIV and XXV) were obtained. These compounds can be considered as analogues of 5-aza-1-deazathymidine (Fig. 6).

These compounds (XXIV and XXV) showed the inhibitory effect on HSV-1 replication in vitro and retained antiherpetic activity in experiments in vivo in rats with herpetic encephalitis. Some of these compounds stimulate the incorporation of labelled pyrimidine precursors into DNA in cancer

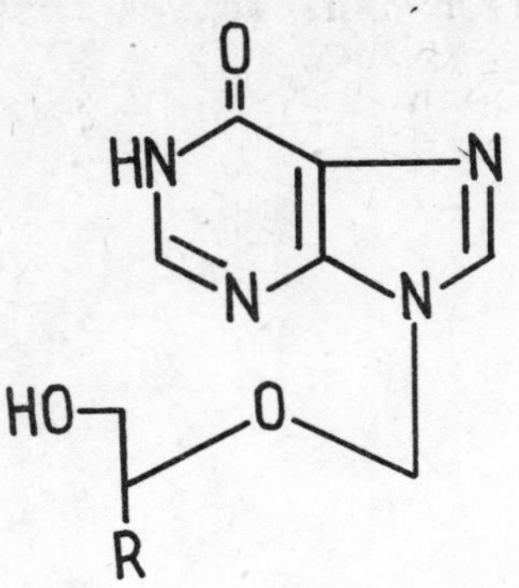

XI R = H

XII R = CH$_2$OH

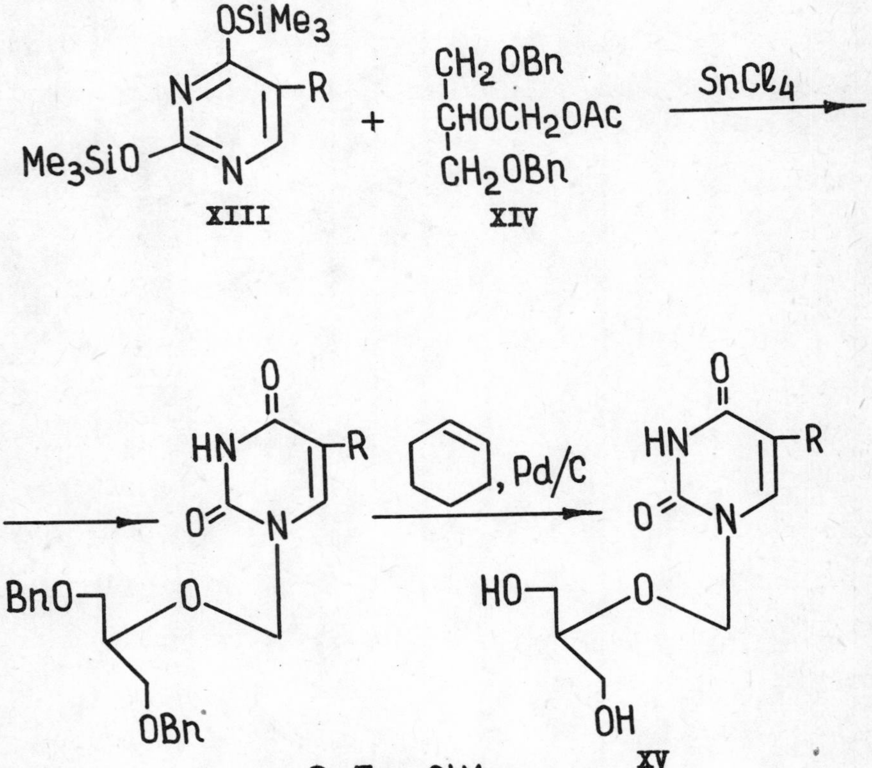

Fig.5. Synthesis of acyclic analogues of 5-FdUrd and 5-Me$_3$SidUrd

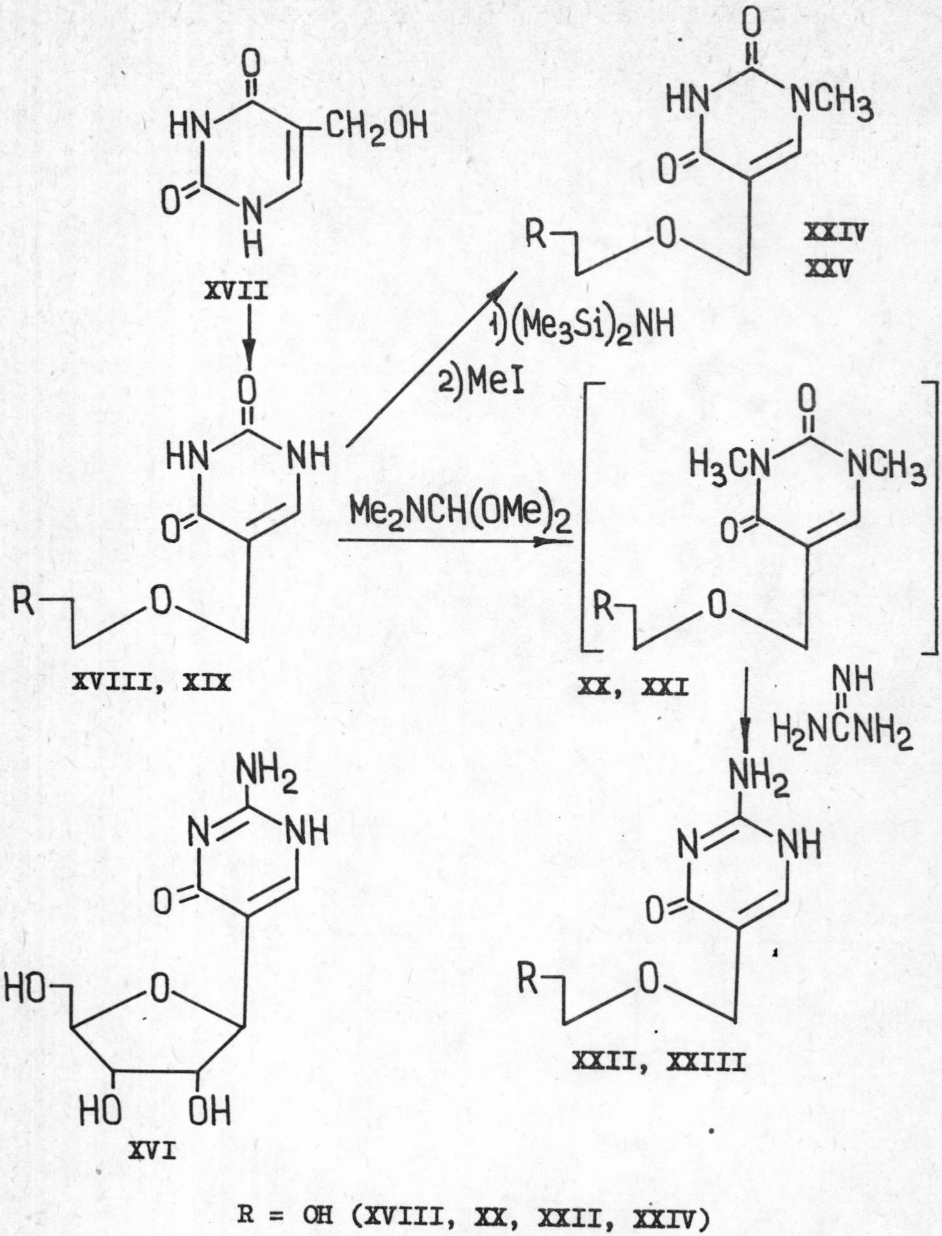

R = OH (XVIII, XX, XXII, XXIV)
R = CH₂OH (XIX, XXI, XXIII, XXV)

Fig.6. Synthesis of acyclic analogues of pyrimidine C-nucleosides

ovarian cell culture and into DNA of mice hepatoma 22A cells in vivo. It was suggested that the stimulation is owing to inhibition of Urd- or dThd phosphorylase. We have shown that (XXIII) and (XXV) inhibit the activity of dThd-phosphorylase isolated from mice hepatoma 22A by 93 and 46%, respectively.

Only a few 9-deazapurine 9-nucleosides are reported. The first such nucleoside 9-ß-D-ribofuranosyl-9-deazaadenine (XXVI) was obtained by multistep process using a synthone already containing carbohydrate moiety and was found to have high cytotoxic activity [24]. This prompted us to synthesize acycloanalogues of 9-deazapurine C-nucleosides (Fig.7).

It is well known that nucleophilic substitution reactions at the skatyl carbon atom are facilitated owing to the possibility of mesomeric indolyl-3-methyl cation formation, the situation being similar to substitution reactions at the uracilyl-5-methyl center. For example, erythro and threo-1(indolyl-3)-2-alkylaminopropanols-1 under the action of alcohols in the presence of acids have been transformed into threo-1-(indolyl-3)-1-alkoxy-2-alkylamino-propanes [25]. For the synthesis of purine-like C-nucleoside analogues we used 4-oxo-3,4-dihydropyrrolo(3,2-d)pyrimidine-7-aldehyde (9-formyl-9-deazahypoxanthine)(XXVII); preparation of the latter from 4-oxo-5-amino-6-methyl-3,4-dihydropyrimidine was described [26]. The reduction of aldehyde (XXVII) with $NaBH_4$ gave 9-hydroxymethyl derivative (XXVIII), which was transformed into 9-chloromethyl compound (XXIX). By condensation of (XXIX) with 1,2-ethanediol or MeOH 9-[(2-hydroxyethoxy)methyl]-9-deazahypoxanthine (XXX) and 9-methoxymeth-

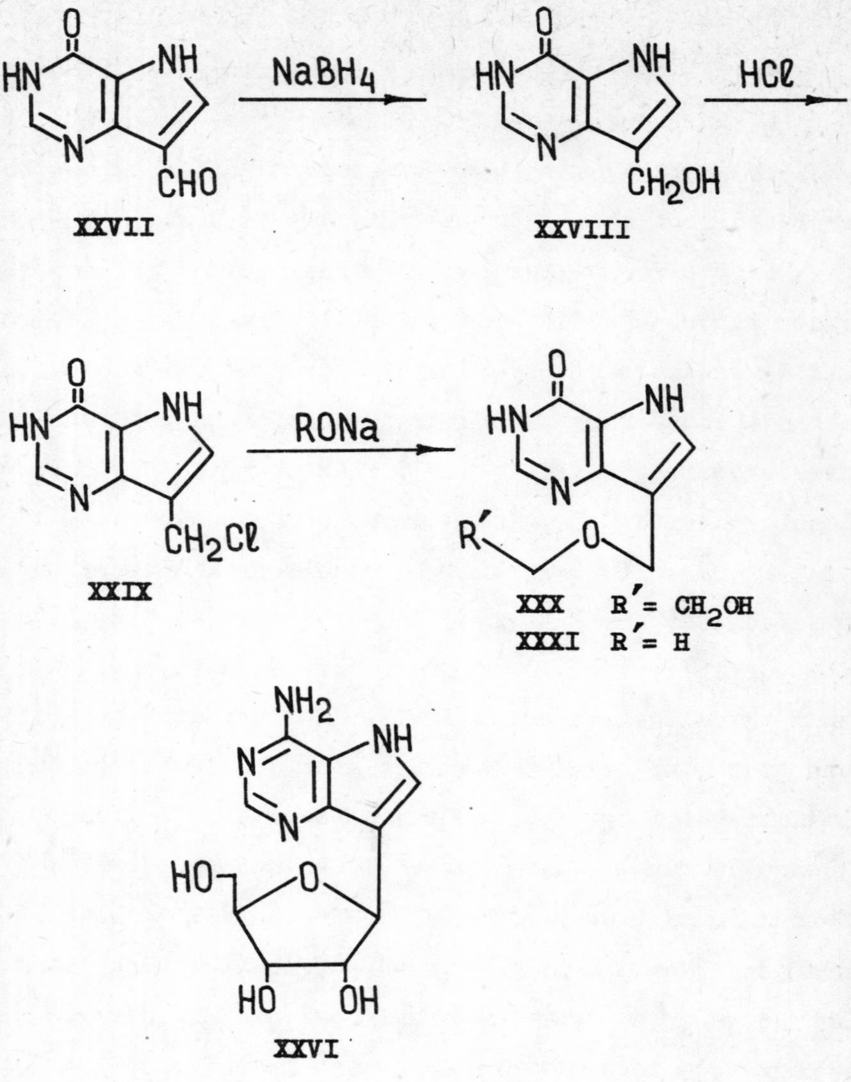

Fig.7. Synthesis of acyclo-9-deazainosine

yl - 9-deazahypoxanthine (XXXI) were prepared, which are the first acycloanalogues of purine C-nucleosides.

1-ß-D-Ribofuranosylpyrazolo(3,4-d)pyrimidines are close analogues of natural purine nucleosides. In the last few years we have developed a regioselective and stereospecific method of synthesis of these compounds and have investigated their biological activity and biotransformation pathways. It was demonstrated previously that in the case of glycosylation of fused pyrazoles [indazoles, pyrazolo(3,4-b)pyridines, pyrazolo(3,4-b)pyrazines] by fusion with per-O-acylated sugars increased reaction temperature and the use of acid catalyst led to predominant or even exclusive formation of the thermodynamically controlled 1-nucleosides as a result of N-2 $\rightarrow$ N-1 isomerization [27]. Similar results were obtained with pyrazolo(3,4-d)pyrimidines, whereas fusion of 4-methyl-thiopyrazolo(3,4-d)pyrimidine (XXXII) with tetra-O-acetyl-ribofuranose at 160° in the presence of 1-5% of bis(p-nitro-phenylphosphate) gave a mixture of per-O-acetylated 1- and 2-ribofuranosides of 4-methylthiopyrazolo(3,4-d)pyrimidine [28]; increase of reaction temperature to 180° and use of 20% of the catalyst permitted to obtain the O-acetylated 1-nucleoside (XXXIII) with the yield of 70%. Deacetylation led to 1-ß-D-ribofuranosyl-4-methylthiopyrazolo(3,4-d)pyrimidine (XXXIV), which produced after ammonolysis 1-ß-D-ribofurano-syl-4-aminopyrazolo(3,4-d)pyrimidine (XXXV). 1-Nucleoside of allopurinol (XXXVI) was obtained by deamination of amino-derivative (XXXV) [29]. These chemical transformations represent the most efficient ways to biologically impor-

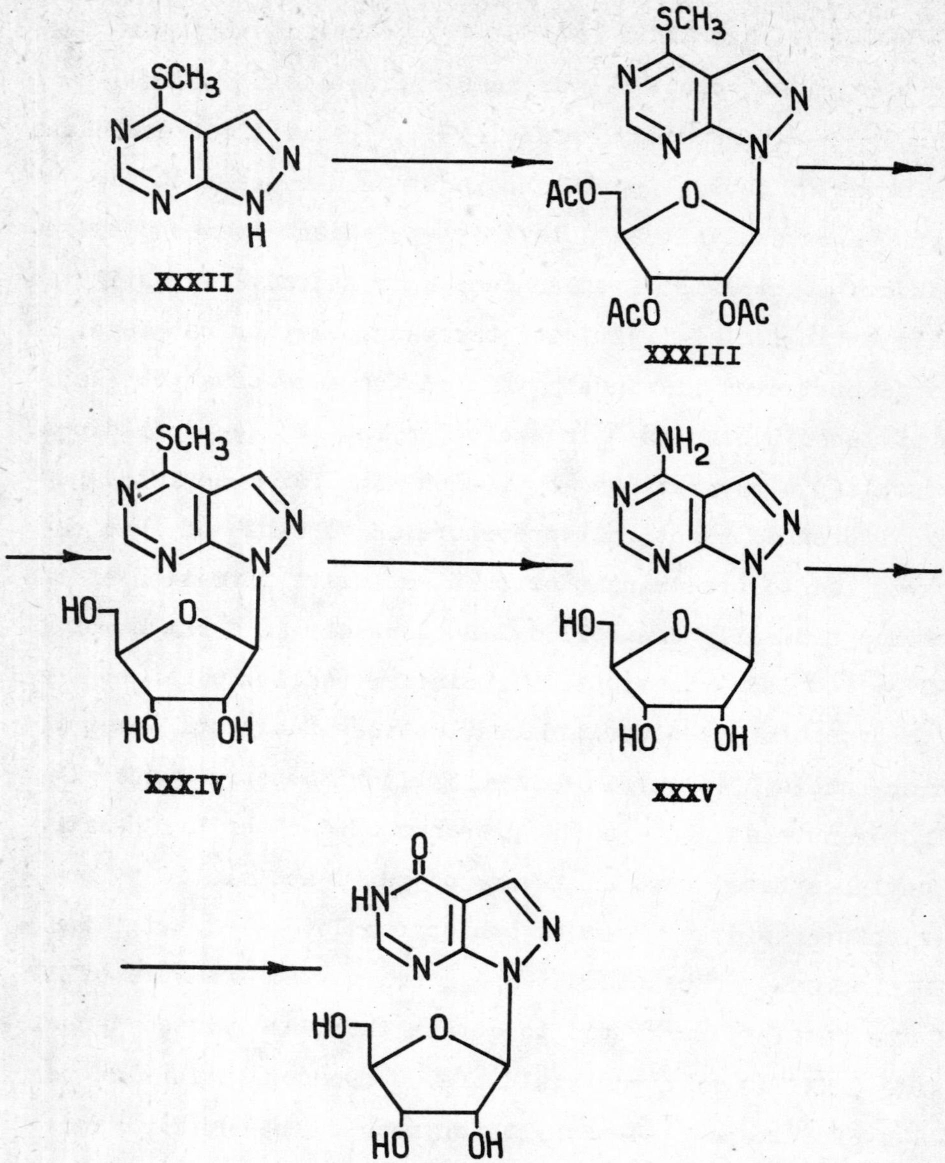

Fig.8. Synthesis of 4-substituted pyrazolo(3,4-d)-
pyrimidine 1-nucleosides

tant pyrazolo(3,4-d)pyrimidine nucleosides (XXXIV-XXXVI)
(Fig. 8) [29-31]. By phosphorylation of (XXXIV) the nucleotide (XLIII) was obtained, which was used in synthesis of
the corresponding 5'-triphosphate [42].

By the method of regioselective glycosylation, series of
1-β-D-ribofuranosides of 4-mono-, 3,4-di-, and 3,4,6-tri-
substituted pyrazolo(3,4-d)pyrimidines (XXXIX, R = Rib)
were obtained from substituted heterocycles (XXXVII or
XXXVIII) as versatile parent compounds (Fig. 9) [32-36].

For preparation of heterocycles (XXXVII or XXXVIII)
3,4-dicyano-5-aminopyrazole (XL, R=CN) was treated with triethyl orthoformate to give 5-ethoxymethylene-aminopyrazole
(XLI, R=CN, R'=OEt) (at 70-100°C), at elevated temperature
(140-150°C) ethylation of pyrazole nitrogen atoms with formation of N-1 and N-2 ethyl-derivatives (XLII, R=CN, R'=OEt)
also takes place [37]. Similarly condensation of 4-cyano-5-
aminopyrazole or 3,4-dicyano-5-aminopyrazole (XL, R=CN or H)
with DMF-diethylacetal in boiling methanol led to dimethylaminomethylene derivatives (XLI, R=NMe$_2$)and boiling the reaction mixture without solvent produced mixtures of (XLI) and
corresponding N-1 and N-2-ethyl compounds (XLII, R=NMe$_2$)
(Fig. 10). Dimethylaminomethylene derivatives of cyanoamino-
pyrazoles proved to be valuable precursors for preparation of
substituted pyrazolo(3,4-d)pyrimidines [38].

Several substituted pyrazolo(3,4-d)pyrimidines and their
1-β-D-ribofuranosides (XXXIX) exhibit cytotoxic, antitumour or
antiviral properties [39,40]. Thioanalogues of allopurinol or
alloxanthine were found to be potent inhibitors of xanthine

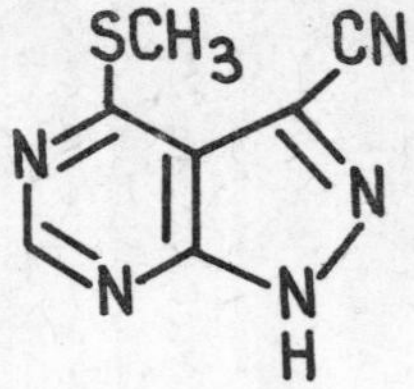

XXXVII

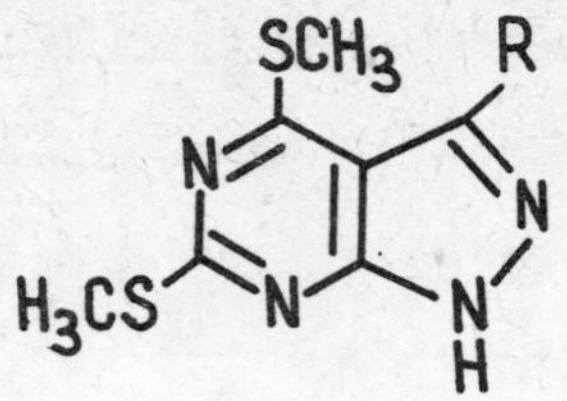

XXXVIII

R = H, CN, CH$_2$CN

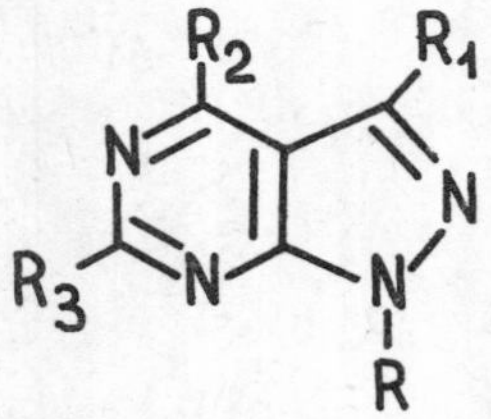

XXXIX

R = H or β-D-ribofuranosyl

R$_1$= H, CN, CH$_2$CN, CSNH$_2$,
C(=NOH)NH$_2$, C(=NH)OMe,
C(=NH)NH$_2$, CH$_2$CONH$_2$,
CH$_2$COONH$_4$, C(=NH)NHNH$_2$,
C(=NH)N⟨⟩, C(=NH)N⟨O⟩

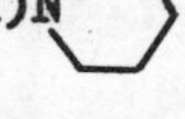

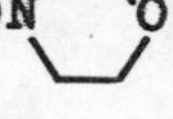

R$_2$= SH, SMe, NH$_2$, N⟨⟩,
N⟨O⟩, OH, OMe

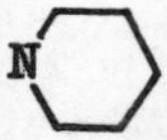

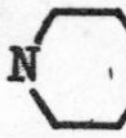

R$_3$= H or SH, SMe

Fig. 9. The substituted pyrazolo(3,4-d)pyrimidines
and their 1-β-D-ribofuranosides

oxidase [41].

1-ß-D-Ribofuranosyl-4-methylthiopyrazolo(3,4-d)pyrimidine (XXXIV) produces 50% inhibition of labelled dThd incorporation into DNA of cultured cancer ovarian cells (CaOv) in vitro at the concentration 1.7×10^{-5} M, the corresponding nucleic base analogue (XXXII) being inactive. Heterocycle (XXXII) has no antitumour properties, whereas nucleoside (XXXIV) as well as nucleotide (XLIII) increase lifespan of mice with leukemia L 1210 or Ehrlich tumour by 40-60%.

The intracellular level of nucleotide (XLIII) was followed by HPLC after administration of an analogue of nucleic base (XXXII), nucleoside (XXXIV) or nucleotide (XLIII) to mice with tumours as well as after incubation of CaOv cells with the compounds in vitro and was found to correlate with anticancer or cytotoxic activity of the nucleoside (XXXIV) or nucleotide (XLIII).

In L1210 cells 2 hrs after i.v. administration of the nucleoside (XXXIV) to leukemic mice in doses 100 mg/kg, the content of the nucleotide (XLIII) was 523 ± 160 pmol/10^6 cells; in similar experiments with the base (XXXII) the phosphate (XLIII) was not detected [42]. In cells of animal tumours resistant to the nucleoside or nucleotide (Ca 755, La, LLC) the intracellular level of the nucleotide (XLIII) was very low. The low level of the phosphate (XLIII) after administration of this compound to animals with nucleoside/nucleotide resistant tumours demonstrates that the nucleotide penetrates cell membranes in very small extent. Our data suggest that bioactivation of the nucleoside (XXXIV) to its

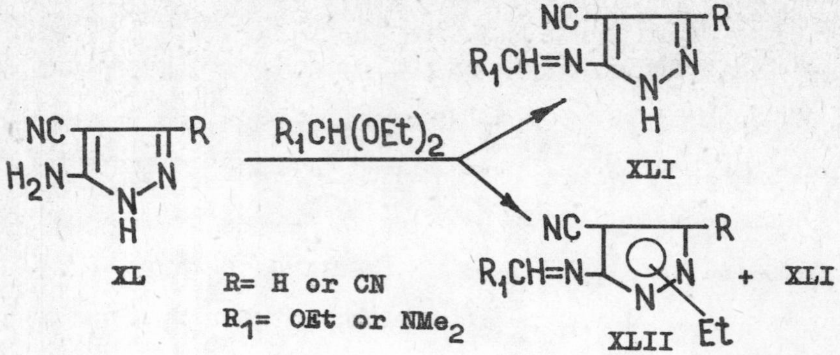

Fig.10.   Interaction of substituted 3-aminopyrazoles
with $HC(OEt)_3$ or $Me_2NCH(OEt)_2$

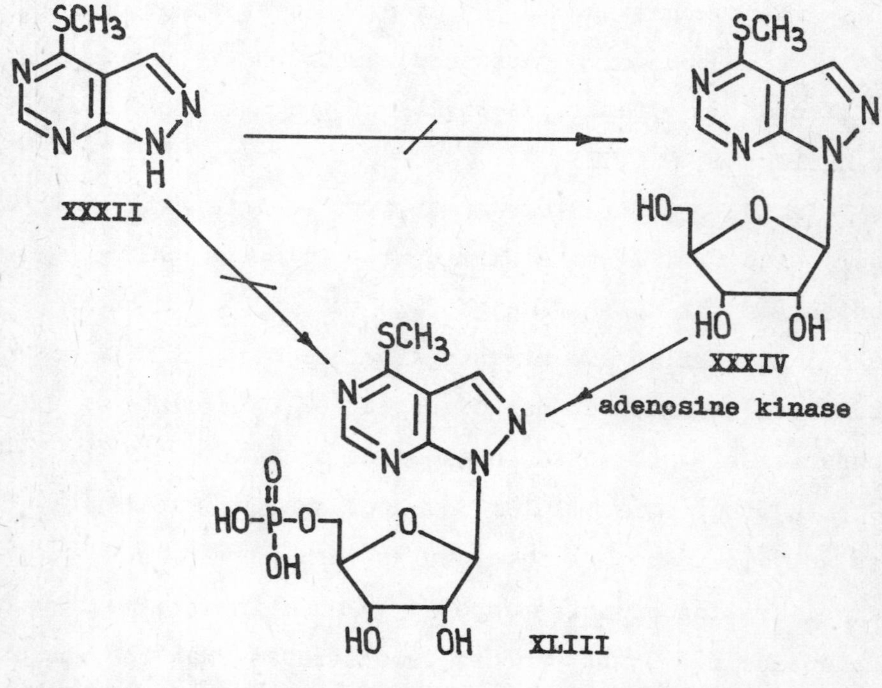

Fig.11.   Biotransformations of 4-methylthiopyrazolo-
(3,4-d)pyrimidine and its nucleoside and nucleotide

phosphate (XLIII) proceeds via direct phosphorylation (by adenosine kinase); transformation of the base (XXXII) to the nucleotide (XLIII) or nucleoside (XXXIV) does not occur (Fig. 11) [42]. Low intracellular level of active metabolite (XLIII) in cells resistant to the investigated compounds may be due to low activity of phosphorylating enzymes or due to rapid biodegradation of the nucleoside (XXXIV) to 1-nucleoside of allopurinol (XXXVI) under the action of adenosine deaminase. The search for metabolically stable substituted pyrazolo(3,4-d)pyrimidine nucleosides is in progress.

REFERENCES

1. **Preobrazhenskaya M.N., Melnik S.Y.** (1984) Analogues of nucleic acids components - inhibitors of nucleic metabolism. Itogi nauki i tekhniki, ser. Bioorganicheskaya khimia, 1, 77-120.

2. Melnik S.Y., Bakhmedova A.A., Vornovitskaya G.I., Dobrynin Y.V., Nikolaeva T.G., Ivanova T.P., Yartseva I.V., Preobrazhenskaya M.N. (1979) Synthesis and investigation of 5-polyfluoroalkyl and 5-polyfluoroalkoxymethyl-2'-deoxypyrimidine nucleosides. Bioorgan. Khimia, 5, 41-46.

3. Melnik S.Y., Bakhmedova A.A., Miniker T.D., Preobrazhenskaya M.N., Zagulyaeva O.A., Mamaev V.P. (1981) Synthesis and investigation of antimetabolite properties of anomeric 5-substituted 2'-deoxyuridines. Nucleic Acids Res., Symp. Ser., № 9, 53-55.

4. Melnik S.Y., Bakhmedova A.A., Nedorezova T.P., Vornovitskaya G.I., Preobrazhenskaya M.N., Avetisyan E.A., German L.S., Polyshchuk V.R., Chekunova E.V., Bektemirov T.A., Andzhaparidze O.G. (1981) Synthesis of anomeric 5-(2-hydroxyhexafluoroisopropyl)-2'-deoxyuridines and investigation of their antimetabolite and antiviral properties. Bioorgan. Khimia, 7, 1047-1053.

5. Bektemirov T.A., Chekunova E.V., Andzhaparidze O.G., Melnik S.Y., Bakhmedova A.A., Preobrazhenskaya M.N. (1979) Investigation of the antiviral activity of anomeric 5-substituted 2'-deoxyuridines, Voprosy virusologii, № 6, 603-606.

6. German L.S., Polyshchuk V.R., Preobrazhenskaya M.N., Melnik S.Y., Kirshchenya L.M. (1974) 5-Polyfluoroalkyluracils. Khimia Geterocycl. Soed., 859.

7. Zagulyaeva O.A., Skorokhodova L.V., Mamaev V.P. (1976) Pyrimidines. LVII. 5-Trimethylsilyluracils. Izv. Sib. Otd. Akad. Nauk SSSR, ser. khim. nauk, № 6, 92-96.

8. Melnik S.Y., Bakhmedova A.A., Preobrazhenskaya M.N., Platonova G.N., Lesnaya N.A., Sof'ina Z.P. (1976) Synthesis and investigation of 5-polyfluoroalkoxymethyluracils and their 1-ß-D-ribosides. Zhurnal organ. chimii, 12,

652-655.

9. Melnik S.Y., Bakhmedova A.A., Yartseva I.V., Nedorezova
   T.P., Vornovitskaya G.I., Preobrazhenskaya M.N., German
   L.S., Polyshchuk V.R., Avetisyan E.A. (1981) Synthesis
   and investigation of 5-perfluoroisopropyl-2'-deoxyuridi-
   ne. Bioorgan. Khimia, 7, 1711-1717.

10. Melnik S.Y., Bakhmedova A.A., Yartseva I.V., Preobra-
    zhenskaya M.N., Zagulyaeva O.A., Mamaev V.P., Chekunova
    E.V., Bektemirov T.A., Andzhaparidze O.G. (1982) Synthe-
    sis and investigation of thymidine analogs with branched
    substituent at 5-position of pyrimidine ring. Bioorgan.
    Khimia, 8, 1102-1107.

11. Preobrazhenskaya M.N., Melnik S.Y., Bakhmedova A.A., Me-
    zhevich Z.M., Mamaev V.P., Zagulyaeva O.A., Bektemirov
    T.A., Chekunova E.V., Andzhaparidze O.G., Pozdnyakov
    V.I., Maychuk Y.F., Shchipanova A.I. (1978) Avt. svid.
    USSR, № 671287, Bull. isobr., № 32, 225 (1983).

12. Santi D. (1980) Perspectives on the design and biochemi-
    cal pharmacology of inhibitors of thymidylate syntheta-
    se, J. Med. Chem. 23, 103-111.

13. Melnik S.Y., Miniker T.D., Yartseva I.V., Preobrazhens-
    kaya M.N. (1982) Synthesis of α-D-arabinofuranosyl py-
    rimidines and their conversion to 2'-deoxy-α-D-ribofu-
    ranosides. Bioorgan. Khimia, 8, 1094-1101.

14. Volodin Y.Y., Kikotj B.S., Preobrazhenskaya M.N. (1982)
    Circular dichroism studies of 5-substituted 2'-deoxyuri-
    dines. Bioorgan. Khimia, 8, 1375-1380.

15. Krajewska E., Shugar D. (1975) Alkylated cytosine nucleo-
    sides: substrate and inhibitor properties in enzymatic
    deamination. Acta Biochim. Pol., 22, 185-194.

16. Schaeffer H.J., Beauchamp L., de Miranda P., Elion G.B.,
    Bauer D.J., Collins P. (1978) 9-(2-Hydroxyethoxymethyl)gu-
    anine activity against viruses of the herpes group. Na-
    ture, 272, 583-585.

17. Elion G.B., Furman P.A., Fyfe J.A., De Miranda P., Beau-
    champ L., Schaeffer H.J. (1977) Selectivity of action of
    an antiherpetic agent, 9-(2-hydroxyethoxymethyl)guanine.

Proc. Natl. Acad. Sci. USA, 74, 5716-5720.

18. De Clercq E., Holy A. (1979) Antiviral activity of aliphatic nucleoside analogues: structure-function relationship. J. Med. Chem., 22, 510-513.

19. Martin J.C., Dvorak C.A., Smee D.F., Matthews T.R., Verheyden J.P. (1983) 9-[(1,3-Dihydroxy-2-propoxy)methyl]guanine: a new potent and selective antiherpes agent. J. Med. Chem., 26, 759-761.

20. Melnik S.Y., Miniker T.D., Yartseva I.V., Nedorezova T.P., Potapova G.I., Preobrazhenskaya M.N. (1983) Synthesis and properties of pyrimidine C-nucleoside acyclic analogues. Bioorgan. Khimia, 9, 1395-1400.

21. Matsuda A., Pankiewicz K., Marcus B.K., Watanabe K.A., Fox J.J. (1982) Synthesis of 3-methylpseudouridine and 2'-deoxy-3-methylpseudouridine. Carbohydr. Res., 100, 297-302.

22. Hirota K., Watanabe K.A., Fox J.J. (1977) Pyrimidines. XIII. Novel pyrimidine to pyrimidine transformation reactions. J. Heterocyclic Chem., 14, 537-538.

23. Hirota K., Watanabe K.A., Fox J.J. (1978) Pyrimidines. 14. Novel pyrimidine to pyrimidine transformation reactions and their application to C-nucleoside conversion. A facile synthesis of pseudoisocytidine. J. Org. Chem., 43, 1193-1197.

24. Lim M., Klein R. (1981) Synthesis of "9-deazaadenosine", a new cytotoxic C-nucleoside isostere of adenosine. Tetrahedron Lett., 22, 25-28.

25. Orlova L.M., Preobrazhenskaya M.N., Turchin K.F., Starostina L.G., Suvorov N.N. (1969) Synthesis and determination of configuration of indole-containing aminoalcohols. Zhurnal Organ. Khimii, 5, 738-746.

26. Sizova O.S., Brytikova N.E., Novitskii K.Y., Shcherbakova L.I., Pershin G.N. (1980) Use of Vielsmeier reaction in the synthesis of pyrrolo(3,2-d)pyrimidine-7-aldehyde derivatives. Khim. Farm. Zhurnal, N 7, 63-66.

27. Preobrazhenskaya M.N., Korbukh I.A., Tolkachev V.N., Dobrynin Y.V., Vornovitskaya G.I. (1979) Studies of

purine nucleoside analogues. <u>Les Colloques de l'INSERM</u>
<u>Nucleosides, Nucleotides and Their Biological Applica-</u>
<u>tions.</u> Eds. Barascut J.-L., Imbach J.-L., vol. 81,
INSERM, Paris, 85-116.

28. Blanco F.F., Korbukh I.A., Preobrazhenskaya M.N. (1976)
    Ribosylation of 4-methylthiopyrazolo(3,4-d)pyrimidine.
    Zhurnal Organ. Khimii, <u>12</u>, 1132-1133.

29. Korbukh I.A., Yakunina N.G., Preobrazhenskaya M.N. (1980)
    Synthesis and reactions of the nucleosides of 4- and
    4,6-substituted pyrazolo(3,4-d)pyrimidines. Bioorgan.
    Khimia, <u>6</u>, 1632-1638.

30. Korbukh I.A., Bulychev Yu.N., Yakunina N.G., Preobrazhen-
    skaya M.N. (1981) The nucleosides of substituted pyrazo-
    lo(3,4-d)pyrimidines. Nucleic Acids Res., Symp. Ser.,
    № 9, 73-75.

31. Korbukh I.A., Yakunina N.G., Blanco F.F., Preobrazhenska-
    ya M.N. (1984) 4-Methylthio-, 4,6-di(methylthio)-, and
    4-amino-1-ß-D-ribofuranosylpyrazolo(3,4-d)pyrimidine.
    <u>Nucleic Acid Chemistry, Improved and New Synthetic Pro-</u>
    <u>cedures, Methods and Techniques in Nucleic Scid Chemist-</u>
    <u>ry.</u> Eds. Townsend L.B., Tipson R.S., vol. 3, John Wiley
    & Sons, New York, contr. № 56.

32. Korbukh I.A., Bulychev Yu.N., Preobrazhenskaya M.N.
    (1979) Synthesis of 3-cyano-4,6-di(methylthio)pyrazo-
    lo(3,4-d)pyrimidine 1-riboside. Khimia Geterocycl. Soed.,
    1687-1692.

33. Bulychev Yu.N., Korbukh I.A., Preobrazhenskaya M.N.
    (1980) Chemical reactions of trisubstituted pyrazolo(3,4-
    -d)pyrimidines and their 1-ribosides. Khimia Geterocycl.
    Soed., 243-250.

34. Bulychev Yu.N., Korbukh I.A., Preobrazhenskaya M.N.
    (1981) Synthesis of pyrazolo(3,4-d)pyrimidine-3-acetic
    acid derivatives and their nucleosides. Khimia Getero-
    cycl. Soed., 536-545.

35. Bulychev Yu.N., Korbukh I.A., Preobrazhenskaya M.N.
    (1984) 3-Cyano-4,6-di(methylthio)-1-ß-D-ribofuranosyl-
    pyrazolo(3,4-d)pyrimidine. <u>Nucleic Acid Chemistry, Imp-</u>

roved and New Synthetic Procedures, Methods and Techniques in Nucleic Acid Chemistry. Eds. Townsend L.B., Tipson R.S., vol. 3, John Wiley & Sons, New York, contr. № 55.

36. Bulychev Yu.N., Korbukh I.A., Preobrazhenskaya M.N. (1984) Synthesis of 3-substituted 4-methylthio- and 4-aminopyrazolo (3,4-d)pyrimidines and their ribosides. Khimia Geterocycl. Soed., 253-258.

37. Bulychev Yu.N., Korbukh I.A., Preobrazhenskaya M.N. (1982) Interaction of 3,4-di(cyano)-5-aminopyrazole with triethyl orthoformate. Khimia Geterocycl. Soed., 1682-1685.

38. Bulychev Yu.N., Korbukh I.A., Preobrazhenskaya M.N., Chernyshev A.I., Esipov S.E. (1984) Interaction of 4-cyano-5-aminopyrazole and 3,4-di(cyano)-5-aminopyrazole with N,N-dimethylformamide diethyl acetal. Khimia Geterocycl. Soed., 259-264.

39. Dobrynin Ya.V., Bektemirov T.A., Ivanova T.P., Chekunova E.V., Andzhaparidze O.G., Korbukh I.A., Bulychev Yu.N., Yakunina N.G., Preobrazhenskaya M.N. (1980). Cytotoxic and antiviral activity of 4- and 3,4-substituted 6-methylthiopyrazolo(3,4-d)pyrimidines and their ribosides. Khim.-Farm.Zhurnal, 5, 10-15.

40. Bektemirov T.A., Chekunova E.V., Korbukh I.A., Bulychev Yu.N., Yakunina N.G., Preobrazhenskaya M.N. (1981) Antiviral activity of substituted 6-methylthiopyrazolo(3,4-d)pyrimidines and their ribosides. Acta virol. 25, 326-329.

41. Vartanyan L.S., Rashba Yu.E., Kazachenko A.I., Korbukh I.A., Bulychev Yu.N., Preobrazhenskaya M.N. (1982) New xanthine oxidase inhibitors of pyrazolo(3,4-d)pyridine series. II. Comparative evaluation of their effectiveness. Khim.-Farm. Zhurnal, № 6, 655-660.

42. Korbukh I.A., Goryunova O.V., Stukalov Yu.V., Ivanova T.P., Dobrynin Ya.V., Preobrazhenskaya M.N. (1984) Biotransformation of 1-ß-D-ribofuranosyl-4-methylthiopyrazolo(3,4-d)pyrimidine and its 5'-monophosphate. Bioorgan. Khimia, 10, № 6.

DISCUSSION

JENEY A. (Budapest, Hungary)
a, What is the policy for the development of new drugs in
your Institute? The approach you used to apply is
chemical, biochemical or pharmacological?
b, Have you observed bone-marrow toxicity after treatment
of animals with your new MNU-analogues?

PREOBRAZHENSKAYA M.N. (Moscow, USSR)
a, The chemical approach was demonstrated when I spoke about
1-glycosyl-3-methyl-3-nitrosoureas which are capable to
trap carbamoylating fragments of the molecules and which
are therefore less toxic than their non-glycosylated
counterparts. We try to use biochemical approaches: try
to understand the mechanism of action and metabolic
pathways of our compound. It is very difficult for a
chemist to use the pharmacological approach because, in
most cases, it is problematic to bridge the pharmacological and the chemical language.
b, 1-Glycosyl-3-methyl-3-nitrosoureas, which have low carbamoylating activity, are less toxic than MNU.

RÉVÉSZ L. (Basle, Switzerland)
Did you try to use monoclonal antibodies as carriers for
your drugs?

PREOBRAZHENSKAYA M.N. (Moscow, USSR)
No, we did not.

PISKALA A. (Praha, Czechoslovakia)
Have you proved the phosphorylation of α-nucleosides in
both infected and uninfected cells and if so, by which
method?

PREOBRAZHENSKAYA M.N. (Moscow, USSR)

Not yet, but we plan to do this.

ÖTVÖS L. (Budapest, Hungary)

Is there a possibility to combine in one molecule the antimetabolite property and alkylating activity?

PREOBRAZHENSKAYA M.N. (Moscow, USSR)

Nucleosides containing nitrosomethyl fragments were synthesized by Prof. Montgomery and co-workers. Some of them demonstrated valuable properties.

Bio-Organic Heterocycles
van der Plas H.C., Ötvös L., and Simonyi M. eds

CHEMICAL STRUCTURE AND BIOCHEMICAL TRANSFORMATION OF PYRIMIDINE DERIVATIVES

DÉNES G.

Central Research Institute for Chemistry,
The Hungarian Academy of Sciences,
Budapest, Pf. 17, H-1525, Hungary

An overwhelming body of evidence suggests, that the genetic information of all cells and many viruses is stored in double-stranded DNA. In some group of viruses the genetic information is stored in single-stranded DNA, single-stranded RNA, or very rarely double-stranded RNA. The double-strandedness of DNA has consequences. 1.) It provides a mechanism of self-replication, since each strand serves as a template on which its complement is formed (replication) 2.) Messenger RNA (mRNA) is formed on one strand of DNA (transcription) 3.) The informa-

---

† Abbreviations used: EU, 5-ethyluracil; iPU, 5-isopropyl-uracil; BVU, 5-bromvinyluracil; FU, 5-fluoro-uracil; UdR, 2'-deoxyuridine; TdR, thymidine; EUdR, 5-ethyl-2'-deoxy-uridine; PUdR, 5-propyl-2'-deoxyuridine; IPUdR, 5-isopro-pyl-2'-deoxyuridine; BUdR, 5-butyl-2'-deoxyuridine; tBUdR, 5-tertiary-butyl-2'-deoxyuridine; HUdR, 5-hexyl-2'-deoxy-uridine; FUdR, 5-fluoro-2'-deoxyuridine; BVUdR, 5-bromo-vinyl-2'-deoxyuridine, diBVUdR, 5-dibromovinyl-2'-deoxy-uridine; UR, uridine; EUr, 5-ethyluridine; EaraU, arabino-furanosil-5-ethyluracil; FUR, 5-fluorouridine;

tion contained in mRNA determines the amino acid sequence in
the polypeptide chains of protein (translations). Any change
in the structure of purine or pyrimidine base - spontaneous or
induced - influences the structure and coding properties of DNA.

It is now well established fact that natural pyrimidine
deoxyribonucleosides - thymidine and cytidine - are incorporated into the DNA of both animal cells and microbial systems.
The metabolic derivatives of these nucleosides in the form of
nucleoside-5'-triphosphates are very important in the life of
cells. Since viruses are critically dependent on these compounds
for their replication, it is not surprising, that synthetic
nucleoside analogues which interfere with either the synthesis
or the function of DNA, do exert antiviral and/or anticancer
activity. The theoretical basis for development of nucleoside
analogues as antiviral or anticancer agents is related to
their role - as precursors of DNA synthesis - in cellular or
viral replication.

Over the past twenty years hundreds of pyrimidine
nucleoside analogues have been synthesized as potential
anticancer or antiviral agents. Some of them were active in
cell culture but only a few in experimental animals or in
man. During these years it has become evident, that double-
stranded DNA viruses - like herpes viruses and others -
after having invaded the cells, induce the synthesis of
their own enzymes. These virus coded enzymes often differ

116

from their cell coded counterparts in substrate specificity
and other properties and as a consequence of these they could
serve as target for the action of antiviral agents.

BIOCHEMICAL TRANSFORMATION PATHWAYS

There are two possibilities for biochemical transformation of pyrimidine nucleosides or nucleoside analogues, as
Fig. 1. shows. They are either inactivated (catabolic pathway)
or activated (anabolic pathway) by different enzyme systems.
The first enzymes in the inactivation process are the cellular nucloside phosphorylases, while the key enzymes in

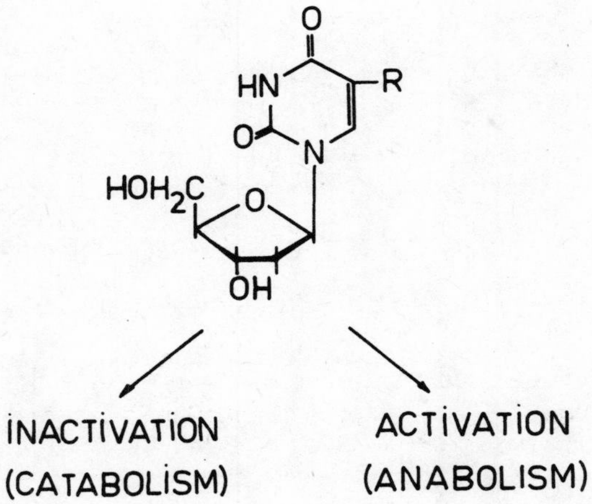

INACTIVATION   ACTIVATION
(CATABOLISM)   (ANABOLISM)

Fig. 1. Metabolic pathways for biochemical transformation of
          nucleoside analogues

the biochemical activation process of pyrimidine nucleosides
or nucleoside analogues are the cellular or virus coded

thymidine kinases. These activated nucleoside analogues are either substrates or inhibitors of the cellular or virus coded DNA-polymerases inhibiting cell division or replication of viruses or both of them.

1. Catabolic pathway

The first step in the catabolism of pyrimidine nucleosides or nucleoside analogues is the phosphorolysis of the ß-N- -glycosyl bond between the pentose and the pyrimidine bases. There are two different pyrimidine nucleoside phosphorylases in the mammalian cells. Uridine phosphorylase catalyzes the

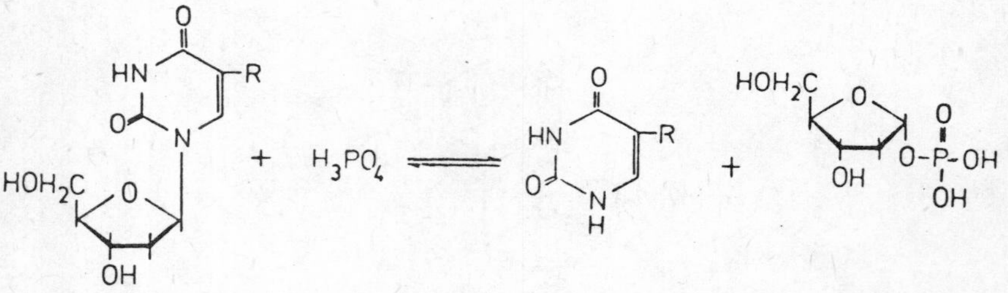

Fig.2. Phosphorolytic cleavage of the ß-N-glycosyl bond of nucleosides by nucleoside phosphorylase

reversible phosphorolysis of pyrimidine ribonucleosides and of 2'-deoxyribonucleosides, while thymidine phosphorylase is specific for 2'-deoxyribonucleosides. The products of these reactions are pyrimidine and ribose-1-phosphate or 2'-deoxy-ribose-1-phosphate, as Fig.2 shows [1, 2, 3, 4]. The rate of phosphorolysis of natural nucleosides and synthetic 2'-deoxyribonucleoside analogues catalysed by uridine phosphoryl-ase are shown in Fig. 3. The studied compounds were synthe-sized by earlier described

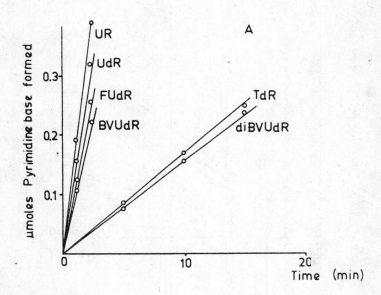

Fig. 3. The rate of phosphorolysis of nucleosides and their
        synthetic analogues by uridine phosphorylase.
        Substrate concentration 3.3 mM

methods in our Institute [5, 6]. It is surprising that the rate of phosphorolysis of uridine, 2'-deoxyuridine, 5-fluoro-2'-deoxyuridine and 5-bromovinyl-2'-deoxyuridine are very similar to each other. Among the tested compounds only thymidine and 5-dibromovinyl-2'-deoxyuridine are phospho-rolysed at lower rate than the others. The rates of phosphorolysis of the same nucleosides are more different from each other in the reaction catalysed by thymidine phosphorylase, as Fig.4 shows. It is important to note that 5-bromovinyl-

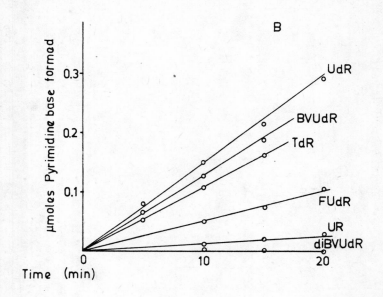

Fig.4. Thymidine phosphorylase catalyzed rate of phosphorolysis of the same nucleosides as shown in Fig.3.
Substrate concentration 3.3 mM

2'-deoxyuridine is one of the most potent antiviral agents, but only in its nucleoside form. 5-bromovinyl-uracil does not have antiviral activity. The rate of phosphorolysis of 5-alkyl-2'-deoxyuridines in the presence of uridine phospho-rylase and thymidine phosphorylase is shown in Fig. 5 and Fig.6. The rates of reactions catalysed by both enzymes are decreasing with increasing number of carbon atoms in the substituents. In this group of compounds, 5-ethyl-2'-deoxy-

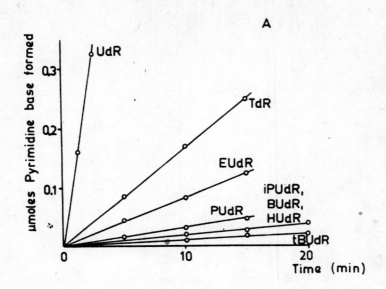

Fig. 5. The rate of phosphorolytic cleavage of 5-alkyl
        substituted uridine derivatives by uridine phospho-
        rylase.   Substrate concentration 3.3 mM.

uridine and 5-isopropyl-2'-deoxyuridine are good antiviral
agents but only in their nucleoside form, since 5-ethyluracil
and 5-isopropyluracil do not have antiviral activity. In more
general terms nucleoside phosphorylases are inactivating most
pyrimidine nucleoside analogues having antiviral activity,
because they are catalysing the phosphorolytic cleavage of the
N-glycosyl bond and the formation of the corresponding substituted pyrimidine bases without antiviral activity. The

initial step of catabolism of citidine and its synthetic
analogues is the deamination of the compounds by cellular
or virus coded citidine deaminase to form uridine or

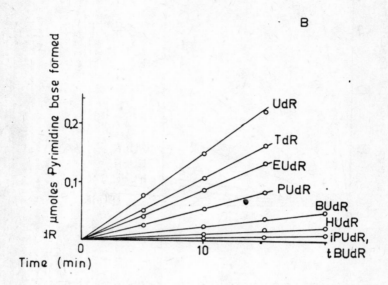

Fig.6. Thymidine phosphorylase catalyzed rate of phospho-
       rolysis of the same 5-alkyl-substituted uridine
       derivatives as in Fig. 5. Substrate concentration
       3.3 mM.

the corresponding uridine derivatives, which are either
substrates or not for uridine phosphorylase or thymidine
phosphorylase. This enzyme is responsible for the inactivation of cytosine arabinoside, one of the most useful
chemotherapeutic agent available for the tretament of acute
myeloblastic leukemia. Since pyrimidine nucleoside analogues

are quite unstable in biological systems because of the
activity of pyrimidine nucleoside phosphorylases, it was
reasonable to look for inhibitors for these enzymes. A series

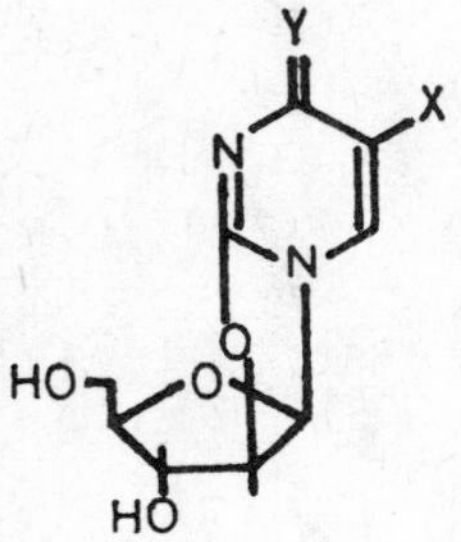

		Y	X
a.		-NH	-H
b.		-O	-F
c.		-O	-H
d.		-O	-CH$_3$
e.		-O	-CH$_2$-CH$_3$
f.		-O	-(CH$_2$)$_2$-CH$_3$
g.		-O	-CH$\begin{smallmatrix}CH_3\\CH_3\end{smallmatrix}$

Fig.7. 2,2'-anhydrouridine and its 5-substituted derivatives
       prepared and used for experiments

of substituted-2,2'-anhydrouridines have been synthetized
in our Institute, as shown in Fig. 7. In enzymological studies
it turned out, that these 2,2'-anhydrouridine derivatives
are not substrates of uridine phosphorylase or of thymidine
phosphorylase. These compounds inhibit the activity of
uridine phosphorylase but do no affect the activity of
thymidine phosphorylase. As Table 1 shows, the most potent
inhibitor of the activity of uridine phosphorylase is
5-ethyl-2,2'-anhydrouridine which inhibits the activity of
enzyme in nM concentration range and the inhibition is
competitive, as shown in Fig.8. The natural pyrimidine bases,
uracil and thymine and some of their substituted analogues

Table 1. Apparent K_i values for inhibition of uridine phosphorylase by 2,2'-anhydropyrimidine nucleosides[+]

Compound	appK$_i$ (nM)
2,2'-Anhydrouridine (c)	2268
2,2'-Anhydro-5-fluorouridine (b)	740
2,2'-Anhydro-5-methyluridine (d)	182
2,2'-Anhydro-5-ethyluridine (e)	20
2,2'-Anhydro-5-propyluridine (f)	79
2,2'-Anhydro-5-isopropyluridine (g)	1461
2,2'-Anhydrocytidine (a)	§

[+]Apparent K_i values were determined from Lineweaver-Burk plots of the data by a computer program with least-squares fitting. §, there is no detecable inhibition.

are catabolysed in the pyrimidine catabolic pathway. The rate litmiting step of degradation is the initial reduction of pyrimidine ring to 5,6-dihydro compound catalysed by dihydro-uracil dehydrogenase. The stubstrate specificity of dihydro-uracil dehydrogenase is narrow and only the natural pyrimidine bases and the 5-halogen substituted uracils as 5-fluorouracil, 5-iodouracil are substrates for the enzyme. The two 5-halogen substituted uracil derivatives 5-iodo-2'-deoxyuridine and 5-fluorouracil are among the few antiviral and cytostatic agents which are currently used in clinical medicine. The reduction of the pyrimidine ring is

followed by a series of hydrolytic reactions catalysed by
enzymes as shown inFig 9. The end-products of uracil and
thymine degradation are β-alanine and β-amino-isobutyric

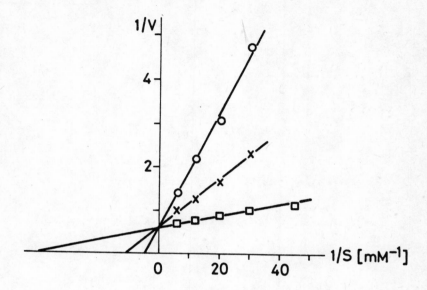

Fig.8. Inhibition of activity of uridine phosphorylase with
2,2'-anhydro-5-ethyluridine. Double reciprocal plot
of initial velocity vs. substrate concentration.
Key: (□) uridine alone; (x) 65.5 nM 2,2'-anhydro-5-
-ethyluridine; (o) 25o nM 2,2'-anhydro-5-ethyluridine.

acid, respectively. From 5-halogen substituted uracils the
corresponding α-halogen substituted β-alanine derivatives
are formed. Most of the 5-substituted uracils derived from
5-substituted-2'-deoxyuridines by phosphorolytic cleavage are
not metabolised in the pyrimidine catabolic pathway. We have

found 5-ethyluracil to be hydroxylated in its ethyl group at the α-position.   5-Isopropyl-2'-deoxyuridine(since it is remarkably resistant to uridine phosphorylase)is hydroxylated as nucleoside in its isopropyl group at β-position.

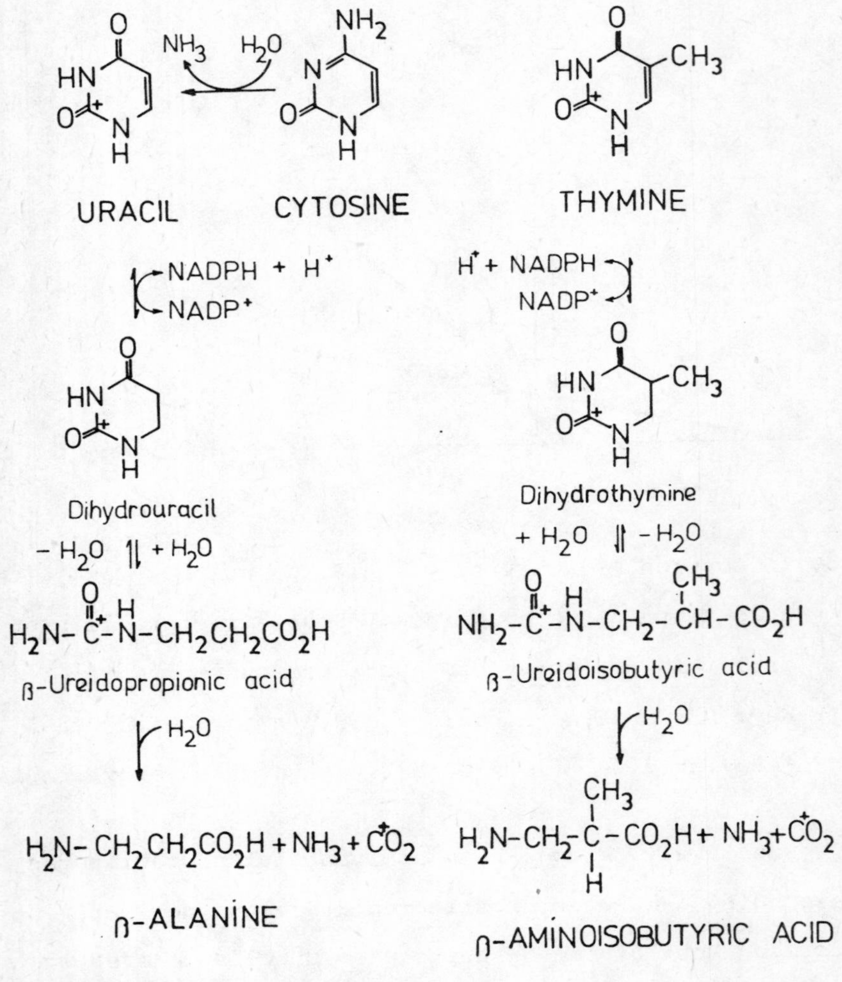

URACIL   CYTOSINE   THYMINE

Dihydrouracil

$-H_2O \parallel +H_2O$

$H_2N-\overset{O}{\overset{\parallel}{C}}-\overset{H}{\overset{+}{N}}-CH_2CH_2CO_2H$

β-Ureidopropionic acid

$\downarrow -H_2O$

$H_2N-CH_2CH_2CO_2H + NH_3 + \overset{+}{C}O_2$

β-ALANINE

Dihydrothymine

$+H_2O \parallel -H_2O$

$NH_2-\overset{O}{\overset{\parallel}{C}}-\overset{H}{\overset{+}{N}}-CH_2-\overset{CH_3}{\overset{|}{C}H}-CO_2H$

β-Ureidoisobutyric acid

$\downarrow -H_2O$

$H_2N-CH_2-\overset{CH_3}{\underset{H}{\overset{|}{C}}}-CO_2H + NH_3 + \overset{+}{C}O_2$

β-AMINOISOBUTYRIC ACID

Fig.9. Catabolic pathways for degradation of natural pyrimidine bases and some of their synthetic analogues.

In general, all natural and some substituted pyrimidine bases
produced by phosphorolytic cleavage of ribo- or deoxyribonucleosides enter into the pyrimidine catabolic pathway for
degradation. The degradation products of the pyrimidine

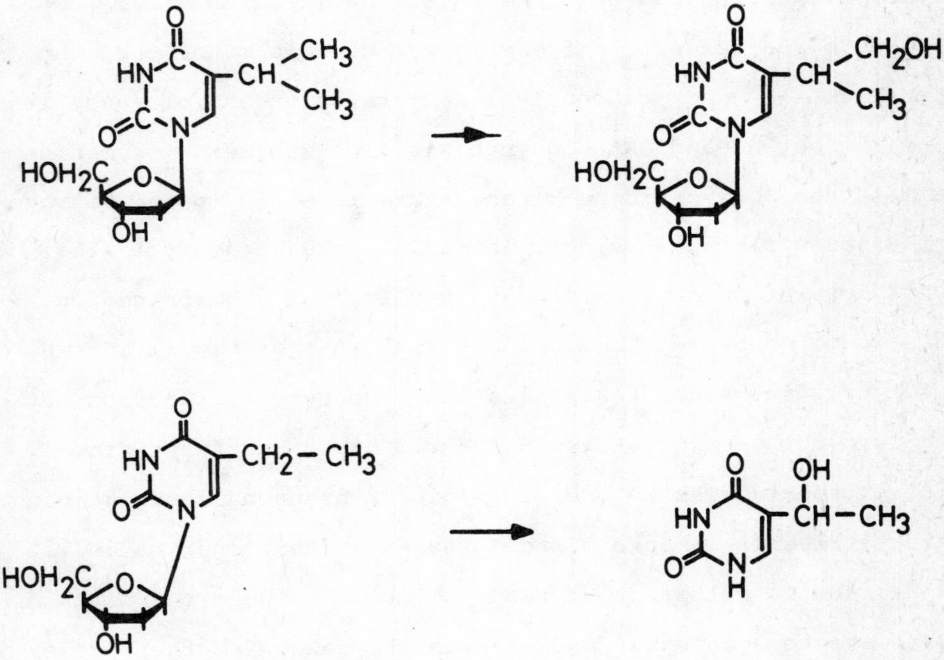

Fig.10. Hydroxylated derivatives of 5-ethyl-2'-deoxyuridine
and 5-isopropyl-2'-deoxyuridine as products of
catabolism.

catabolic pathway (together with pyrimidine derivatives not
metabolized because of structural reasons) are transported by
blood from the cells, where they are produced to the kidneys,
where they are excreted.

2. Anabolic pathway.

It is a well known fact that all kind of foods contain
nucleic acids. These nucleic acids are degraded by hydrolytic
enzymes in the gastrointestinal tract of man or animals and,
among others, pyrimidine-2'-deoxyribonucleosides - thymidine
and 2'-deoxycytidine - are formed. These 2'-deoxyribonucleosides
after being absorbed from the intestinal tract are transported by
blood to the cells of different organs. This means that the
pyrimidine-2'-deoxyribonucleosides are salvaged for reuse in
the nucleotide and DNA biosynthesis. Two enzymes - thymidine
kinase and deoxycytidine kinase - are involved in the salvage
of 2'-deoxyribonucleosides. Thymidine kinase catalyzes the
phosphorylation of thymidine to thymidine-5'-phosphate and
the compound is required for DNA synthesis, as Fig.11 shows.
Deoxycytidine kinase catalyzes the same type of reaction but
the product of the catalyzed reaction is 2'-deoxycytidine-
-5'-phosphate. The enzymes thymidine kinase and deoxycytidine
kinase are not members of the previously mentioned catabolic
and anabolic pathways but they connect the two types of
pathways to each other and this is the reason of their name
as amphibolic enzymes. These enzymes are playing crucial
role in the activation of synthetic nucleoside analogues.
The substrate specificity of the two amphibolic enzymes are
quite narrow. The cellular thymidine kinase catalyzes the
phosphorylation of thymidine and a few thymidine analogues

as 5-fluoro-, 5-chloro-, 5-iodo-, and 5-trifluoromethyl-2'-
deoxyuridine to 5'-phosphates. The 5'-phosphates are phosphorylated by other cellular enzymes to nucleoside-5'-tri-
phosphates and they are incorporated into the DNA of the
cells [7,8]. The specificity of cellular deoxycytidine kinase

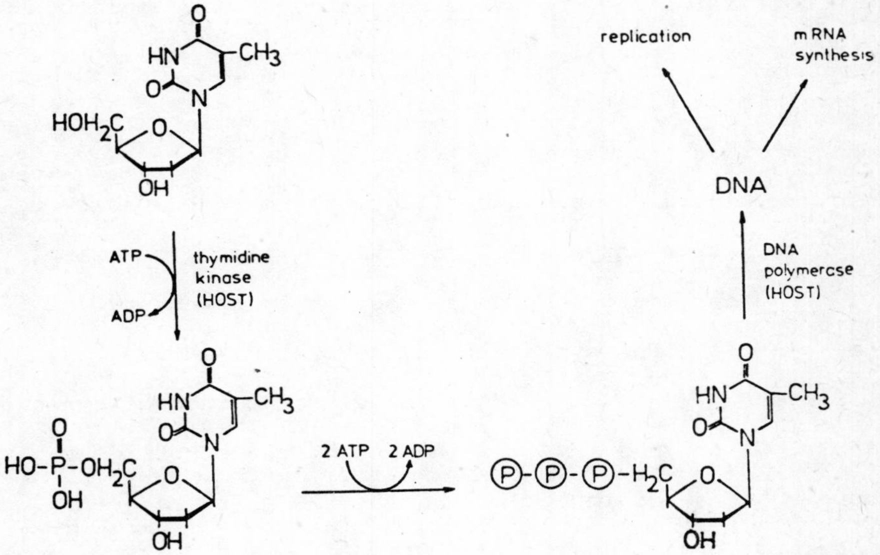

Fig.11. Activation of thymidine by cellular thymidine kinase

is also narrow and it phosphorylates 5-halogen substituted
deoxycytidine derivatives [9]. Since the above mentioned
2'-deoxyribonucleoside analogues are activated and incorporated
into the cellular DNA by enzymes of the cells, they are
cytotoxic and some of them are used as anticancer agents.

It was discovered in the early sixties that herpes simplex virus induces an increase of thymidine kinase activity in the infected cells [1o]. It was later shown that herpes simplex virus induces the formation of both thymidine kinase and deoxycytidine kinase and the two enzyme activities belong to the same protein molecule [11]. The cellular thymidine kinase and deoxycytidine kinase are separate enzymes but these two enzyme activities induced by herpes simplex virus belong to one single protein molecule, therefore the enzyme was termed deoxypyrimidine kinase [11] It was proved experimantally that the herpes simplex virus type 1 and type 2 coded deoxypyrimidine kinase enzymes are different from each other and from the cellular enzymes [12, 13]. Among the herpes simplex virus coded enzymes, one which is absolutely necessary for the herpes virus DNA synthesis is the DNA polymerase [14, 15, 16]. Since double-stranded DNA viruses are coding the synthesis of their own enzymes involved in the activation of deoxyribonucleosides and in the synthesis of DNA, it was a rational approach to develop analogues of thymidine and deoxycytidine which could be phosphorylated only by virus-coded enzymes but not by human or animal cellular thymidine or deoxycytidine kinase. Because these analogues would not be phosphorylated in the healthy cells, it is not probable, that they would exert any adverse effect on the normal cell growth. The most likely group of deoxyribonucleoside analogues having specific antiviral effect without influence on cellular processes were the 5-alkyl-2'-deoxyuridines [17, 18, 19, 2o, 21]. It was a surprise, that 5-iodo-5'-amino-2',5'-dideoxyuridine is

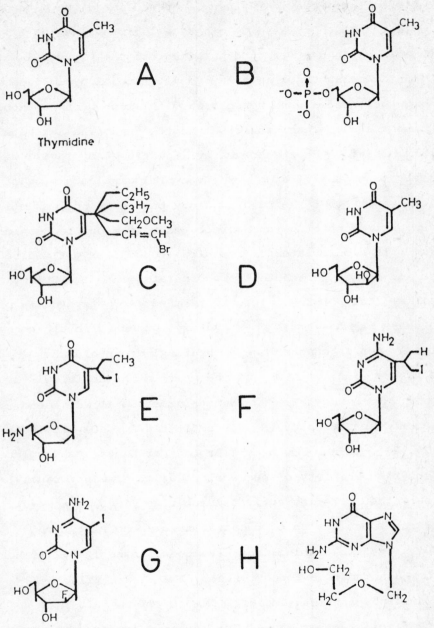

Fig.12. Substrates for herpes simplex virus type 1 coded thymidine kinase

a substrate for the virus coded deoxypyrimidine kinase and it catalyzes the formation of a phosphoroamidate. The phospho-roamidate is phosphorylated by other cellular enzymes to tri-phosphoamidate, then it is incorporated into the virus DNA [22]. This finding means that the 5'-hydroxyl group of pyrimidine deoxyribonucleosides can be replaced with amino group without affecting the substrate nature of the compound. The antiviral activity of 9-(2-hydroxyethoxymethyl)-guanine (Acyclovir) was the most unexpected finding [23]. The biological activity of Acyclovir shows that the sugar moiety per se is not essential and the pyrimidine ring can be replaced by the purine, guanine. Acyclovir is a substrate for the virus-coded deoxypyrimidine kinase and its triphosphate is incorporated into the DNA of the virus. The incorporated Acyclovir terminates the synthesis of DNA because it does not have a hydroxyl group which stereo-chemically corresponds to the 3'-hydroxyl group of the ribofuranose ring [24]. One of the most potent antiviral agent is 5-(2-bromovinyl)-2'-deoxyuridine and it represents the second generation of the 5-substituted-2'-deoxyuridines [25, 26, 27]. It was found earlier that arabinose can replace the deoxyribose moiety of nucleosides as in arabinofuranosyl-thymine without loosing antiviral activity [28]. As a result of further substitution of the arabinofuranose ring, two second generation arabinosides have been synthesized: 2'-fluoro-2'-deoxy-arabinosyl-5-iodocytosine (FIAC) and 2'-fluoro-2'-deoxy-arabinosyl-5-methyluracil (FMAU) [29, 3o]. It is important to note that the 2' "up" arabino configuration of the fluoro substitutent of FIAC is important from the point of view of antiviral activity. If the 2'-fluoro substituent of FIAC is in "down" position or in ribo configuration, the compound is at least thousand times less active than the arabino isomer. Another important feature of FIAC is its resistance to phosphorolytic cleavage catalyzed by nucleoside phosphorylases. We may assume some kind of interaction e.g. a hydroge: bond between functional groups of the pyrimidine ring and the substituted sugar moiety or between different groups of sugar moiety which stabilizes the N-glycosidic bond of the nucleoside. It is possible also that either the conformation of FIAC or an enzyme induced conformational change (syn or anti) stabilizes the compound.

The results concerning the chemical structure - biological activity relationship in connection with the problems of biotransformation of nucleosides are convincing enough to see that this field of research just entered into an exciting phase of drug development.

REFERENCES

1. Krenitshy, T.A., Mellors, J.W., and Barclay R.K. Pyrimidine Nucleosidases. Their classification and relationship to uric acid ribonucleoside phosphorylase. J.biol.Chem. 240, 1281-1286 (1965).

2. Ishitsuka, H., Miwa, M., Takemoto, K., Fukuoka, K., Itoga, A., and Maruyama, H.B. Role of uridine phosphorylase for antitumor activity of 5'-deoxy-5-fluorouridine. Gann, 71, 112-123 (1980).

3. Woodman, P.W., Sarrif, A.M., and Heidelberger, C.
    Specificity of pyrimidine nucleoside phosphorylases and
    the phosphorolysis of 5-fluoro-2'-deoxyuridine.
    Cancer Res. 40, 507-511 (1980).

4. Nakayama, C., Wataya, Y., Meyer, R.B., Santi, D.V.
    Thymidine Phosphorylase. Substrate specificity for
    5-substitued 2'-deoxyuridines. J.med.Chem. 23,
    962-964 (1980).

5. Szabolcs, A., Sági, J., Ötvös, L. Synthesis of 5-alkyl-
    -2'-deoxyuridines. J.Carbohydrates, Nucleosides., Nucleo-
    tides. 2, 197-211 (1975).

6. Szabolcs, A., Kruppa, G., Sági, J., Ötvös, L. Unnatural
    nucleosides and nucleotides. III. Preparation of $2-^{14}C$
    and $4-^{14}C$ labelled 5-alkyluracils and 5-alkyl-2'-
    -deoxyuridines. J.of Labelled Comp. and Radiopharm.
    14, 713-726 (1978).

7. Prusoff, W.H. Synthesis and biological activites of
    iododeoxyuridine, an analog of thymidine. Biochim.
    biophys. Acta 32, 295-296 (1959).

8. Bresnick, E., Williams, S.S. Effects of 5-trifluoro-
    methyldeoxyuridine upon deoxythymidine kinase. Biochem.
    Pharmacol. 16, 503-507 (1967).

9. Cooper, G.M., and Greer, S. Phosphorylation of 5-
    -halogenated deoxycytidine analogues by deoxycytidine
    kinase. Mol.Pharmacol. 9, 704-710 (1973).

10. Kit, S., and Dubs, D.R. Acquisition of thymidine kinase
    activity by herpes simplex infected mouse fibroblast
    cells. Biochem. Biophys. Res. Commun. 11, 55-59 (1963).

11. Jamieson, A.T., Gentry, G.A., and Subak-Sharpe, J.H. J.gen.Virol. 24, 465-480 (1974).

12. Ogino, T., and Rapp, F. Differences in thermal stability of deoxythymidine kinase activity in extracts from cells infected with herpes simplex virus type 1 and 2. Virology. 46, 953-955 (1971).

13. Thouless, M.E. Serological properties of thymidine kinase produced in cells infected with type 1 or type 2 herpesvirus. J.gen.Vird. 17, 307-315 (1972).

14. Keir, H.M. DNA polymerases from mammalian cells. Prog.nucl.Acid.Res.molec.Biol. 4, 81-129 (1965).

15. Müller, W.E.G., Falke, D., and Zahn, R.K. DNA dependent DNA polymerase pattern in noninfected and herpesvirus infected rabit kidney cells. Arch.ges.Virusforsch. 42, 278-284 (1973).

16. Weissbach, A., Hong, S.L., Aucker, J., and Müller, R. Characterization of herpes simlex virus-induced deoxy-ribonucleic acid polymerase. J.biol.Chem. 248, 6270-6277 (1973) .

17. Swierkowsky, M., and Shugar, D. A nonmutagenic thymidine analog with antiviral activity, 5-ethyldeoxyuridine. J.med.Chem. 12, 533 (1969).

18. De Clercq, E., and Shugar, D. Antiviral activity of 5-ethyl pyrimidine deoxynucleosides. Biochem.Pharmacol. 24, 1073-1078 (1975).

19. Cheng, Y.C., Domain, B.A., Sharma, R.A., Bobek, M. Antiviral action and cellural toxicity of four thymidine analogues; 5-ethyl-, 5-vinyl-, 5-propyl-, and 5-alkyl--2'-deoxyuridine. Antimicrob.Agents Chemother. 10, 119-122 (1976).

20. Sagi, J., Novak, R., Zumundzka, B., Szemző, A., and Ötvös, L. Study of the substrate specificity of mamalian and bacterial DNA polymerases with 5-alkyl--2'-deoxyuridine 5'triphosphates. Biochim.biophys. Acta. 606, 196-201 (1980).

21. Kovaluik, L., Gauri, K.K. Spadari, S., Pedrali-Nay, G., Kühne, J., and Koch, G. Differntial incorporation of thymidylate analogues into DNA by DNA polymerase α and by DNA polymerases specified by two herpes simplex viruses. J.gen.Virol. 62, 29-38 (1982).

22. Chen, M.S., Ward, D.C., Prusoff, W.H. Specific herpes simplex virus induced incorporation of 5-iodo-5'--amino-2',5' dideoxyuridine into into deoxyribonucleic acid. J.biol.Chem. 251, 4833-4838 (1976).

23. Elion, G.B., Furman, P.A., Fyfe, J.A., De Miranda, P., Beauchamp, L., and Schaeffer, H.J. Selectivity of action of an antiherpetic agent, 9-(2-hydroxyethoxymethyl) guanine. Proc.natl.Acad.Sci. 74, 5716-5720 (1977).

24. Derse, D., Cheng, Y-C., Furman, P.A., St.Clair, M.H., and Elion, G.B. Inhibition of purified human and herpes simplex virus-induced DNA polymerases by 9-(hydroxyethoxymethyl) guanine triphosphate. Effects on primer--template function. J.biol.Chem. 256, 11447-11451 (1981)

25. De Clercq, E., Descamps, J., De Somer, P., Barr, P.J., Jones, A.A., and Walker, R.T. E-5-(2-bromovinyl)-2'--deoxyuridine: a potent and selective anti-herpes agent. Proc.natl.Acad.Sci. 76, 2947-2951 (1979).

26. Sagi, J., Szabolcs, A., Szemző, A., and Ötvös, L. /E/--5-(2-bromovinyl)-2'-deoxyuridine-5'-triphosphate as a DNA polymerase substrate. Nucleic Acids Res. 9, 6985-6994 (1981).

27. Mancini, W.R., De Clerq, E., and Prusoff, W.H. Relation-
     ship between incorporation of E-5-(2-bromovinyl)-2'-
     -deoxyuridine into herpes simplex virus type 1 DNA with
     virus infectivity and DNA integrity. J.biol.Chem. _258_,
     792-795 (1983).

28. Gentry, G.A., and Aswell, J.F. Inhibition of herpes
     virus replication by ara T. Virology _65_, 294-296 (1975).

29. Watanabe, K.A. Reichman, U., Hirota, K., Lopez, C., and
     Fox, J.J. Nucleosides. 110. Synthesis and antiherpes
     virus activity of some 2'-fluoro-2'-deoxyarabino-
     furanosylpyrimidine nuckeosides. J.med.Chem. _22_, 21-24
     (1979).

30. Wanatabe, K.A., Su, T.-L., Klein, R.S., Chu, C.K.,
     Matsuda, A., Chun, M.W., Lopez, C. Fox, J.J. Nucleosides.
     123. Synthesis of antiviral nucleosides: 5-substituted
     1-(2'-deoxy-2'-halogeno-β-D-arabino-furanosyl)-cytosines
     and -uracils some structure activity relationship.
     J.med.Chem _26_, 152-156 (1983).

DISCUSSION

PREOBRAZHENSKAYA M.N. (Moscow, USSR)

In connection with the stability of FIAC against enzymatic phosphorolysis, do you have a CD spectrum of the compound? It may prove whether your hypothesis concerning a hydrogen bond stabilizing the molecule in syn conformation is correct.

DÉNES G. (Budapest, Hungary)

As I mentioned before, FIAC has been synthesized in Prof. Fox's laboratory and we do not have the compound or its CD spectrum. I agree with you that pyrimidine-2'-deoxy-ribonucleoside molecules are found in anti conformation in the crystalline state or in solution. In the case of an enzyme-catalyzed reaction, the conditions are completely different. The interaction of a substrate with the binding site of the enzyme may induce conformational change in the substrate molecule. It is also possible that a molecule may have two different conformations and they are in equilibrium with each other. If the proportion of the two conformers is 100:1 or 1000:1 and only the minor conformer is taking part in enzyme-substrate complex formation, you would think that the conformation of the substrate is the same as the conformation of the majority of molecules. I should like to mention that the half saturation concentration of uridine phosphorylase with 5-ethyl-2,2'-anhydrouridine having syn conformation is in the nM concentration range, while the half saturation concentration of the enzyme with 5-ethyl-2'-deoxy-uridine having anti conformation is in the µM concentration range. I think, this difference of three orders of magnitude in the concentrations is in connection with syn or anti conformation of the compounds studied. Theoretically, the N-glycoside bond may be stabilized by weak interactions between the pyrimidine and the arabinofuranose rings, or the conformational change of the substituted arabinofuranose ring in itself protects the glycosidic bond. I hope, we can get a sample of FIAC and we shall be able to clarify the structural reason for this mysterious stability of the compound in the enzyme-catalyzed phosphorolytic reaction.

PISKALA A. (Praha, Czechoslovakia)
I agree with Professor Preobrazhenskaya that hydrogen bonding is questionable. Instead, my explanation on stability is rather the increased acidity of oxygen in the furanose ring caused by the fluorine atom.

DÉNES G. (Budapest, Hungary)
This is a possible explanation but we do not have any experimental evidence.

PISKALA A. (Praha, Czechoslovakia)
In my opinion, the inactivity of the fluorine epimer is the consequence of its ribose configuration being not a substrate of HSV-1 coded thymidine kinase.

DÉNES G. (Budapest, Hungary)
I agree with you.

PANDIT U.K. (Amesterdam, The Netherlands)
Hydrogen bond can be very significant in the mechanism of stability and I suggest you to think about a hydrogen bond between 5'-OH and 2'-F when the latter is positioned above the sugar ring.

DÉNES G. (Budapest, Hungary)
Thank you for your suggestion.

PREOBRAZHENSKAYA M.N. (Moscow, USSR)
My second remark is connected with the enzymatic phosphorolysis reaction. Transformation the of nucleoside into sugar-α-1-phosphate is tne consequence of different stabilities: α-1-phosphates are more stabile than β-1-phosphates.

DÉNES G. (Budapest, Hungary)

In reply I should like to mention that in the case of nucleosides prepared by synthetic methods, a mixture of the α- and β-anomers is formed. After separation, the α- and β-anomer nucleosides are equally stabile chemically and no spontaneous inversion occurs between them. In α-D-ribose-1-phosphate formation from β-N-pyrimidine nucleosides and inorganic phosphate catalyzed by pyrimidine nucleoside phosphorylase, the anomeric carbon atom certainly undergoes inversion from β-to α-configuration in the catalytic pathway. In case of 5-phospho-β-D-ribosyl-1-amine formation from 5-phospho-α-D-ribose-1-pyrophosphate and glutamine catalized by aminophosphoribosyl transferase, the anomeric carbon atom undergoes inversion again, but in this case from the α-to the β-configuration. It seems that the above inversions of the anomeric carbon atom of ribose are important events in the catalytic pathway of the mentioned enzymes. I agree with you, however, if we use the term "stability" from the point of view of thermodynamics.

Bio-Organic Heterocycles
van der Plas H.C., Ötvös L., and Simonyi M. eds

STEREOCHEMICAL ASPECTS OF THE INTERACTION OF HETEROCYCLIC SUBSTRATES AND DRUGS WITH ENZYMES

YOUNG D.W.

School of Chemistry and Molecular Sciences,
University of Sussex,
Falmer, Brighton, BN1 9QJ, United Kingdom

The nucleic acids DNA and RNA are 'coded' by four heterocyclic bases. Two of these are purines and two are pyrimidines. Three of these bases are common to both DNA and RNA but the base thymine (1, R' = H) is present only in DNA, being replaced in RNA by the base uracil (2, R' = H). Cell division (mitosis) occurs by fission of the two paired helical strands of DNA. Each of the developing single strands then acts as a coded template on which the new second strand is built. If there is an insufficiency of thymidine monophosphate (1, R' = deoxyribosemonophosphate) in the cell during this process to pair with the adenine of the template then a second strand cannot develop and mitosis will not occur.

(1) (2) (3)

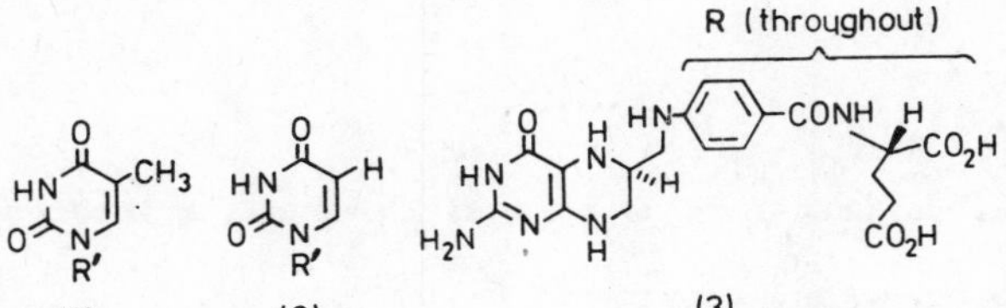

One difference between some cancer cells and their healthy counterparts is that the cancer cells divide more rapidly. Compounds which interfere with cell division will therefore be more likely to interact with a cancer cell than with a healthy cell and so may exhibit selective

toxicity to the cancer cells. Since control of cellular thymidine levels
is a way of interfering with mitosis, compounds which interfere with the
biosynthesis and breakdown of thym(id)ine are of interest as potential
anticancer drugs.

The biosynthesis of thymidine monophosphate (1) involves methylation
of deoxyuridine monophosphate (dUMP) (2). This is a one-carbon transfer
and is mediated by the coenzyme tetrahydrofolic acid (3) [1]. This coenzyme can react chemically as a 1,2-diamine with formaldehyde to form
the adduct (4) which may also be formed, as we shall see later, in an
enzymic reaction. There are enzymes which will oxidise the adduct (4)

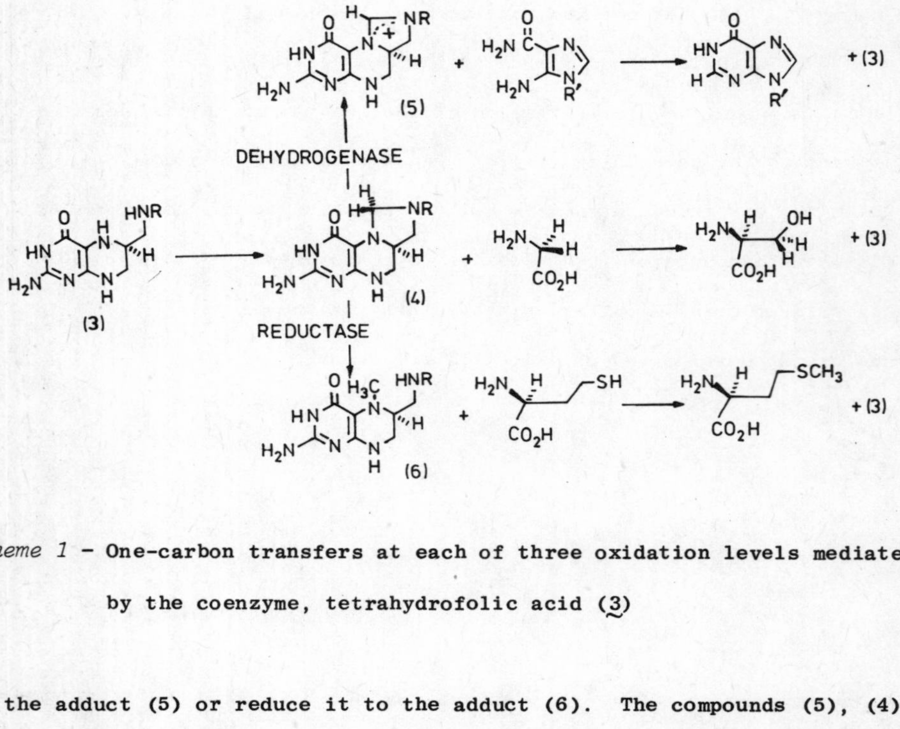

Scheme 1 - One-carbon transfers at each of three oxidation levels mediated
by the coenzyme, tetrahydrofolic acid (3)

to the adduct (5) or reduce it to the adduct (6). The compounds (5), (4)
and (6) are adducts containing tetrahydrofolic acid covalently bound to
a one-carbon unit in the formic acid, formaldehyde and methanol oxidation
levels respectively and these are used to transfer the one-carbon moiety

at the appropriate oxidation level to a biological substrate. Three representative transfers are shown in Scheme 1 above and in each case the coenzyme (3) is regenerated.

The one-carbon transfer involved in the biosynthesis of thymidine monophosphate (1) from deoxyuridine monophosphate (2) is of particular interest to the chemist since, although the total transfer is at the methanol oxidation level, the formaldehyde oxidation level adduct (4) is used to affect the one-carbon transfer. This transfer is catalysed by the enzyme *thymidylate synthetase* and the change in oxidation level required in the process is achieved by the hydrogen at C-6 (H_A) of tetrahydrofolate being transferred to the resultant methyl group of the thymine moiety [2]. Thus the coenzyme is oxidised to dihydrofolic acid (7) in the process and a second enzyme, *dihydrofolate reductase* is required to regenerate the coenzyme and thus make thymidine biosynthesis a continuous process. This is shown in Scheme 2 below.

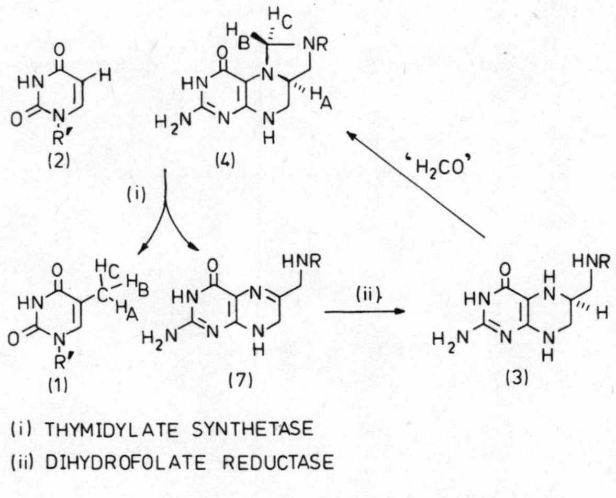

(i) THYMIDYLATE SYNTHETASE
(ii) DIHYDROFOLATE REDUCTASE

Scheme 2 − **Biosynthesis of thymidine monophosphate**

Inhibition of the two enzymes involved in thymidine biosynthesis will evidently reduce cell thymidine levels and so inhibitors may be useful anti-cancer drugs.   It has been shown [3] that inhibitors of *dihydrofolate reductase* such as methotrexate (8), and of *thymidylate synthetase* such as a metabolite of 5-fluorouracil (9) are indeed clinically useful for cancer treatment.   These compounds structurally resemble the substrates of the enzymes.

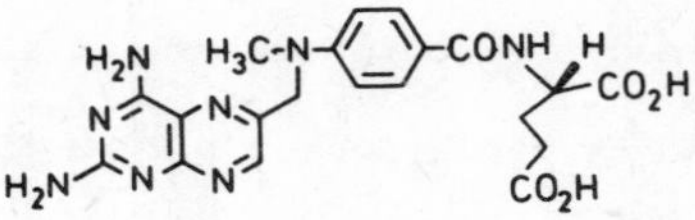

(8) 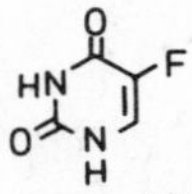   (9)

We have been interested in the mechanism of methylation of dUMP with its fascinating hydrogen migration and, as part of a programme to study the process, have been looking at the overall stereochemistry of the two enzymic reactions involved.   Our first target was the reaction catalysed by *dihydrofolate reductase*.   This enzyme reduces the vitamin folic acid (10) as well as dihydrofolic acid (7) and so it was of interest to assess the stereochemistry of both reductive steps, not only with respect to the centres at C-7 and C-6 of the pteridine ring but also with respect to C-4 of the reducing coenzyme NADPH (11).

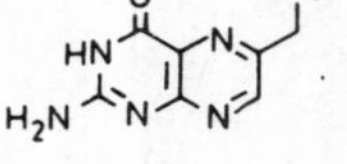

(10) 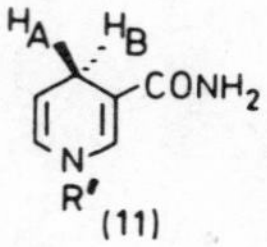   (11) 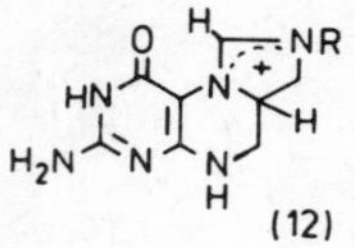   (12)

We had originally intended to assess the stereochemistry at C-6 of the active coenzyme by total stereospecific synthesis of both C-6 epimers.   In 1979, however, Fontecilla-Camps *et al.* [4] published an X-ray structural analysis of both diastereoisomers of 5,10-methenyltetra-

hydrofolic acid (12). Since the absolute stereochemistry of these

compounds was known, formal assessment of the stereochemistry

of the active coenzyme (3) would be achieved by its

chemical conversion to one of these compounds. We therefore prepared

dihydrofolic acid (7) by dithionite reduction of folic acid (10) and

converted this to the coenzyme by reduction using *dihydrofolate reductase*

and NADPH. The active coenzyme was then reacted with formic acetic

anhydride to yield the diformyl derivative (13). This was selectively

deformylated using sodium hydroxide to yield that isomer of folinic acid

(14) which had been converted to the 6R diastereoisomer of compound (12)

in the X-ray studies. The stereochemistry of the coenzyme was therefore

(6S) and so the enzyme had reduced dihydrofolic acid from the *re*-face at

C-6.

(13)   (14)

The next objectives were to assess the stereochemistry of the

reduction at C-7 of folic acid (10) and at C-4 of NADPH. We were able to

achieve these ends in collaboration with Dr. J. Feeney and his colleagues

at the National Institute for Medical Research. $(4R)[4-^2H_1]$-NADPH

$(11, H_A = {}^2H)$ was prepared by reducing $[4-^2H_1]$NADPH [5] using the

'B specific' enzyme *glucose-6-phosphate dehydrogenase* [6]. This was then

used with the enzyme *dihydrofolate reductase* to reduce the vitamin folic

acid (10) to the coenzyme (3) in an n.m.r. tube. The ^{1}H-n.m.r. spectrum

of the resultant coenzyme is shown in Figure 1B together with the

spectrum (Figure 1A) from a similar experiment where unlabelled NADPH was used as reductant. These spectra show that deuterium rather than hydrogen has been transferred to both C-6 and C-7 of the coenzyme in the process so that it is the (4-*pro R*) hydrogen of NADPH which is transferred in *both* reduction steps. Further, of the C-7 hydrogens, it is the one which exhibits the smaller vicinal coupling constant which has been replaced by deuterium. This has been assigned as the C-7 hydrogen which is *cis* to the hydrogen at C-6 by careful analogy [7] and so must be the (7-*pro S*) hydrogen.

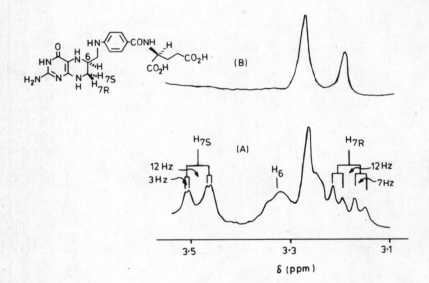

Figure 1 - Part of the 270 MHz ^{1}H-n.m.r. spectrum in ^{2}H$_2$O of
(A) tetrahydrofolic acid; (B) [6,7-^{2}H$_2$]-tetrahydrofolic acid from enzymic reduction of folic acid.

Reduction of the vitamin folic acid therefore involves the (4-*pro R*)-hydrogen of NADPH in both steps and it is the *re*-face with respect to C-6 and the *si*-face with respect to C-7 of the pteridine which

146

is reduced as shown in Scheme 3 below. In spite of the confusion caused
by nomenclature, this represents addition *at the same face* of the
pteridine at C-6 and C-7.

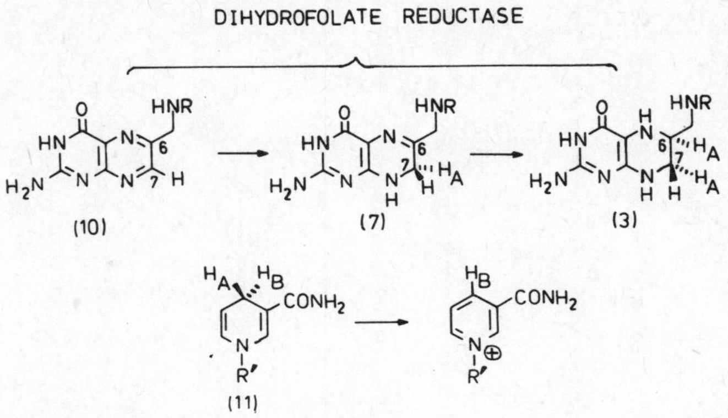

DIHYDROFOLATE REDUCTASE

Scheme 3 - Reduction of folic acid by NADPH

These results have particular significance when considered in the
light of an X-ray study of a ternary complex of *dihydrofolate reductase*
with the coenzyme NADPH and the anti-cancer drug methotrexate (8) [8].
This work shows that although it is the (4-*pro R*) hydrogen of NADPH which
approaches the pteridine ring of methotrexate (8) at C-6 and C-7, it is
in fact the opposite face of the pteridine (8) which is approached to
that face of the pteridine ring which we have shown to be reduced in the
substrate folic acid (10). Thus the substrates (10) and (7) must be bound
to the enzyme in a different way from the anti-cancer drug (8) in spite
of their apparent structural similarity. Use of computer graphics [9] has
suggested to us that this may be due to the fact that a strong hydrogen
bond between the 4-amino group of methotrexate and the enzymic
Leu-4/Trp-5 amide carbonyl is not possible when methotrexate is replaced
by compounds (10) and (7). The carbonyl group at C-4 of these pteridines

may however hydrogen bond to the carboxyl group of Asp 26 of the enzyme while keeping most other interactions intact. This would require the pteridine ring to be rotated through 180° compared to methotrexate and so would account for the stereochemistry of the reduction.

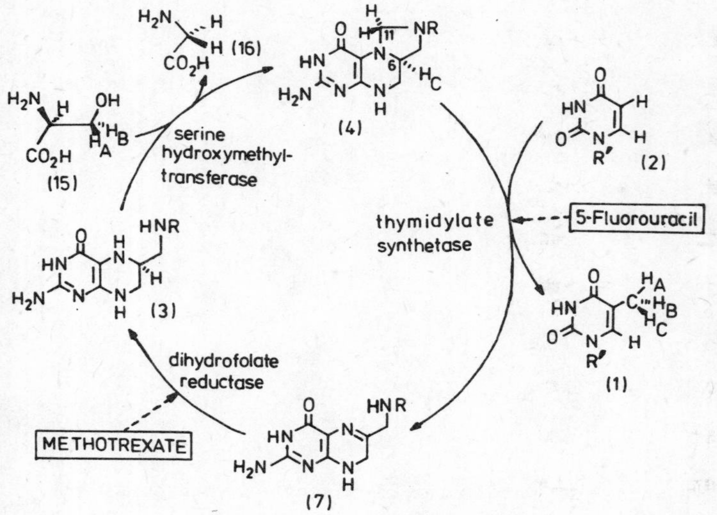

Scheme 4 - Synthesis of thymidine in nature

Having assessed the overall stereochemistry of one of the enzymic reactions involved in thymidine biosynthesis, we now turned to the second of these reactions, the one-carbon transfer reaction catalysed by the enzyme *thymidylate synthetase*. In the simplistic description of thymidine synthesis of Scheme 2, nature in fact converts the coenzyme (3) to the adduct (4) by coupling the reaction to a reverse aldol process in which serine (15) is converted to glycine (16). This reaction is catalysed by the enzyme *serine hydroxymethyltransferase* and so a more accurate description of the cycle involved in thymidine biosynthesis is as shown in Scheme 4. The overall stereochemistry of the combined reactions catalysed by *serine hydroxymethyltransferase* and *thymidylate synthetase* has been shown by Floss and Benkovic [10] to be as depicted in Scheme 4 so that the discovery of the steric course of either of these enzymic reactions would reveal the stereochemistry of the other reaction.

We have prepared the adduct (4) stereospecifically labelled at the bridge carbon, C-11, with deuterium [11] and so a synthesis of serine (15) labelled stereospecifically at C-3 with deuterium was required for a study of the stereochemistry of the enzymic reactions. We achieved this synthesis by adapting a synthesis of C-3 labelled glutamic acids which we had completed in our laboratory [12].

The commercially available enzyme *aspartase* is known [13] to add ammonia across the double bond of fumaric acid with *trans* stereospecificity. Thus we were able to obtain synthetically useful amounts of $(2S,3S)[2,3-^2H_2]$-aspartic acid (17, $H_A = {}^2H$) from $[2,3-^2H_2]$-fumaric acid using this enzyme whilst use of unlabelled fumaric acid and ${}^2H_2O$ in the incubation gave $(2S,3R)[3-^2H_1]$-aspartic acid (17, $H_B = {}^2H$). These samples of aspartic acid were used separately in the synthesis of labelled serine.

Reaction of labelled aspartic acid (17) with trifluoroacetic anhydride gave the intermediate N-trifluoroacetylanhydride which, on reaction with methanol, gave a mixture containing 80% of the α-monomethyl ester (18) and 20% of the regioisomeric β-monomethyl ester. The mixture of esters was converted to a mixture of acid chlorides from which the acid chloride (19) could be obtained by recrystallisation from benzene. Reaction of the acid chloride with diazomethane gave the corresponding diazoketone and reduction of this with hydrogen iodide in chloroform gave the methyl ketone (20). Baeyer Villiger oxidation and hydrolysis gave a mixture from which serine (15) (19%) and aspartic acid (17) (32%) could be isolated by ion-exchange chromatography.

11 Plas

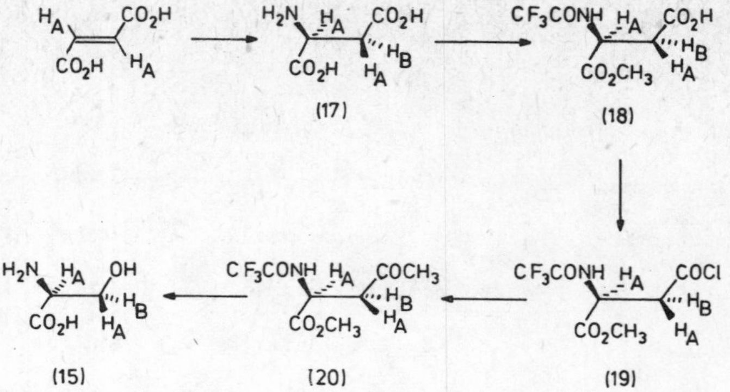

H_A CO_2H / CO_2H H_A → H_2N H_A CO_2H / CO_2H H_B (17) → CF_3CONH H_A CO_2H / CO_2CH_3 H_B H_A (18)

H_2N H_A OH / CO_2H H_B H_A (15) ← CF_3CONH H_A $COCH_3$ / CO_2CH_3 H_B H_A (20) ← CF_3CONH H_A $COCl$ / CO_2CH_3 H_B H_A (19)

Scheme 5 – Synthesis of serine stereospecifically labelled with
deuterium at C-3

The synthesis, outlined in Scheme 5 has as its key step the Baeyer
Villiger reaction which would be expected [14] to occur with retention of
chirality at the labelled chiral migrating group. The [1]H-n.m.r. spectra
(Figure 2, following page) of the resultant serines indicated chiral
purity but the yield of serine was low due to a lack of regiospecificity
in the Baeyer Villiger step. We had expected migration of the labelled
centre rather than of the methyl group because of the well-known low
migratory aptitude of methyl groups [14]. It would seem, however, that
the effect of distant electron withdrawing groups, noted recently in some
other work [15], has decreased the migratory aptitude of the methylene
moiety to a level lower than that of the methyl group.

Having prepared samples of serine (15) stereospecifically
deuteriated at C-3 and samples of the stereospecifically deuteriated
adduct (4) we are at present in the position of investigating the final
stages in assessing the steric course of the one-carbon transfer

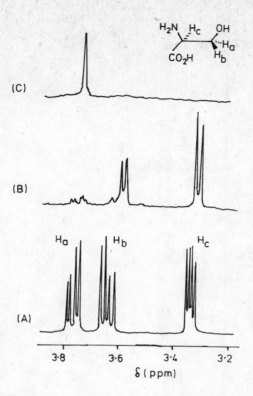

Figure 2 - 360 MHz ^{1}H n.m.r. spectra in 10% $NaO^2H/^2H_2O$ of (A) L-serine; (B) (2S,3R)[3-$^2H_1$]-serine; and (C) (2S,3S)[2,3-$^2H_2$]-serine

reactions catalysed by *serine hydroxymethyltransferase* and by *thymidylate synthetase*. During this phase of our programme we had started work aimed at assessing the steric course of the catabolism of thymine (1, R' = H). Interestingly the enzyme system which degrades thymine to α-methyl-β-alanine will also degrade uracil (2, R' = H) and 5-fluorouracil (9) as shown in Scheme 6. Since 5-fluorouracil is metabolised to a suicide inhibitor for *thymidylate synthetase*, this fact is doubly interesting.

11*

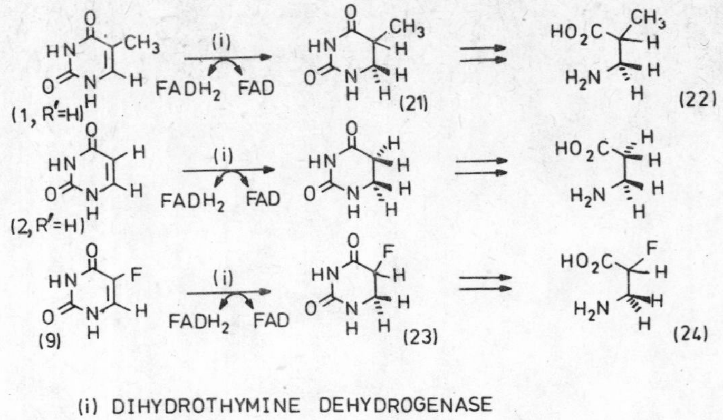

(i) DIHYDROTHYMINE DEHYDROGENASE

Scheme 6 – Catabolism of thymine, uracil and 5-fluorouracil by
mammalian liver enzymes

We have been able to prepare a partially purified enzyme system from
beef liver which will perform the catabolic processes summarised in
Scheme 6. The first and rate limiting step in these processes, the
reduction of the pyrimidines, is catalysed by *dihydrothymine dehydrogenase*,
an enzyme which has been shown to contain FAD [16]. The FAD is reduced
to $FADH_2$ by the coenzyme NADPH. Owing to the high cost of this latter
coenzyme, we economised by using *glucose-6-phosphate dehydrogenase* to
re-reduce the NADP produced in the reaction to NADPH by the oxidation of
glucose-6-phosphate to 6-phosphogluconate. By linking this latter system
to our partially purified catabolic system we were able to examine the
catabolic reactions. These were performed in parallel, one set of
reactions using labelled substrates where necessary and aqueous medium,
and the other set using unlabelled substrates and ^2H_2O. The involvement
of $FADH_2$ in the process meant that labelling could be achieved using ^2H_2O
although there was an isotope effect of *ca.* 4 in these experiments. In

152

our first studies with this system we were able to deduce the steric

course of the catabolism of uracil [17] and now turned to a study of the

emotive pairing of thymine (1, R' = H) and 5-fluorouracil (9).

Catabolism of thymine (1, R' = H) yields 3-amino-2-methylpropanoic

acid (22) and, since the 2R-isomer (22a) of this acid is excreted in

human urine [18], it is likely that the metabolic product has this

configuration. The 2S-enantiomer (22b) is, however, found in some peptides

[19] and it has been reported that the 2S-isomer is itself metabolised

whilst the 2R-isomer is not [20]. It is therefore essential that the

stereochemistry of the metabolic product be confirmed. We had been

working on the enzyme *glutamate mutase* [21] from *Clostridium tetanomorphum*

and so had the enzyme β-*methylaspartase* from this source. We were

therefore able to use this enzyme to prepare a sample of (2S,3S)-3-methyl-

aspartic acid (25) of known [22] absolute stereochemistry. This compound

could be decarboxylated by pyrolysis in a melt with *p*-methoxyacetophenone

followed by hydrolysis in 3M aqueous hydrochloric acid. The (2S)-3-amino-

2-methylpropanoic acid (22b) obtained in this experiment had $[\alpha]_D + 12.7^o$

compared to $[\alpha]_D - 15.4^o$ for the isomer obtained by catabolism of thymine

using our bovine liver enzyme system. That these compounds were

enantiomeric could be further proved by preparing their camphanic acid

amides and that of commercial (2RS)-3-amino-2-methylpropanoic acid. The

360 MHz [1]H-n.m.r. spectra of these compounds showed clearly that the C-3

protons had different chemical shifts in the 2R- and 2S-isomers and that,

although decarboxylation of (2S,3S)-3-methylaspartic acid (25) had been

accompanied by a small amount of racemisation at C-2, the product from

catabolism of thymine was clearly (2R)-3-amino-2-methylpropanoic acid

(22a). This would imply that the product of the first step in thymine

catabolism was (5R)-dihydrothymine and that reduction of thymine by

(21a)   (22a)   (22b)   (25)

In order to assess the steric course of the reduction of thymine at C-6, it was necessary to have a sample of (2R)- or (2RS)-3-amino-2-methylpropanoic acid which was unambiguously stereospecifically labelled at C-3 with deuterium. This would permit an assay for the stereochemistry of the labelled catabolic products to be developed. Witkop [23] has shown that catalytic hydrogenation of thymidine (1, R' = deoxyribose) takes place with asymmetric induction so that, on hydrolysis, (5S)-dihydrothymine (21a) is produced. Since hydrogenation should involve *cis* addition of hydrogen, we repeated Witkop's procedure but used $^2H_2$ and ^2H_2O in the reduction step. Hydrolysis of the resultant dihydrothymine with hydrochloric acid, known [24] to result in racemisation at C-2, gave (2RS,3S)[3-$^2H_1$]-3-amino-2-methylpropanoic acid. This was converted into an amide with (1S,4R)-camphanic acid and the ^1H-n.m.r. spectrum of this compound (Figure 3(B)) confirmed the stereochemical purity of the 3S label in both the 2S- and 2R-isomers of 3-amino-2-methylpropanoic acid and provided an assay for products from the catabolic experiments. The ^1H-n.m.r. spectra of the camphanic acid amides of the samples of (2R)-3-amino-2-methylpropanoic acid (22a) obtained from catabolism of (a) [6-$^2H_1$]-thymine in H_2O, and (b) thymine in ^2H_2O are shown in Figures 3(C) and 3(D) respectively. These clearly show that enzymic reduction of thymine must have occurred from the *si*-face at C-6. The reduction process therefore involves *anti* addition of hydrogen at C-5 and C-6.

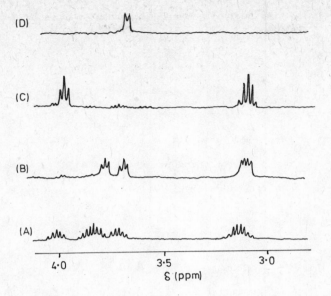

Figure 3 - 360 MHz ^{1}H n.m.r. spectra in C^2H_5N of (A) (2RS)-3-

camphanoylamino-2-methylpropanoic acid; (B) (2RS,3S)-

[3-$^2H_1$]-3-camphanoylamino-2-methylpropanoic acid (synthetic);

(C) (2R,3R)[3-$^2H_1$]-3-camphanoylamino-2-methylpropanoic acid;

(D) (2R,3S)[2,3-$^2H_2$]-3-camphanoylamino-2-methylpropanoic acid.

The samples in (C) and (D) were derived ultimately from the

enzymic experiments.

Catabolism of 5-fluorouracil (9) yields 3-amino-2-fluoropropanoic

acid (24). One stereoisomer of this compound had been prepared [25] by

reaction of N,N-dibenzyl-L-serine benzyl ester (26, R' = CH$_2$Ph) with

diethylaminosulphur trifluoride (DAST) followed by hydrogenolysis. When

we prepared the methyl ester (26, R' = CH$_3$) and treated it with DAST

followed by hydrogenolysis and hydrolysis, we obtained 3-amino-2-fluoro-

propanoic acid (24), $[\alpha]_D$ = + 28.4°. This was identical to a sample which we obtained by catabolism of 5-fluorouracil. Since the low anomalous dispersion of fluorine made use of Bijvoet's method difficult, we converted (24) into its methyl ester and thence into the camphanoyl amide (27). X-ray structure analysis of this compound showed it to have the stereo-chemistry as in Figure 4. Since the (−)camphanic acid used was derived from (+)camphor of known [26] absolute stereochemistry, the absolute stereochemistry of the catabolic product must be (2R) as in (24a). The enzymic reduction of 5-fluorouracil had therefore occurred from the *si*-face at C-5.

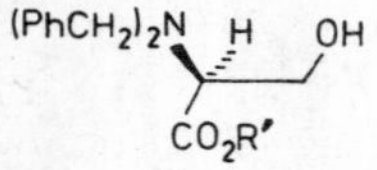

(26) 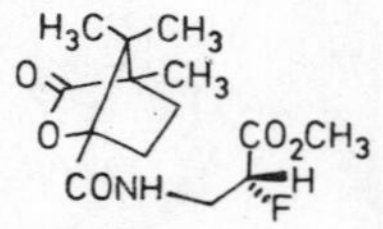   (27) 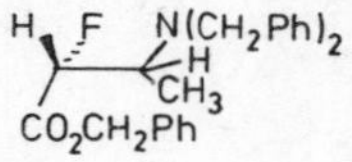   (28) 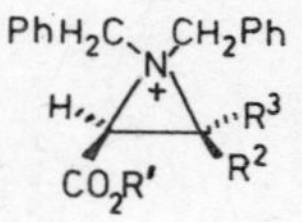   (29)

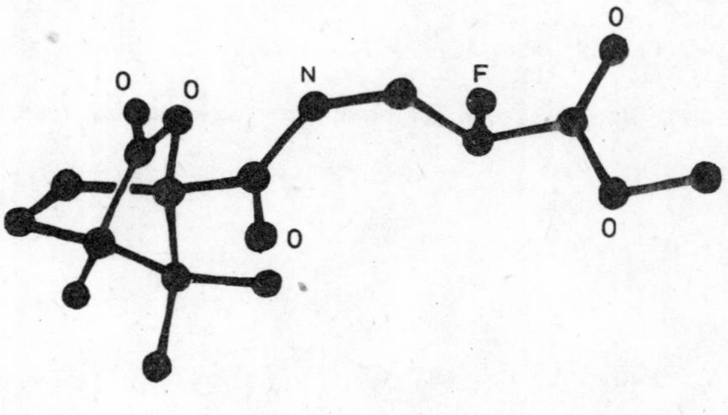

Figure 4 – Molecular structure of the camphanoyl amide (27)

When L-threonine and L-allothreonine replaced L-serine in the DAST synthesis of α-fluoro-β-amino acids [25] the products were the *threo*- and *erythro*-isomers respectively of (28). Thus the *relative* stereochemistry was the same as in the starting amino acid. Our X-ray analysis of the

156

L-serine product suggested that both centres would be inverted in the process and this would be compatible with the intermediacy of an aziridinium ion (29). Since we had already prepared samples of L-serine stereospecifically labelled at C-3, we were now in a position to synthesise samples of the catabolic product (24) stereospecifically labelled at C-3 with deuterium. The absolute stereochemistry at C-3 of these compounds

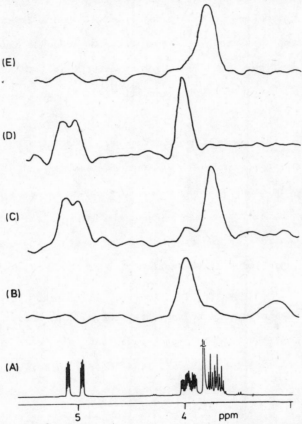

Figure 5 – ^{1}H and ^{2}H n.m.r. of (A) (2R) 3-amino-2-fluoropropanoic acid;

(B) (2R,3S) $[3-^2H_1]$-3-amino-2-fluoropropanoic acid (synthetic);

(C) (2R,3R) $[2,3-^2H_2]$-3-amino-2-fluoropropanoic acid (synthetic);

(D) (2R,3S) $[2,3-^2H_2]$-3-amino-2-fluoropropanoic acid (from

catabolism of 5-fluorouracil; (E) (2R,3R) $[3-^2H_1]$-3-amino-2-

fluoropropanoic acid (from catabolism of $[6-^2H_1]$-5-fluorouracil) would be epimeric with the stereochemistry at C-3 in the labelled samples of L-serine. When the samples of labelled L-serine (15) were converted to the deuteriated fluoro-β-alanines (24) and these were further converted to the N-camphanoyl methyl esters (27), the ^{2}H-n.m.r. spectra (Figure 5(B) and (C)) showed the stereochemical purity of the label and provided an assay for chirality of labelling at C-3. We now used our bovine liver enzyme system to catabolise [6^{2}H]-5-fluorouracil in H_2O and 5-fluorouracil in ^2H_2O. The N-camphanoyl methyl esters of these compounds had ^{2}H-n.m.r. spectra as shown in Figure 5(D) and (E) respectively and clearly showed that the enzymic reduction of 5-fluorouracil has occurred from the *si*-face at C-6.

These experiments show that the catabolism of *both* the DNA base thymine and the anti-cancer drug 5-fluorouracil involves *anti* addition of hydrogen to the pyrimidines at the *si*-face at C-5 and the *si*-face at C-6. This suggests that both compounds are bound similarly at the active site of the enzyme in contrast to the case of the substrate folic acid and the drug methotrexate which bind at the active site of *dihydrofolate reductase* with quite different geometries.

The work discussed in this lecture has been done at Sussex by my colleagues Peter Charlton, Steven Field, David Gani and Peter Hitchcock and I am grateful to them for their skill and energy in bringing it to a successful conclusion. I am also grateful to Drs. Feeney, Roberts and Birdsall at the National Institute for Medical Research, Mill Hill, for their collaboration on part of the work and to the Science and Engineering Research Council (U.K.) for their financial support. Some of the work has been published in full and preliminary form [26-28].

References

1. For a review of tetrahydrofolate mediated one-carbon transfers see Young, D.W. (1983) 'Chemistry and Biology of Pteridines - Pteridines and Folic Acid Derivatives', ed. Blair, J.A., Walter de Gruyter, Berlin, p321.

2. (a) Pastore, E.J. and Friedkin, M. (1962) *J. Biol. Chem.*, **237**, 3802; (b) Blakley, R.L., Ramasastri, B.V. and M^{C}Dougall, B.M. (1963) *J. Biol. Chem.*, **238**, 3075; (c) Lorenson, M.Y., Maley, G.F. and Maley, F. (1967) *J. Biol. Chem.*, **242**, 3332.

3. (a) Blakley, R.L. (1969) 'The Biochemistry of Folic Acid and Related Pteridines, North Holland, Amsterdam; (b) Danenberg, P.V. (1977) *Biochim. Biophys. Acta*, **473**, 73.

4. Fontecilla-Camps, J.C., Bugg, C.E., Temple, C., Rose, J.D., Montgomery, J.A., and Kisliuk, R.L. (1979), *J. Am. Chem. Soc.*, **101**, 6114.

5. San Pietro, A. (1955) *J. Biol. Chem.*, **217**, 579.

6. Young, D.W. (1978) 'Isotopes in Organic Chemistry', vol.4, eds. Buncel, E. and Lee, C.C., Elsevier, Amsterdam, p184.

7. (a) Poe, M. and Hoogsteen, K. (1978) *J. Biol. Chem.*, **253**, 543; (b) Furrer, H-J, Bieri, J.H. and Viscontini, M. (1978) *Helv. Chim. Acta*, **61**, 2744; (c) Armarego, W.L.F. and Schou, H. (1977) *J. Chem. Soc., Perkin Trans. I*, 2529.

8. Matthews, D.A., Alden, R.A., Bolin, J.T., Filman, D.J., Freer, S.T., Hamlin, R., Hol, W.G.J., Kisliuk, R.L., Pastore, E.J., Plante, L.T., Xuong, N., and Kraut, J. (1978) *J. Biol. Chem.*, **253**, 6946.

9. We thank Mr. E.A. Piper, National Institute for Medical Research, London, U.K. for the computer graphic work.

10. Tatum, C., Vederas, J., Schleicher, E., Benkovic, S.J., and Floss, H.G. (1977) *J. Chem. Soc., Chem. Commun.,* 218.

11. Charlton, P. and Young, D.W., *unpublished work* - see also

    (a) Tatum, C.M., Benkovic, P.A., Benkovic, S.J., Potts, R., Schleicher, E. and Floss, H.G. (1977) *Biochemistry,* **16**, 1093;

    (b) Farina, P.R., Farina, L.J. and Benkovic, S.J. (1973) *J. Am. Chem. Soc.,* **95**, 5409; (c) *ibid* (1975) *J. Am. Chem. Soc.,* **97**, 4151.

12. (a) Field, S.J. and Young, D.W. (1979) *J. Chem. Soc., Chem. Commun.,* 1163; (b) *ibid* (1983) *J. Chem. Soc., Perkin Trans. I,* 2387.

13. See ref.6, p231.

14. Smith, P.A.S. (1967) 'Molecular Rearrangements', ed. de Mayo, P., Wiley Interscience, New York, part 1, p577.

15. (a) Noyori, R., Sato, T. and Kobayashi, H. (1980) *Tetrahedron Lett.,* **21**, 2569; (b) Grudzinski, Z., Roberts, S.M., Howard, C. and Newton, R.F. (1978) *J. Chem. Soc., Perkin Trans. I,* 1182.

16. Shiotani, T. and Weber, G. (1981) *J. Biol. Chem.,* **256**, 219.

17. Gani, D. and Young, D.W. (1983) *J. Chem. Soc., Chem. Commun.,* 576.

18. Balenovic, K. and Bregant, N. (1959) *Tetrahedron,* **5**, 44.

19. Kakimoto, Y., Kanazawa, A. and Sano, I. (1965) *Biochim. Biophys. Acta,* **97**, 376.

20. Armstrong, M.D., Yates, K., Kakimoto, Y., Taniguchi, K. and Kappe, T. (1963) *J. Biol. Chem.,* **238**, 1447.

21. Field, S.J. and Young, D.W., *unpublished work.*

22. Sprecher, M. and Sprinson, D.B. (1966) *J. Biol. Chem.,* **241**, 868.

23. Kondo, Y. and Witkop, B. (1968) *J. Am. Chem. Soc.,* **90**, 764.

24. Rachina, V. and Blagoeva, I. (1982) *Synthesis,* 967, reference 12.

25. Somekh, L. and Shanzer, A. (1982) *J. Am. Chem. Soc.,* **104**, 5836.

26. Charlton, P.A., Young, D.W., Birdsall, B., Feeney, J., and
    Roberts, G.C.K. (1979) *J. Chem. Soc., Chem. Commun.*, **922**.

27. Gani, D. and Young, D.W. (1983) *J. Chem. Soc., Perkin Trans. I*, **2393**.

28. Gani, D., Hitchcock, P.B. and Young, D.W. (1983) *J. Chem. Soc., Chem.
    Commun.*, **898**.

DISCUSSION

PANDIT U.K. (Amsterdam, The Netherlands)
I wonder, whether the reaction catalyzed by thymidylate
synthetase can be practically reversed, because it might
contribute to the elucidation of the stereochemistry of the
enzymic reaction.

YOUNG D.W. (Brighton, United Kingdom)
According to my knowledge the reaction is not reversible
in practice.

HERNÁDI F. (Debrecen, Hungary)
Could methotrexate be used to inhibit bacterial dihydrofolate reductase, too?

YOUNG D.W. (Brighton, United Kingdom)
Yes it seems to have good affinity for all dihydrofolate
reductases. This lack of specificity causes toxicity to
mammals and so it would be unlikely to be a useful antibacterial agent and is only practical in cancer therapy.
The use of folinic acid (5-formyltetrahydrofolate) as a
cancer "rescue agent" can now allow quite large doses to
be applied but one would only use such a dangerous compound
for this most serious disease and not far bacterial
diseases. The specificity of drugs such as trimethoprim
for the bacterial enzyme is, however, extremely hopeful.

OTTENHEIJM H.C.J. (Nijmegen, The Netherlands)
On one slide you showed the methylation of the α-carboxyl
group of aspartate, a reaction during which the β-CO_2H was
esterified in part, too. Could not you use your "formaldehyde equivalent" to protect the α-CO_2H function as well
as the amino group to form a five membered ring? The stage
would then be set to work on the β-carboxyl group.

YOUNG D.W. (Brighton, United Kingdom)

In fact we have synthesized several compounds of biological importance labelled with deuterium from labelled aspartate and the method which I have discussed here is not the only one used. It is, however, one of the better ones since, although there is an 80:20 mixture, it is easy to separate and the total recovery is high so that our isolated yield must approach 80 % which we are happy with. Aesthetically I like your idea of using the proximity of the α-carboxyl group to the amine in an attempt to obtain a ring which would act both as α-amino and α-carboxyl protector. The $NH-CH_2O-C=O$ functionality would not be extremely stable, however. This may in fact have an analogy in one of our methods where we achieve specifically the β-ester from the reaction of thionyl chloride and methanol on aspartate, since it could be that an $\alpha-NH-SO-O-C=O$ ring is an inter- mediate. Other methods have used the greater acidity of the α-carboxyl to affect regiospecificity. In practice, however, we are well satisfied with our yield.

van der PLAS H.C. (Wageningen, The Netherlands)

To use the term "adduct" to indicate the reaction of formal- dehyde and the coenzyme is not correct; it should be called a condensation reaction.

YOUNG D.W. (Brighton, United Kingdom)

I would agree that, strictly speaking, my use of the word adduct is wrong. I would, however, plead "poetic licence" in my usage here as I wish to convey the impression of an ephemeral compound in which the important one-carbon unit is attached to the coenzyme. That is to say, I focus on a "one carbon adduct" at one of three oxidation levels rather than a condensation product of formic acid, formaldehyde or methanol, or their equivalents.

SNATZKE G. (Bochum, German Federal Republic)

I would like to make a general comment in connection with
all the lectures presented up till now.
Chirality is a geometrical property, so it should be used
only in connection with molecular shape. Optical activity
is a property of a substance, i.e. of an ensemble of many,
many molecules. Although frequently used, a term like
"a chiral solvent" is meaningless. What actually is meant
by this phrase is an "optically active solvent" and you
need this e.g. in PIRKLE's method, although you are not
interested in optical properties at all when you apply this.
There is still a need of a short term for a substance whose
molecules are chiral, and are all of the same "handedness"
(this would be one of the two pure enantiomers), or where
at least one type preponderates.
Mind that in a racemate all molecules are chiral and still
there can no optical activity be measured; in that sense
also substances like tetralin are racemates, as each of the
two possible half-chair conformations are chiral, but they
are interconverted into each other extremely quickly.

YOUNG D.W. (Brighton, United Kingdom)

Professor Snatzke's comment on the fact that he would like
a new name to determine the ensemble of molecules of the
same handedness obviously does not need reply, but it might
be worth recording his plea for suggestions for a new name
to be sent to Prof. Eliel, Notre Dame University.

Bio-Organic Heterocycles
van der Plas H.C., Ötvös L., and Simonyi M. eds

BIOSYNTHESIS OF ANTHRANILIC ACID DERIVED ALKALOIDS

GRÖGER D.

Institute of Plant Biochemistry,
Academy of Sciences of the GDR
GDR-4010 Halle (Saale), Weinberg 3, GDR

INTRODUCTION

A number of protein amino acids and some other closely related compounds e. g. ornithine, nicotinic acid and anthranilic acid may serve as essential building blocks in alkaloid biosynthesis. Usually, amino acids are decarboxylated in the formation of N-heterocycles. In anthranilic acid derived alkaloids the carboxyl group is retained in the course of alkaloid formation. Anthranilic acid is the first intermediate in the tryptophan branch of the aromatic amino acid pathway. Anthranilate synthetase catalyzes the formation of anthranilate from chorismic acid and L-glutamine as N-donor (in higher plants) (Fig. 1). Interestingly, anthranilate and 3-hydroxyanthranilate may also be formed in the catabolic pathway of tryptophan. Kynureninase is responsible for the degradation of kynurenine and 3-OH-kynurenine, respectively (Fig. 2).

Anthranilate derived secondary metabolites are found in microorganisms, plants and animals. Among them there are "true"alkaloids and a number of antibiotics (Figs. 3 and 4).

12 Plas

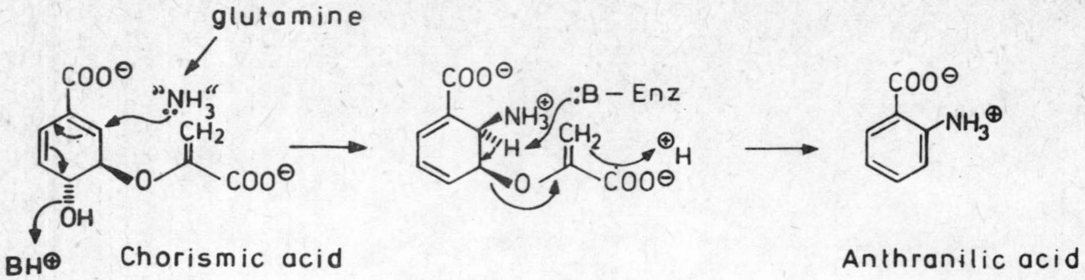

Fig. 1 Possible mechanism of anthranilate synthetase reaction

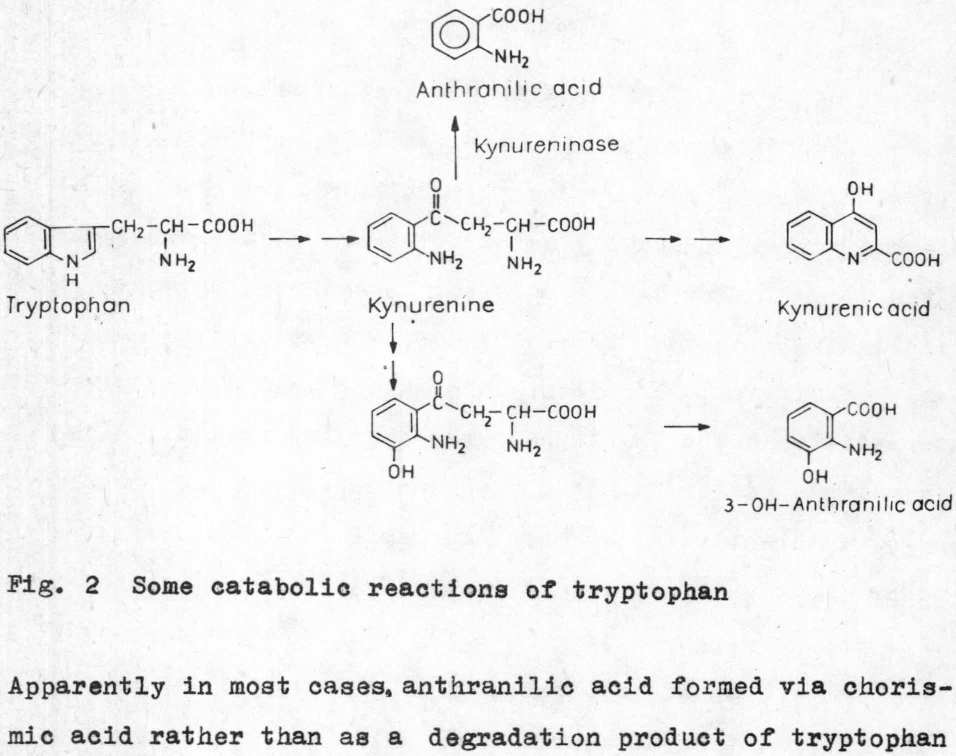

Fig. 2 Some catabolic reactions of tryptophan

Apparently in most cases, anthranilic acid formed via chorismic acid rather than as a degradation product of tryptophan serves as precursor for natural product biosynthesis.

COOCH3
NHCH3
OCH3
Damascenine
CH3O
O
OH
N
O
OH
DIMBOA
OCH3
N
O
Dictamnine
COOH
NH2
O
OH
OCH3
N
OCH3
CH3
Arborinine
O
N
N
CH3
CH3
Glomerine
N
N
OH
Vasicine
N
N
O
H
N
CH3
Evodiamine

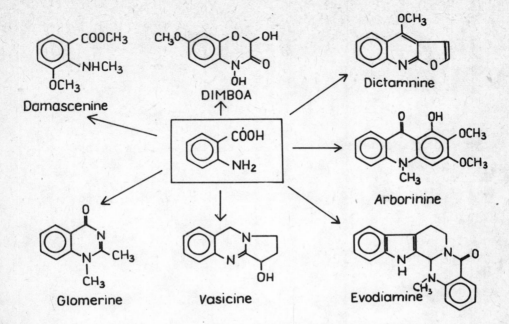

Fig. 3 Alkaloids in animals and higher plants related to anthranilic acid

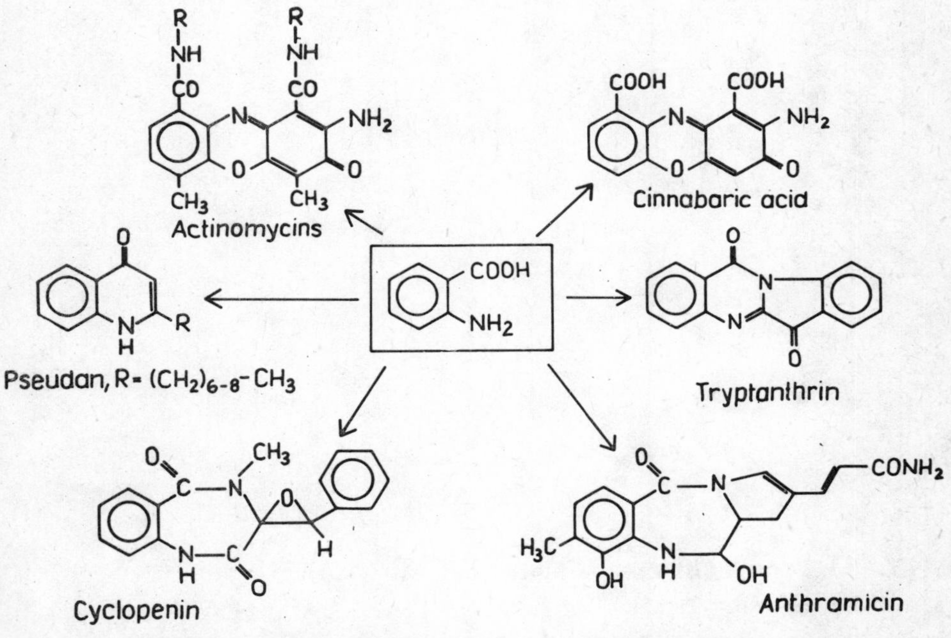

R
NH
CO
R
NH
CO
N
NH2
O
O
CH3
CH3
Actinomycins
COOH
COOH
N
NH2
O
O
Cinnabaric acid
O
N
H
R
Pseudan, R= (CH2)6-8-CH3
COOH
NH2
O
N
N
O
Tryptanthrin
O
CH3
C
N
O
N
C
H
O
H
Cyclopenin
O
C
N
CONH2
H3C
N
H
OH
OH
Anthramicin

Fig. 4 Microbial products derived from anthranilic acid

This review will focus on some aspects of anthranilate derived metabolites. For broader coverage other compilations are available /1,2,3/.

PROTOALKALOIDS

Minor constituents of many volatile oils are the methyl esters of anthranilic acid as well as its N-methyl derivative (e.g. citrus oil). Damascenine is a trimethylated derivative of 3-hydroxyanthranilic acid, hitherto found only

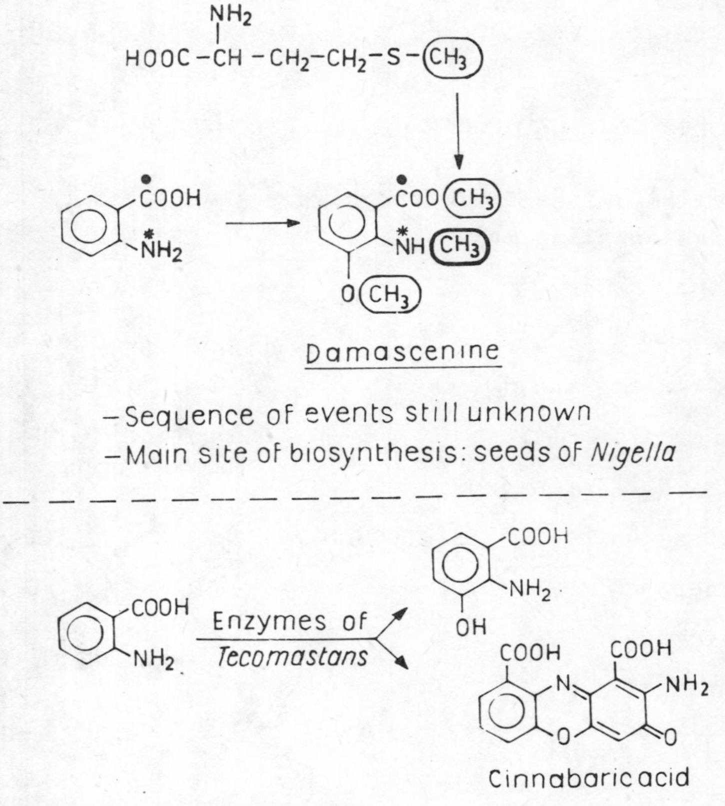

$$HOOC-\overset{\overset{NH_2}{|}}{CH}-CH_2-CH_2-S-\boxed{CH_3}$$

Damascenine

– Sequence of events still unknown
– Main site of biosynthesis: seeds of *Nigella*

Cinnabaric acid

Fig. 5 a) Biosynthesis of damascenine
        b) Hydroxylation of anthranilate by extracts of
           leaves of <u>Tecoma stans</u>

in the genus <u>Nigella</u> (Ranunculaceae). ($^{14}CH_3$)-Methionine is
an efficient precursor for all three variously bound methyl
groups but labels preferentially the N-methyl group
(Fig. 5). The aromatic ring arises from the shikimate pathway and some feeding experiments indicate that anthranilate
is directly hydroxylated /4/. Such an anthranilic acid oxidase system was detected in leaves of <u>Tecoma stans</u> /5/.

QUINOLINES

From the biogenetic point of view the quinolines represent a
rather heterogeneous group of natural compounds. The <u>Cinchona</u>
alkaloids and the anticancer agent camptothecine originate
from tryptophan and the monoterpene loganin. The different
skeletons are formed due to various rearrangements via
strictosidine. Kynurenic acid and related compounds which
are found in mammals, insects and <u>Pseudomonas</u> species arise
as products of tryptophan catabolism (Fig. 2). A third pathway leading to the quinoline nucleus comprises anthranilate
and acetate as major precursors. Especially rich in anthra-
nilate-derived quinolines are the Rutaceae. They are found
in many genera of the sub-families Rutoideae, Toddaliodeae
and Aurantioideae /6/. The biosynthesis of furoquinolines
has been studied most thoroughly. Anthranilic acid, including its amino and carboxyl groups, is well incorporated. The
carbon atom 10 stems from the carboxylic group of acetate
/7,8/. The carbon atoms 2 and 3 of the furane ring were
specifically labeled after feeding of (4-^{14}C)- and (5-^{14}C)-
mevalonic acid, respectively.

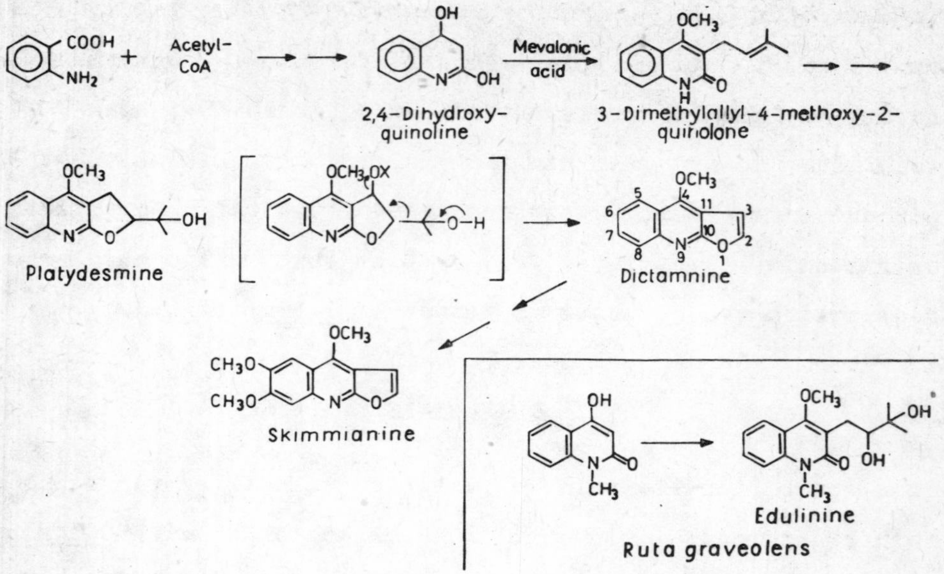

Fig. 6 Biosynthesis of quinoline alkaloids in Rutaceae

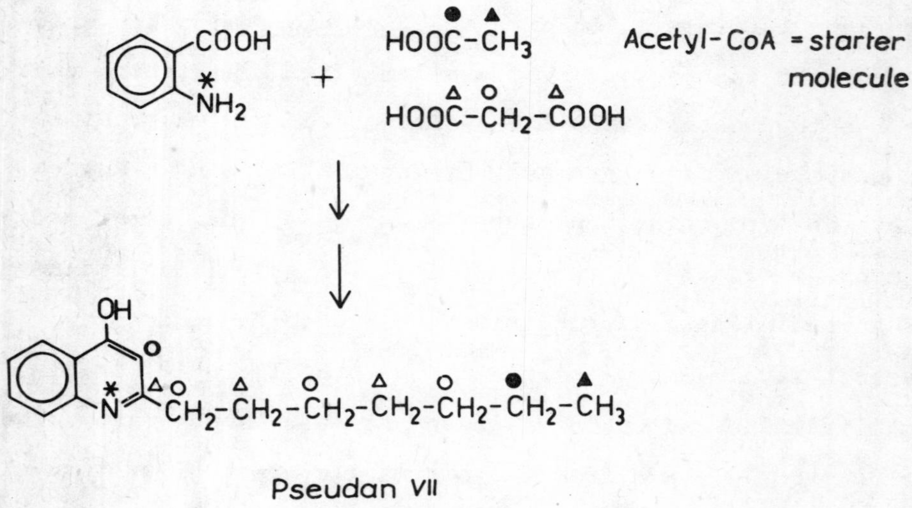

Pseudan VII

Fig. 7 Biosynthesis of pseudanes in Pseudomonas species

Surprisingly, (^{14}C)-acetate was incorporated solely in the quinoline part of dictamnine but not in those carbon atoms which are derived from mevalonic acid /9/. The sequence of furoquinoline biosynthesis was well established by feeding potential intermediates (Fig. 6). On this route 2,4-di-hydroxyquinoline is converted to 3-dimethylallyl-4-methoxy-2-quinolone which in turn gives a 2-isopropyl-dihydrofuro-quinoline. Formation of the furane ring proceeds with loss of the isopropyl group but retention of one hydrogen at C-3 /10,11/. Using cell cultures of Ruta graveolens, the biotransformation of 4-hydroxyquinoline into dictamnine and γ -fagarine as well as 4-hydroxy-N-methyl-2-quinolone into edulinine were observed /12/.

Some strains of Pseudomonas produce 2-n-alkyl-4-hydroxy - quinolines (pseudans) which differ in the length of the aliphatic side chain in position 2. Recently 2-(n-undecyl) quinolone-4 was isolated from flowers of Ptelea trifoliata and roots of Ruta graveolens /13/. In Pseudomonas the pseu-dans are formed from anthranilic acid and acetate /14/. Apparently a ß-ketoacid-CoA-ester corresponding to a given pseudan reacts with "activated" anthranilic acid. Experi-ments with (^{14}C)-malonate revealed that the side-chain is build up from the terminal C-atoms as it is usually found in fatty acid biosynthesis (Fig. 7).

QUINAZOLINES

About 80 natural compounds containing quinazoline skeleton have been found in higher plants, microorganisms and
even animals. A number of structural types occur in taxo-
nomically unrelated plant families e.g. simple substituted
quinazolino-4-ones, pyrroloquinazolines, pyridoquinazolines
and indolopyridoquinazolines. In microorganisms, quinazolines
were isolated from Candida-, Aspergillus-, Pseudomonas- and
Streptomyces species. Most prominent among the quinazolines
found in animals is tetrodotoxin, which is a chemically
very complex alkaloid and a powerful non-protein neurotoxin.
A comprehensive review on natural quinazolines was recently
published /15/. The biosynthesis of some quinazolines were
also summarized /16/.

The millipede alkaloid glomerine (1,2-dimethylquinazoline-
4-one) is derived from (carboxyl-^{14}C)-anthranilic acid /17/.
Arborine, the main alkaloid of the leaves of the Indian
plant Glycosmis arborea is biosynthesized from anthranilic
acid and phenylalanine (Fig. 8). Apparently anthranilate
gets first methylated prior to the condensation with
phenylalanine. The specific incorporation of (3-^{14}C,^{15}N)-
phenylalanine shows clearly that with exception of the
carboxyl group the aliphatic side chain of L-Phe is involved
in arborine formation. Phenylacetic acid can be excluded as
immediate precursor /18/.

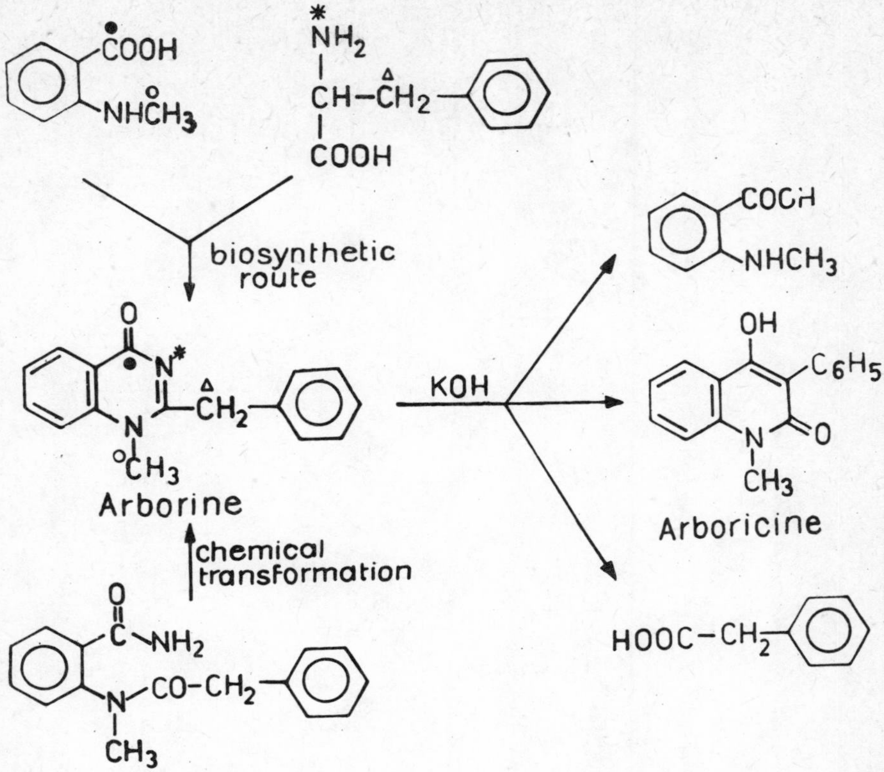

Fig. 8 Biosynthesis of arborine in Glycosmis arborea

The oldest known quinazoline alkaloid is vasicine=peganine. The biosynthesis of this particular pyrrolidino-quinazoline has been intensively investigated. Anthranilic acid including the nitrogen atom is specifically incorporated. Still controversial is the origin of the "non-anthranilic acid moiety" of vasicine. In the authors laboratory it was found that carbon atoms C-3 and C-10 may derive from a C_2-unit, e.g. $(2'-^{14}C, ^{15}N)$N-acetylanthranilic acid was specifically incorporated. Judging from various labeling experiments, it

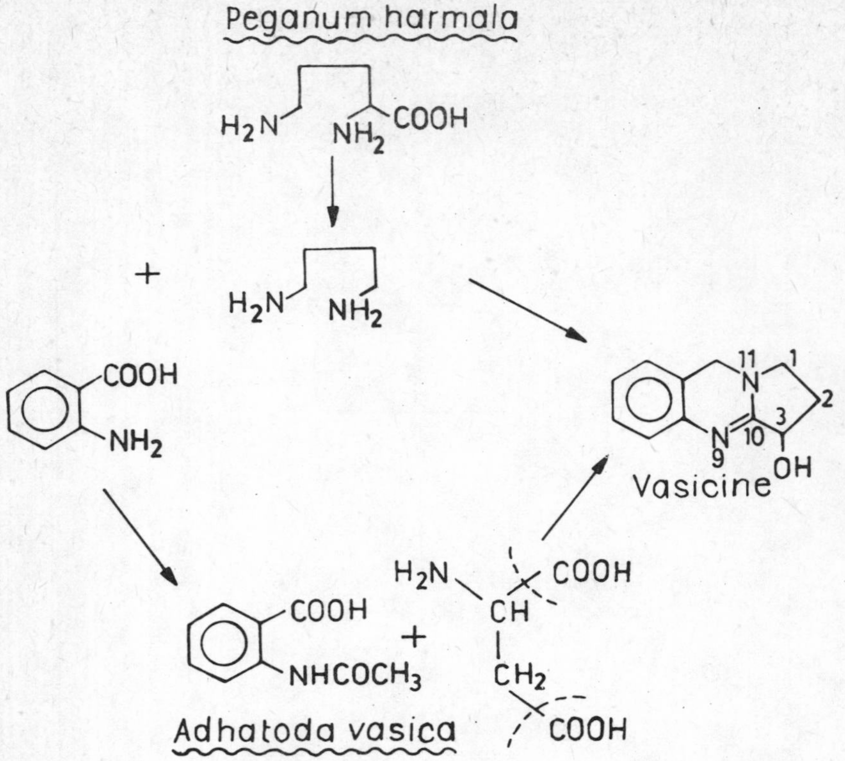

Fig. 9 Biosynthetic routes to vasicine in A. vasica
and P. harmala

might be concluded that in Adhatoda vasica aspartic acid or a
closely related compound - by loss of the carboxyl groups -
may provide C-1, C-2 and N-11 of vasicine. Members of the
glutamic acid family showed random incorporation. On the
other hand Lijegren's group /19,20/ found that in Peganum
harmala labeled ornithine, proline, putrescine and related
compounds are more or less specifically incorporated into the
pyrrolidino-ring system. The existence of two different

pathways leading to vasicine is still perplexing (Fig. 9). The indolopyridoquinazolines evodiamine and rutaecarpine are present in the fruits of **Evodia rutaecarpa**. Japanese authors demonstrated the specific incorporation of tryptophan into the ß-carboline part of both alkaloids. The carbon atom C-3 and in the case of evodiamine also the N-methyl group are derived from methionine. The other aromatic part was provided by anthranilic acid. The relationship between rutae- carpine and evodiamine as well as the possible involvement of a ß-carboline in Evodia alkaloid biosynthesis are still obscure /21,22/ (Fig. 10).

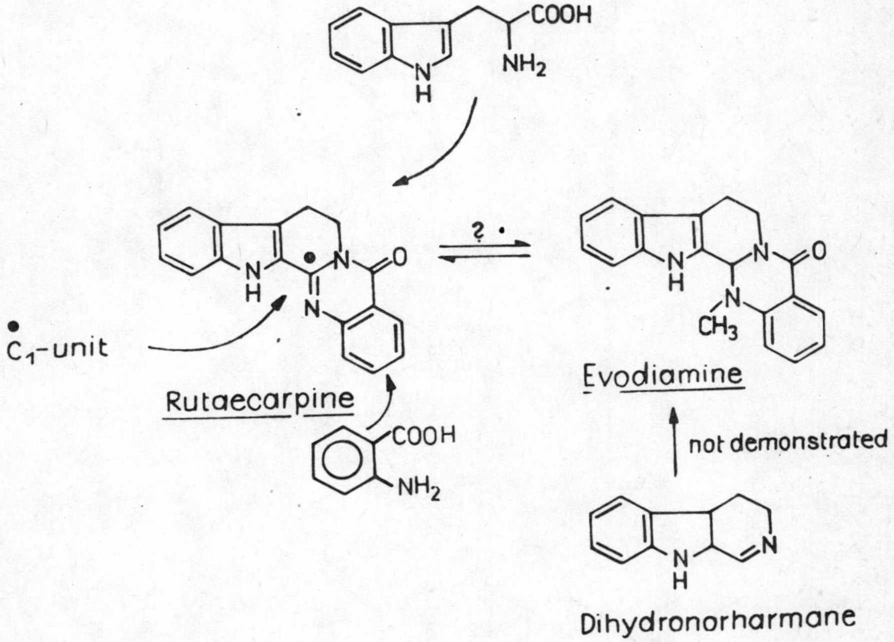

Fig. 10 Biosynthesis of **Evodia** alkaloids

Certain strains of the yeast _Candida lipolytica_ form in tryptophan-supplemented media tryptanthrin. This quinazolone is derived from tryptophan and anthranilic acid which originated via tryptophan degradation. Feeding of anthranilic acid or tryptophan derivatives gave the corresponding new "tryptanthrines". Two representative experiments are shown in Fig. 11 /23/.

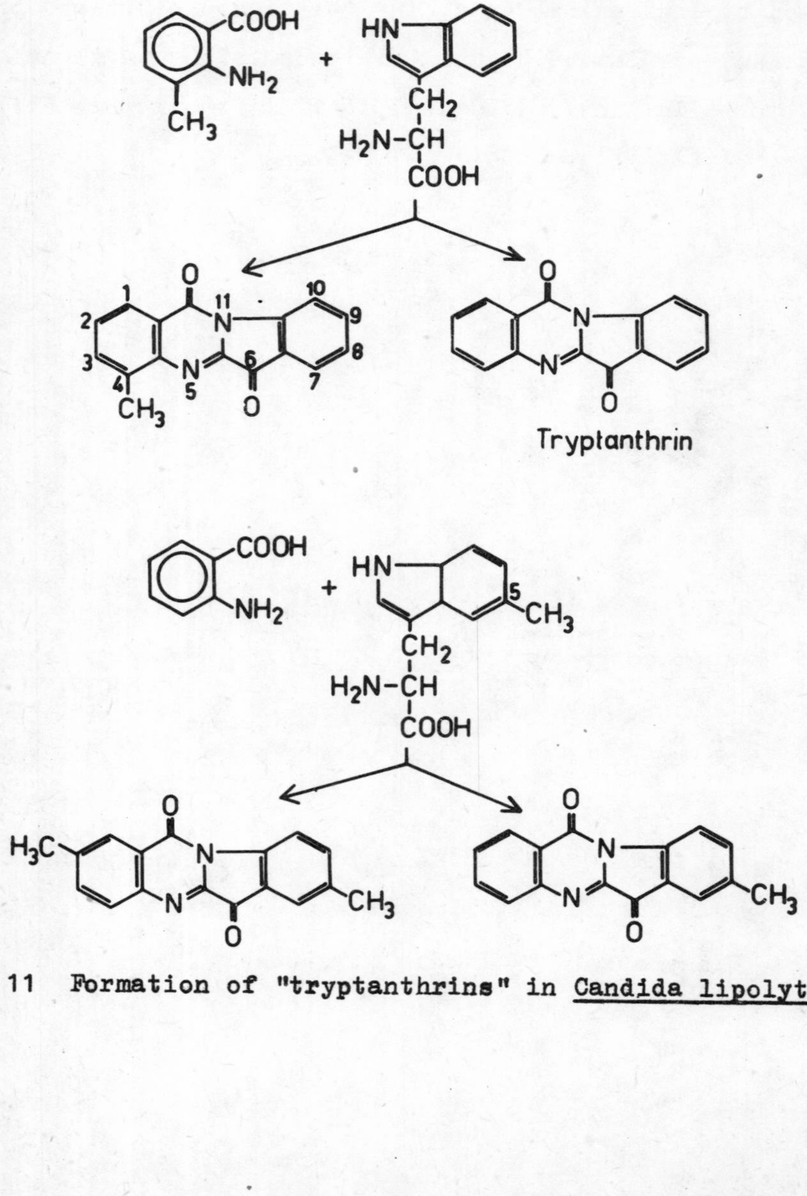

Tryptanthrin

Fig. 11 Formation of "tryptanthrins" in _Candida lipolytica_

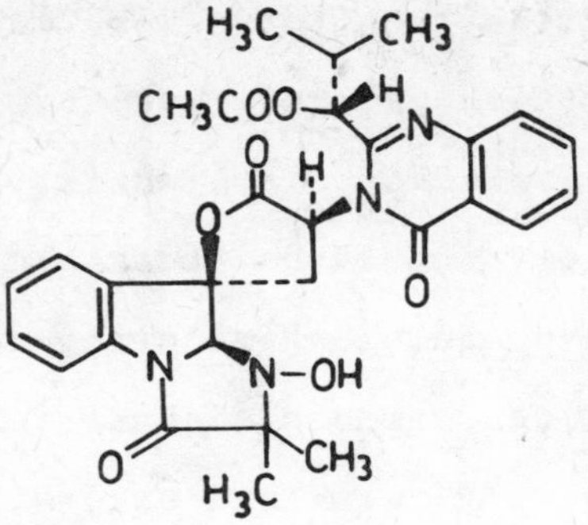

Fig. 12 Tryptoquivaline

From various species of Aspergillus, tremor-producing metabolites were isolated. Tryptoquivaline (Fig. 12) is the proto-
type of this group of neurotropic mycotoxins. As potential
precursors tryptophan, anthranilic acid, alanine and valine
were proposed, but experimental evidence is still lacking
/24/.

ACRIDONE ALKALOIDS

Natural products with an acridone nucleus are very weak bases
which are yellow in colour. About 50 acridone alkaloids have
been found which occur solely in some genera of the Rutaceae
mainly in species of Acronychia, Evodia, Glycosmis, Melicope
and Ruta. Chlorine-containing acridones were isolated and
also dimeric acridone alkaloids have been obtained. A well-
known potent anticancer agent is acronycine. First studies
revealed that anthranilic acid proved to be the precursor
of ring A and the adjacent carbon atom as well as the N-atom
of some acridones /25,26/.

According to Robinson's scheme, ring C may be derived from acetate via a polyketo-acid. As potential intermediates 2-aminobenzophenones have been discussed /27/. Remarkable progress in acridone alkaloid biosynthesis has been made by using tissue cultures of R. graveolens. The precursor role of anthranilate also in the case of rutacridone (dihydrofuro-acridone) was confirmed. Surprisingly, mevalonic acid was a poor and unspecific precursor for the isopropylidene dihydro-furan part of this particular alkaloid /28/. Carbon-13 label-ing experiments have shown unequivocally that ring C of rut-acridone is acetate-derived. After administration of $(1,2-^{13}C_2)$-acetate, six carbons were enriched from the incor-poration of three intact acetate units. Two different arran-gements of acetate units were observed. The rotation of ace-tate-derived ring C seems to occur at the stage of an N-methylated aminobenzophenone intermediate /29/. Labeling experiments and enzymatic work revealed that N-methylation of anthranilic acid is the first pathway-specific step in acri-done biosynthesis. The particular enzyme, anthranilic acid N-methyltransferase was only detectable in rutacridon synthesizing plant cell cultures /30/. It has been speculated that acridone alkaloid biosynthesis might involve a conden-sation of an activated anthranilic acid with 3 molecules ma-lonyl-Co A. Apparently no N-methylanthranilate-CoA ligase but rather an N-methylanthranilate adenyltransferase is respon-sible for activation /31/. The present data on acridone alkaloid biosynthesis are summarized in Fig. 13.

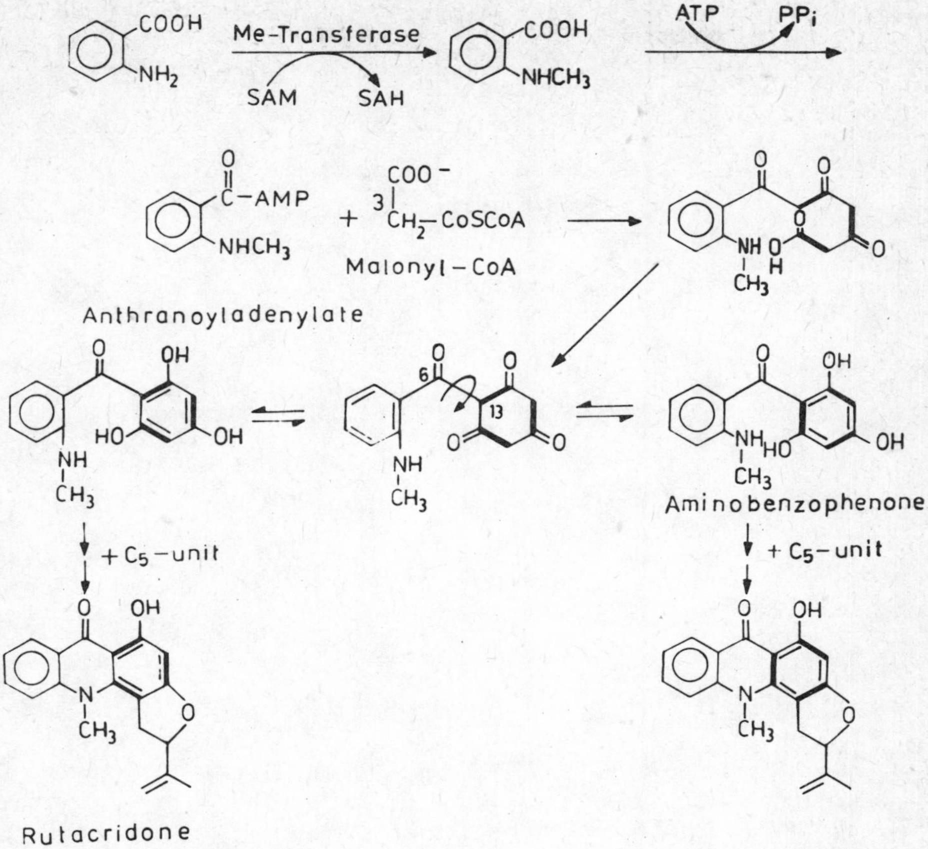

Fig. 13 Biosynthesis of rutacridone

BENZODIAZEPINES

Some Penicillium species accumulate several types of alkaloids. Compounds of the cyclopenin-viridicatin group are
synthesized in species of the section Asymetrica, subsection
Fasciculata. Cyclopeptine is the first intermediate which
originates from anthranilate, S-phenylalanine and the CH$_3$-

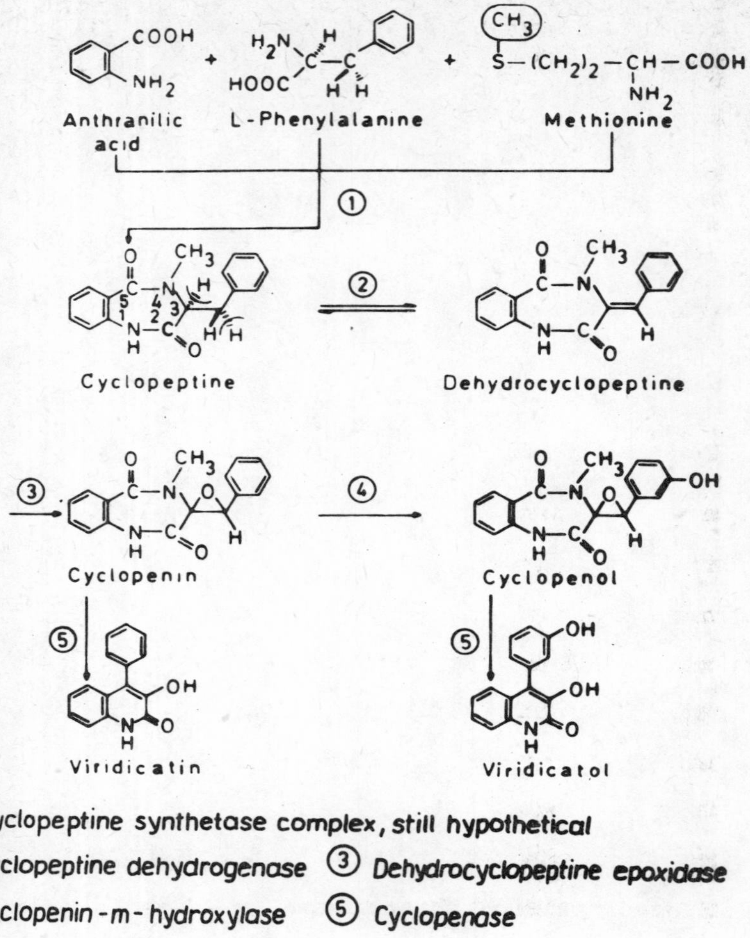

Fig. 14 Formation of the alkaloids of the cyclopenin-
viridicatin group in Penicillium cyclopium

group of methionine. This alkaloid is converted by a dehydrogenase to dehydrocyclopeptine. In the next step an epoxide is
formed. The resulting compound is cyclopenin which in turn
might be hydroxylated by an m-hydroxylase to cyclopenol. The
cyclopenin-m-hydroxylase is a mixed-function oxygenase.

The hydroxylation is accompanied by a NIH-shift. The enzyme cyclopenase which is responsible for the transformation of the benzodiazepines into quinoline alkaloids e.g. viridicatin is solely located in the conidiospores of the fungi. Most work on biosynthesis and regulation of <u>Penicillium</u> alkaloids has been done in Luckner's group /32/ (Fig. 14). Anthramycin, tomamycin and sibiromycin are closely related antibiotics with antitumor activity. These pyrrolo (1,4) benzodiazepine antibiotics are produced by thermophilic <u>Streptomyces</u> species.

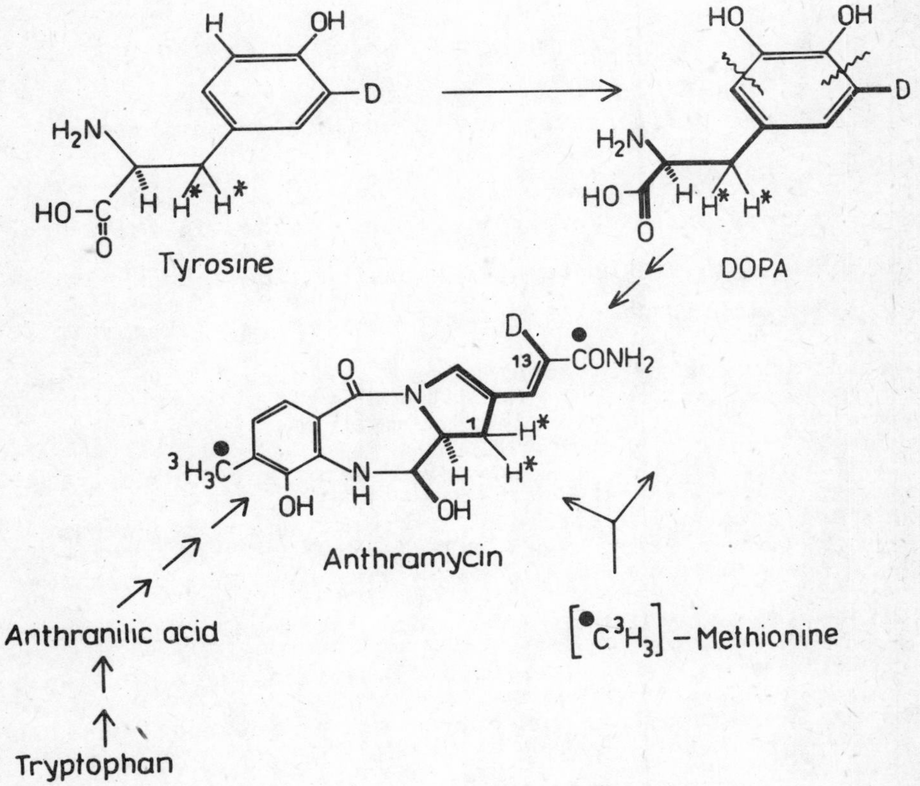

Fig. 15 Biosynthetic route to anthramycin

Essential building blocks are anthranilic acid and tyrosine.
The mode of assembling of these precursors was mainly investigated in Hurley's group /33/. Labeling experiments revealed that tyrosine is degraded via dopa and a ring cleavage product finally giving the "proline" part of anthramycin. The methyl group in the aromatic ring and the carbox -
amide group originate from the S-methyl group of methionine.
Deuterium present at C-_ of tyrosine is ultimately located
at C-13 in anthramycin indicating "meta" cleavage of the
ring of dopa. Furthermore, C-1 in anthramycin retains all the
hydrogens originally present at .C-3' of tyrosine (Fig. 15).

ACTINOMYCINS

The well-known actinomycins are chromopeptide antibiotics
which are produced by a number of Streptomycetes. The various actinomycins differ only from each other in the amino
acids present in the pentapeptide chains. The common chromo-
phor actinocine is a phenoxazinone derivative. Other natural
compounds possessing a phenoxazinone skeleton are the ommo-
chromes produced by arthopodes and molluscs, certain pigments found in fungi, and a small number of antibiotics. Tryptophan and some of its catabolic metabolites are key precursors of phenoxazinone biosynthesis. Phenoxazinone synthetase
catalyzes the formation of actinocin when 4-methyl-3-hydroxy-
anthranilic acid (4-MHA) serves as substrate. Also 4-MHA-pep-
tides are converted into the corresponding actinocyl peptides.
Mutants of Streptomyces parvulus that are blocked in actinomycin biosynthesis accumulate 4-MHA.

4-Methyl-3-hydroxyanthraniloyl peptides $\xrightarrow[\text{synthetase}]{\text{Phenoxazinone}}$ Actinocinylpeptides

Salzmann et al. 1969

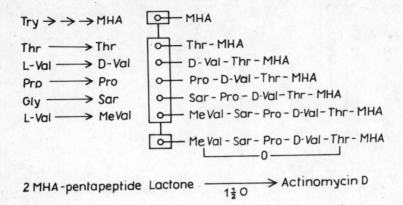

Try → → → MHA   ▢—○— MHA

Thr ——→ Thr   ○— Thr - MHA
L-Val ——→ D-Val   ○— D-Val - Thr - MHA
Pro ——→ Pro   ○— Pro - D-Val - Thr - MHA
Gly ——→ Sar   ○— Sar - Pro - D-Val - Thr - MHA
L-Val ——→ MeVal   ○— MeVal - Sar - Pro - D-Val - Thr - MHA

▢—○— MeVal - Sar - Pro - D-Val - Thr - MHA
⎣————— 0 —————⎦

2 MHA-pentapeptide Lactone $\xrightarrow[1\frac{1}{2}\,O]{}$ Actinomycin D

Tentative scheme according to Troost and Katz 1979

Fig. 16 Tentative scheme of actinomycin biosynthesis on a protein template

Normally this essential precursor of actinocin formation is not excreted during growth or antibiotic synthesis by the parental strains. A possible mechanism of actinomycin D formation on a protein template was recently proposed /34,35/ (Fig. 16).

MISCELLANEOUS

A number of 1,4-benzoxazin-3-ones were isolated from maize, wheat and rye. Of special interest is 2,4-dihydroxy-7-methoxy-1,4-benzoxazin-3-one (DIMBOA) found in maize. DIMBOA acts as feeding deterrent for insects and inhibits spore germination of a phytopathogenic Helminthosporium species. It is bio-

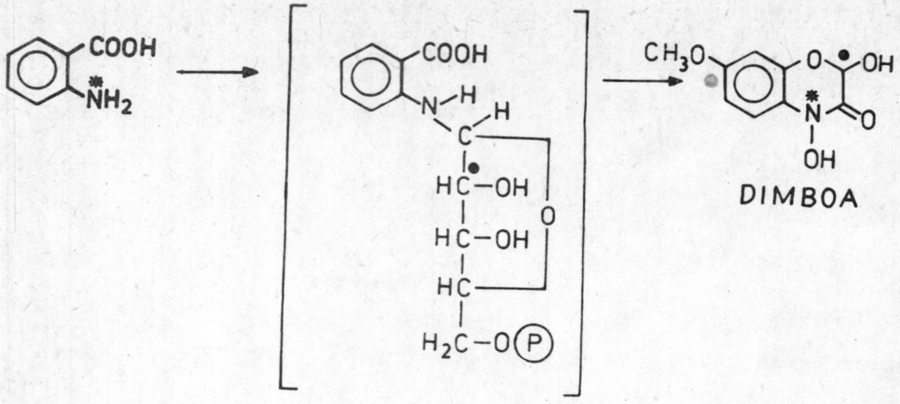

DIMBOA and related compounds occur in cereal grasses

Fig. 17 Biosynthesis of DIMBOA

synthetically derived from anthranilic acid and carbon atoms
2 and 3 of the oxazine ring from carbon 2 and 1 of ribose,
respectively. As intermediate N-(5'phosphoribosyl) anthranilate has been assumed, but no experimental evidence has been
provided /36/ (Fig. 17).

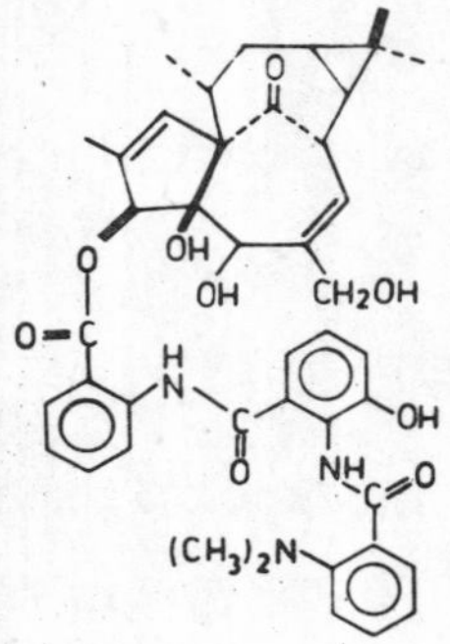

Fig. 18 Structure of
          milliamine C

The milliamines represent unique compounds. These ingenol
esters have been isolated from roots and stems of Euphorbia
milii. Instead of fatty ester groups present in other irritant ingenol derivatives, they contain anthraniloyl tripeptides
in various positions of the diterpene ingenol. For example, milliamine C (Fig. 18) is a highly active irritant which,
however, is not a tumour promoter. It would be of particular
interest to unravel the biosynthesis of these extremely potent secondary metabolites.

References

1. GRÖGER, D., Anthranilic acid as precursor of alkaloids. Lloydia, _32_, 221-246 (1969).

2. GRÖGER, D., Alkaloids derived from tryptophan and anthranilic acid. In: Encyclopedia of Plant Physiology, New Series, Vol. 8., Secondary Plant Products. BELL, E.A. and CHARLWOOD, B.V. (eds.) Springer-Verlag, Berlin-Heidelberg pp. 128-159 (1980).

3. CORDELL, G.A., Introduction to Alkaloids. A Biogenetic Approach. J. Wiley & Sons, New York (1981).

4. MUNSCHE, D., Biosynthese des Damascenins. Abhandl. Deut. Akad. Wiss. Berlin, Kl. Chem. Geol. Biol. _3_, 611-615 (1966).

5. NAIR, P.M. and VAIDYANATHAN, C.S., Anthranilic acid oxidase system of Tecoma stans - II Studies on the properties of a purified anthranilic acid oxidase system and its separation into different enzymic components. Phytochemistry _3_, 513-523 (1964).

6. WATERMAN, P.G., Alkaloids of the rutaceae: their distribution and systematic significance. Biochem. Syst. Ecol. _3_, 149-180 (1975).

7. MONKOVIC, I. and SPENSER, I., Biosynthesis of dictamnine. J. Chem. Soc. Chem. Commun. 205-207 (1966).

8. MATSUO, M., YAMAZAKI, M. and KASIDA, Y., Biosynthesis of skimmianine. Biochem. Biophys. Res. Commun. _23_, 679-682 (1966).

9. COLONNA, A.O. and GROSS, E.G., Biosynthesis of skimmianine in Fagara coco. Phytochemistry _10_, 1515-1521 (1971).

10. COBET, M. and LUCKNER, M., 2,4-Dihydroxychinolin, ein direktes Zwischenprodukt bei der Biosynthese des Furochinolinalkaloids Skimmianin bei Ruta graveolens. Phytochemistry _10_, 1031-1036 (1971).

11. GRUNDON, M.F., HARRISON, D.M. and SPYROPOULOS, C.G., Biosynthesis of aromatic isoprenoids,Part III. Mechanism of formation of the furan ring and origin of the 4-methoxy-group in the biosynthesis of furoquinoline alkaloids. J. Chem. Soc., Perkin I 302-304 (1975).

12. STECK, W., GAMBORG, O.L. and BAILEY, B.K., Increased yields of alkaloids through precursor biotransformation in cell suspension cultures of Ruta graveolens. Lloydia _36_, 93-95 (1973).

13. REISCH, J., ROSZA, Zs., SZENDREI, K. and KÖRÖSI, J.,
    2-(n-Undecyl)-chinolon-(4) aus den Blüten von Ptelea
    trifoliata und den Wurzeln von Ruta graveolens. Phyto-
    chemistry 14, 840-841 (1975).

14. RITTER, Ch. und LUCKNER, M., Zur Biosynthese der 2-n-
    alkyl-4-hydroxychinolinderivate (Pseudane) bei Pseudo-
    monas aeruginosa. Eur. J. Biochem. 18, 391-400 (1971).

15. JOHNE, S., The quinazoline alkaloids. Progress Chem.
    Org. Nat. Products, (1984) in the Press.

16. JOHNE, S. and GRÖGER, D., Natürlich vorkommende China-
    zolin-Derivate. Pharmazie 25, 22-44 (1970).

17. SCHILDKNECHT, H. und WENNEIS, W.F., Über Arthropoden-
    Abwehrstoffe XXV. Anthranilsäure als Precursor der
    Arthropoden-Alkaloide Glomerin und Homoglomerin. Tetra-
    hedron Lett. 1815-1818 (1967).

18. JOHNE, S., WAIBLINGER, K. und GRÖGER, D., Zur Biosyn-
    these des Chinazolinalkaloids Arborin. Eur. J. Biochem.
    15, 415-420 (1970).

19. LILJEGREN, D.R., The biosynthesis of quinazoline alka-
    loids of Peganum harmala. Phytochemistry 7, 1299-1306
    (1968).

20. LILJEGREN, D.R., Biosynthesis of quinazoline alkaloids
    of Peganum harmala. Phytochemistry 10, 2661-2669 (1971).

21. YAMAZAKI, M. and IKUTA, A., Biosynthesis of Evodia alka-
    loids. Tetrahedron Lett. 3221-3224 (1966).

22. YAMAZAKI, M., IKUTA, A., MORI, T. and KAWANA, T., Bio-
    synthesis of Evodia alkaloids, II. The participation of
    C-1 unit to the formation of indoloquinazoline alkaloids.
    Tetrahedron Lett. 3317-3320 (1967).

23. FIEDLER, E., FIEDLER, H.P., GERHARD, A., KELLER-SCHIER-
    LEIN, W., KÖNIG, W.A. und ZÄHNER, H., Stoffwechselpro-
    dukte von Mikroorganismen. 156. Mitt. Synthese und Bio-
    synthese substituierter Tryptanthrine. Arch. Mikrobiol.
    107, 249-256 (1976).

24. YAMAZAKI, M., The biosynthesis of neurotropic mycotoxins
    In: The Biosynthesis of Mycotoxins; A Study of Secondary
    Metabolism. STEYN, P.S. (ed.), Academic Press, New
    York, pp. 193-222 (1980).

25. GRÖGER, D. und JOHNE, S., Zur Biosynthese einiger Alka-
    loide von Glycosmis arborea (Rutaceae). Z. Naturforsch.
    236, 1072-1075 (1968).

26. PRAGER, R.H. and THREDGOLD, H.M., Studies using radio-isotopes. III. The biosynthesis of acridone alkaloids. Aust. J. Chem. 22, 2627-2634 (1969).

27. KHAN SHAFIG, M., LEWIS, J.R. and WATT, R.A., Conversion of tecleanone into 1,3-dimethoxy-10-methylacridone: a possible biosynthetic route to acridone alkaloids. Chem. Ind. 744-745 (1975).

28. BAUMERT, A., KUZOVKINA, I.N., KRAUSS, G., HIEKE, M. and GRÖGER, D., Biosynthesis of rutacridone in tissue cultures of Ruta graveolens L., Plant Cell Reports 1, 168-171 (1982).

29. ZSCHUNKE, A., BAUMERT, A. and GRÖGER, D., Biosynthesis of rutacridone in cell cultures of Ruta graveolens: Incorporation studies with (^{13}C) acetate. J. Chem. Soc., Chem. Commun. 1263-1265 (1982).

30. BAUMERT, A., HIEKE, M. and GRÖGER, D., N-Methylation of anthranilic acid to N-methylanthranilic acid by cell-free extracts of Ruta graveolens tissue cultures. Planta medica 48, 258-262.

31. BAUMERT, A. and GRÖGER, D., Enzymatic activation of anthranilic acid during the course of rutacridone biosynthesis, in the Press.

32. LUCKNER, M., Alkaloid biosynthesis in Penicillium cyclopium - does it reflect general features of secondary metabolism? J. Nat. Products 43, 21-40 (1980).

33. HURLEY, L.H. and SPEEDIE, M.K., Pyrrolo (1,4) benzodiazepine antibiotics: anthramycin, tomamycin and sibiromycin. In: Antibiotics, IV Biosynthesis. CORCORAN, J.W. (ed.) Springer-Verlag, Berlin-Heidelberg, pp. 262-294 (1981).

34. SALZMANN, L., WEISSBACH, H. and KATZ, E., Enzymatic synthesis of actinocinyl peptides. Arch. Biochem. Biophys. 130, 536-545 (1969).

35. TROOST, T. and KATZ, E., Phenoxazinone biosynthesis: Accumulation of a precursor, 4-methyl-3-hydroxyanthranilic acid, by mutants of Streptomyces parvulus. J. Gen. Microbiol. 111, 121-132 (1979).

36. TIPTON, C.L., WANG, M.-C., TSAE, F.H.-C., LIN TU, C.-C. and HUSTED, R.R., Biosynthesis of 1,4-benzoxazin-3-ones in Zea mays. Phytochemistry 12, 347-352 (1973).

37. MARSTEN, A. and HECKER, E., On the active principles of the Euphorbiaceae VI: Isolation and Biological activities of seven milliamines from Euphorbia milii. Planta medica 47, 141-147 (1983).

DISCUSSION

LEWIS J.R. (Aberdeen, Scotland)

I should like to comment on an interesting observation in rutacridone biosynthesis; ^{13}C coupling indicates that the aminobenzophenone intermediate must show free rotation before introduction of the prenyl side chain. This indicates that, contrary to accepted theories, the all keto form of the phloroglucinol moiety must first cyclise and then leave the enzyme surface. Whether it remains in the keto form or enolizes (in its free state) cannot be determined at this point of time. Further feeding experiments may be able to clarify this aspect of the biosynthesis of acridones.

GRÖGER D. (Halle, German Democratic Republic)

I thank you for this valuable remark and I would like to add the following. The second pathway-specific reaction in acridone alkaloid biosynthesis seems to be the "activation" of the N-methylanthranilic acid. According to our cell-free system, no CoA-ester is involved but an AMP-derivative of anthranilic acid.

van der PLAS H.C. (Wageningen, The Netherlands)

In one of your slides I saw a reactor using plant cells for the production of rutacridone. Do you feel that using immobilized plant cells as units for the preparation of certain chemical components, has a great future?

GRÖGER D. (Halle, German Democratic Republic)

There are many efforts around the world to produce biologically active compounds with the aid of tissue cultures. Some success has been achieved on a commercial scale in Japan and in the Soviet Union. The fermentation technology for mass production of plant cells is far advanced. But in many cases the yield of a given secondary metabolite is not high enough to compete with the intact plant.

KISFALUDY L. (Budapest, Hungary)

You referred to benzodiazepine formation starting from anthranilic acid in plants. Is it possible that such a biotransformation could exist in higher living organisms, too? It would relate to the existence of an endogenous ligand not found so far for the benzodiazepine receptor discovered in brain.

GRÖGER D.: We did not perform such experiments.

CHEMISTRY OF PLANT BILE PIGMENTS

GOSSAUER A.

Institut für organische Chemie, Universität
Freiburg i.Ue., Pérolles, CH-1700 Fribourg,
Switzerland

In addition to bile pigments which occur in man and animals as products of degradation of hemoglobin, some coloring substances are found in plants, mainly algae; these structures were recognized about 55 years ago by Lemberg (1928,1933) and related to the bile pigments. Nowadays, it is well recognized that these pigments act as the prosthetic groups of some chromoproteins (so-called biliproteins) which perform important functional roles as light harvesters and light sensors. Thus, blue-green algae (Cyanobacteria) as well as red algae (Rhodophyta) and some Cryptophyta (cryptomonad algae) contain besides chlorophyll, as the essential photosynthetic pigment, some blue and red accessory pigments (phycocyanins and phycoerythins, respectively) which emit a strong fluorescence when illuminated with visible light. This fluorescence vanishes on denaturation of the protein with urea or guanidine hydrochloride and reappears on dilution of the denatured protein solution with water.

Another physiologically active biliprotein, phytochrome, governs plant morphogenesis. In the living plant, phytochrome occurs in two modifications, a red absorbing form (P_r) and a far red absorbing form (P_{fr}) which are interconverted by irradiation with light of λ_{max} 665 and 725 nm, respectively. The regulative function of phytochrome is associated with the P_{fr} form. The biological function as well as the physicochemical properties of biliproteins and phytochrome have been reviewed recently by Scheer (1981) and Rüdiger (1980), respectively.

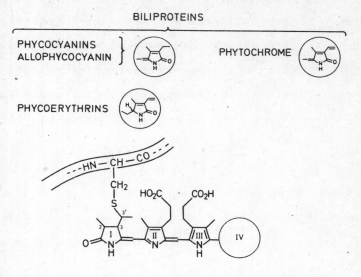

Scheme 1   **Structural relationship between the different types of biliprotein chromophores.**

A peculiar feature of the biliproteins is the covalent linkage between the prosthetic group and the apoprotein (cf. Scheme 1). Therefore, the elucidation of the structure of the corresponding chromophores is hampered both by the rather drastic hydrolytic conditions required for the cleavage of the bile pigment chromophore and by the lability of the latter.

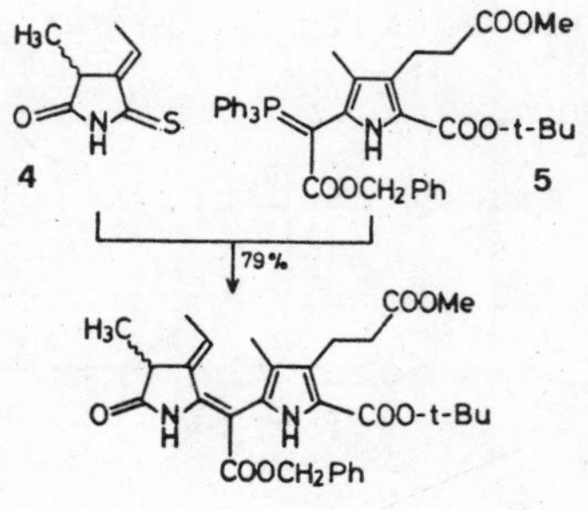

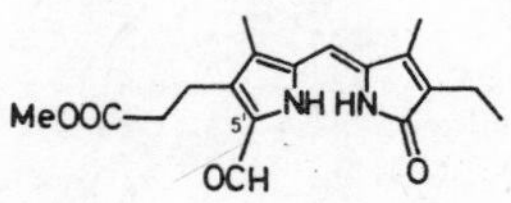

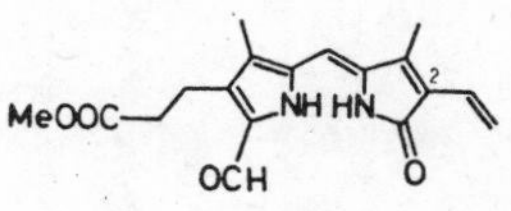

Until now, only three biliprotein chromophores, namely phycocyanobilin (1), phycoerythrobilin (2) and phytochromobilin (3) have been unambiguously characterized. Their structures have been elucidated by spectroscopic and degradation studies, and confirmed by chemical synthesis. Particularly in the case of phycoerythrobilin the total synthesis of its enantiomeric pure dimethyl ester played an important rôle in the assignement of the absolute configuration of the chromophore (Gossauer and Weller 1978). Characteristic of all bile pigments which have been isolated hitherto from biliproteins is the presence of an ethylidene group at the C-3 position of the saturated lactame ring I in the bilindione skeleton (cf. 1-3). As a matter of fact, the conventional methodology used in bile pigment chemistry proved to be impracticable for the synthesis of the biliprotein chromophores and new methods had to be developped for the preparation of the required building blocks. Particularly, the 3,4-dihydro-1(10H)-dipyrrinone 6, which was used as a key compound for the synthesis of the bile pigments 1-3, was obtained by reaction of monothiosuccinimide 4 with the pyrrolic phosphorous ylide 5. This kind of Wittig reaction involving the C=S bond represents a new general method for the synthesis of alkylidene lactams from monothiosuccinimides and resonance-stabilized phosphorous ylides, (Gossauer et al. 1977, 1979, Gossauer and Slopianka 1981 a,b).

Acid catalyzed condensation of **6** with methyl 5'-formylisoneoxanthobilirubinate (**7**) or methyl 5'-formyl-[2-vinyl]-isoneoxanthobilirubinate (**8**), led

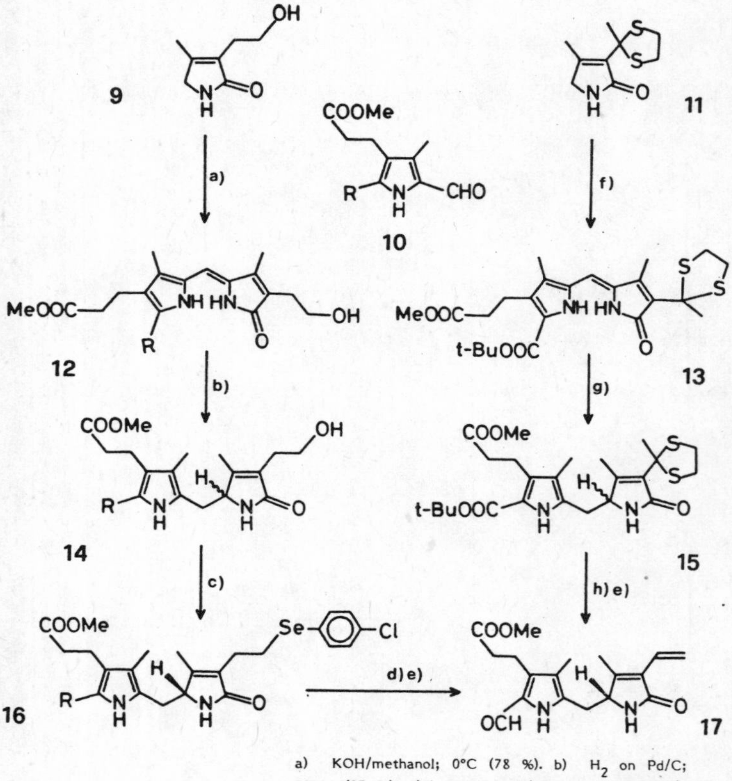

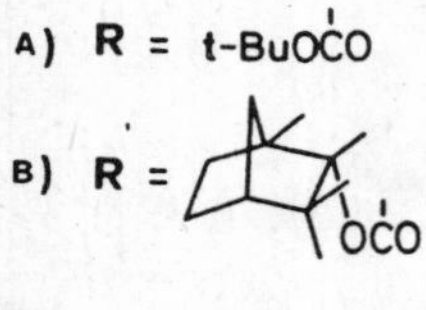

a) KOH/methanol; 0°C (78 %). b) H$_2$ on Pd/C; 20°C (97 %). c) i) H$_3$C-SO$_2$Cl/DMFA; 0-20°C (82 %); ii) p-Cl-C$_6$H$_4$-SeH/THF, 20°C (62 %). d) H$_2$O$_2$/ THF; 0°C (87 %). e) HC(OC$_2$H$_5$)$_3$/TFA; 20°C (65 %). f) KOH/methanol; 0-4°C (58 %). g) i) Na(Hg)/ methanol; -10°C (94 %); ii) NaOCH$_3$/methanol; 0°C (95 %). h) Raney Ni/ethanol; 78°C (44 %). (DMFA = dimethylformamide; THF = tetrahydrofuran).

A) R = t-BuOCO

B) R =

Scheme 2   Synthesis of 5'-formyl[2-vinyl]-isoneobilirubinate.

to phycocyanobilin dimethyl ester (Gossauer and Hirsch 1974) (Gossauer and Hinze 1978) and phytochromobilin dimethyl ester (Gossauer and Weller 1980), respectively. In order to synthesize phycoerythrobilin dimethyl ester, methyl 5'-formyl-[2-vinyl]-isoneobilirubinate (17) was prepared by two independent routes (Gossauer and Weller 1978) (Gossauer und Klahr 1979) (cf. Scheme 2). Moreover, using the α-methylfenchyl ester 10B (instead of the tert-butyl ester 10A) as precursor of the aldehyde 17, the latter could be obtained enantiomeric pure by previous fractional crystallization of the selenide 16B. Condensation of the R-configurated aldehyde 17 with 6 afforded two diastereomeric phycoerythrobilins (as dimethyl esters) which were separated by chromatography. The absolute configuration at the asymmetric C-2-atom was elucidated by oxidative degradation of 2 which afforded laevorotatory (i.e. R-configurated) 2-methyl-3-(E)-ethylidenesuccinimide. As the synthetic compound proved to be identical with phycoerythrobilin dimethyl ester isolated from C-phycoerythrin (from Phormidium persicinum) with boiling methanol, the absolute configuration of the native chromophore at the positions C-2 and C-16 was etablished (Gossauer and Weller 1978).

As mentioned above, biliproteins act as photoregulators in phanerogames as well as photosensitizers of chlorophyll in algae. Just recently, W. Rüdiger et al. (1983) could demonstrate that the photoreversible transformation of the P_r and P_{fr} forms of phytochrome involves a change of configuration at the C-15-C-16 double bond of the chromophore (cf. Scheme 1). As found by H. Falk et al., (1983) synthetic all Z configurated 3,4-dihydrobilindiones photoisomerize in the same way yielding the corresponding Z,Z,E-isomer. It is noteworthy that our own attempts to synthesize E,Z,Z- and E,Z,E-isomers of 3,4-dihydrobilindiones through condensation of the appropriate 1(10H)-dipyrrinone derivatives led, irrespectively of the configuration at the exocyclic double bond of the 3,4-dihydro-1(10H)-dipyrrinone moiety, to Z,Z,E-

or Z,Z,Z-isomers (Blacha-Puller, 1979). Thus, the Z-configuration at the exocyclic double bond adjacent to the saturated lactame ring seems to be thermodynamically more favoured than in the case of the corresponding (unsaturated) 1(10H) dipyrrinone moiety (see Scheme 3).

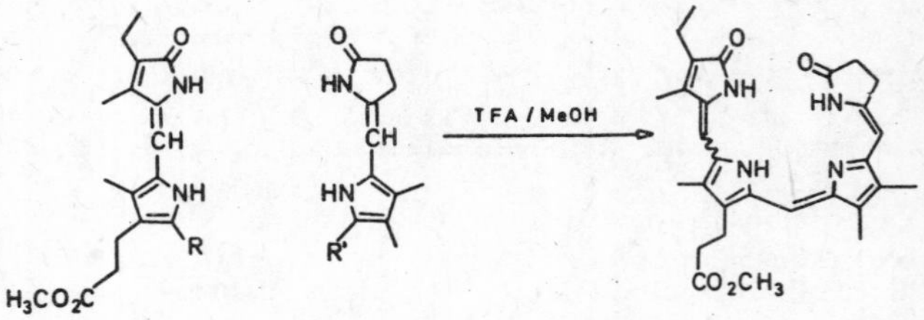

R = CHO	R' = CO$_2$t-Bu		
Z	Z	Z	Z
Z	E	Z	Z
E	E	E	Z
E	Z	E	Z

R = CO$_2$t-Bu	R' = CHO		
Z	E	Z	Z
E	E	Z	Z

Scheme 3   **Synthesis of Z,Z,E-configurated 3,4-dihydro-biliverdins from Z- and E-configurated 1(10H)-dipyrrinone derivatives.**

Until now, no isomerization at the exocyclic double bond of the dipyrrine moiety has been observed in bile pigments. Most likely, the Z-configuration is stabilized in this case by the hydrogen **bond** between the pyrrole and pyrrolenine rings which are mutualy interconverted by proton jump from one nitrogen atom to the other. For this reason, it has been suggested that the above mentioned strong fluorescence of the native biliproteins is associated with a "stretched" conformation of the chromophore molecule inside the hydrophobic environment of the apoprotein (Scheer 1981). In order to investigate the influence of conformation on the desactivation of the excited state of bile pigment molecules, we have synthesized some amphiphilic biliverdine derivatives (e.g. **18**) which should be able to be incorporated into organic membranes (A. Gossauer and P. Nesvadba, unpublished results).

18

Acknowledgments - The work carried out in the author's research group has been supported financially by the Deutsche Forschungsgemeinschaft (Projects Go 209/5 and Go 209/12) and by the Schweizerischen Nationalfonds zur Förderung der wissenschaftlichen Forschung (Project 2.320-0.81). I am grateful to them, and to all the able co-workers whose work is referred to herein.

References

BLACHA-PULLER, M., 1979: PhD Thesis, Technische Universität Braunschweig. cf. A. GOSSAUER, M. BLACHA-PULLER, R. ZEISBERG, and V. WRAY, 1981: Synthese von E,Z,Z-konfigurierten Biliverdinen aus entsprechend konfigurierten 5(1H)-Pyrromethenonen. Liebigs Ann. Chem., 342-346.

FALK, H., GRUBMAYR, G. KAPL, N. MUELLER, U. ZRUNEK, 1983: Phytochrommodellstudien: Diastereomere 2,3-Dihydrobilatriene-abc. Monatsh. Chem. 114, 753-771.

GOSSAUER, A., and W. HIRSCH, 1974: Totalsynthesen des racemischen Phycocyanobilins (Phycobiliverdins) sowie eines "Homophycobiliverdins". Liebigs Ann. Chem., 1496-1513.

GOSSAUER, A., R.-P. HINZE, and H. ZILCH, 1977: Umsetzung cyclischer Imide von Monothiodicarbonsäuren mit Phosphor-Yliden: Eine neue Methode zur Synthese von ω-Alkyliden-lactamen. Angew. Chem. 89, 429-430. Angew. Chem. Int. Ed. Engl. 16, 418.

GOSSAUER, A., and J.-P. Weller, 1978: Chemical Total Synthesis of (+)-(2R,16R)- and (+)-(2S,16R)-Phycoerythrobilin Dimethyl Ester. J. Am. Chem. Soc. 100, 5928-5933.

GOSSAUER, A., and R.-P. HINZE, 1978: An Improved Chemical Synthesis of Racemic Phycocyanobilin Dimethyl Ester. J. Org. Chem. 43, 283-285.

GOSSAUER, A., and E. KLAHR, 1979: Totalsynthese des racem. Phycoerythro bilin-dimethylesters. Chem. Ber. 112, 2243-2255.

GOSSAUER, A., F. ROESSLER, H. ZILCH, and L. ERNST, 1979: Einige präparative Anwendungen der Reaktion von N-(Thioacyl)urethanen und deren Vinylogen mit resonanzstabilisierten Phosphor-Yliden. Liebigs Ann. Chem., 1309-1321.

GOSSAUER, A., and J.-P. WELLER, 1980: Synthese und Photoisomerisierung des racem. Phytochromobilin-dimethylesters. Chem. Ber. 113, 1603-1611.

GOSSAUER, A., and M. SLOPIANKA, 1981a: Thiocarbonyl Olefination III: A new Synthesis of β-Aminoacids from N-Thiocylurethanes. Synthetic Commun. 11, 95-99.

GOSSAUER, A., and M. SLOPIANKA, 1981b: Darstellung von β-Aminosäuren aus N-(Acetyl)thioamiden; Totalsynthese der Iturinsäure, Liebigs Ann. Chem., 2258-2265.

LEMBERG, R., 1928: Die Chromoproteide der Rotalgen I. Liebigs Ann. Chem. 461, 46-89.

LEMBERG, R., 1933: Die Phycobiline der Rotalgen. Ueberführung in Mesobilirubin und Dehydro-mesobilirubin. Liebigs Ann. Chem. 505, 151-177.

RUEDIGER, W., 1980: Phytochrome. A Ligth Receptor of Plant Photomorphogenesis. Structure and Bonding, 40, Springer-Verlag, Heidelberg, 101-140.

RUEDIGER, W., T. THUEMMLER, E. CMIEL, and S. SCHNEIDER, 1983: Chromophore structure of the physiologically active form (P_{fr}) of phytochrome. Proc. Natl. Acad. Sci., USA, 80, 6244-6248.

SCHEER, H, 1981: Biliproteine. Angew. Chem. 93, 230-250; Angew. Chem. Int. Ed. Engl. 20, 241-261.

DISCUSSION

PANDIT U.K. (Amsterdam, The Netherlands)
   a, Is the second amide group essential for the thio-Wittig
      reaction?
   b, How are you preparing your thioimides, especially in the
      case of unsymmetrical systems?

GOSSAUER A. (Fribourg, Switzerland).
   a, Yes, in contrast to Eschenmoser's sulfide contraction
      method, the olefination of the thiocarbonyl group with
      resonance-stabilized phosphorous ylides does not work
      in the case of thiolactams. Nevertheless, it is known
      that some thioketones which do not occur as dimers,
      react with Wittig reagents.
   b, In order to prepare monothioimides we used three
      different approaches: i/ reaction of imides with P_4S_{10}
      or, alternatively, with Lawesson's reagent. ii/ Thiolysis
      of the corresponding iminoester with hydrogen sulfide,
      and iii/ thiolysis of the corresponding imidines with
      hydrogen sulfide in the presence of pyridine. As imidines
      can be obtained by cyclisation of the corresponding
      ω-cyanoamides, which are available by regioselective
      methods, the latter approach enables the synthesis of
      unsymmetrical monothioimides.

ÖTVÖS L. (Budapest, Hungary)
   You have mentioned that the reason of the extremely high
   optical rotation of urobilins is the helicity of the
   chromophore. Is it known which is the helicity of the
   natural compounds? I think that this is important in
   relation to the binding of the chromophore to the
   apoprotein.

GOSSAUER A. (Fribourg, Switzerland)
   According to Moscowitz's model for optical active urobilins,
   the helicity of the chromophore is determined by the absolute

configuration at the chiral atoms C-4 and C-16. Thus, if the absolute configuration of these are opposite, the chromophore itself belongs to the symmetry group C_s and is, therefore, achiral. Symmetrically substituted urobilins which have R or S configuration at both chiral atoms belong to the symmetry group C_2 and are chiral. Molecular models point out that the S,S configured chromophore has probably M helicity. Until now, however, this point has not been clarified experimentally.

SNATZKE G. (Bochum, German Federal Republic)

a, What is the E/Z-stereochemistry in your Wittig-reaction?

b, May the "stretched" form be possible in a model compound by avoiding hydrogen bonding between the 2- and 3-pyrrole units, e.g. by making the N-methyl or N-acetyl derivative?

GOSSAUER A. (Fribourg, Switzerland)

a, The olefination of monothioimides with resonance-stabilized phosphorous ylides is Z stereoselective. Most likely, the reason for this selectivity is the formation of a hydrogen bond between the NH group and the carbonyl group of the Wittig reagent at an earlier stage of the reaction (e.g. in the betaine intermediate).

b, The stereochemistry of bilidiones bearing an N-methyl group in the dipyrrine moiety of the molecule has been investigated recently by H. Falk (University of Linz). Even in these cases, the all-Z configuration of the exocyclic double bonds seems to be preferred. Certainly, as the β-positions of pyrrole rings in the investigated molecules were substituted also by methyl groups, steric hindrance may favour the Z configuration in the dipyrrine moiety of the molecule despite of the absence of a hydrogen bond between the two nitrogen atoms. Therefore, we are trying to "stretch" bile pigment molecules by intermolecular interactions similarly to those which govern the structure of the chromophore in native biliproteins.

MEDZIHRADSZKY K. (Budapest, Hungary)

On one of your slides I saw a reaction between a
diazomethylketone and a pyrrole derivative. Could you
comment on the conditions of this reaction? Is this reaction
pH-dependent?

GOSSAUER A. (Fribourg, Switzerland)

Usually, we carried out the reaction adding dropwise a
solution of the diazolaevulinic ester in benzene to a
mixture of copper dust and the pyrrole derivative at the
melting temperature of the latter. Although the yield of
the reaction is generally not high (about 30 %), more than
60 % of unchanged pyrrole can be recovered. Actually, we
tried to increase the yield using other catalysts (e.g.
copper(II) acetylacetonate) but other modifications of the
reaction conditions have not been investigated up to now.

POSTERS

Bio-Organic Heterocycles
van der Plas H.C., Ötvös L., and Simonyi M. eds

DIARYL-N$_4$-14-MEMBERED MACROCYCLES
A NEW CLASS OF ANTIARRHYTHMIC AGENTS

HANKOVSZKY H.O., HIDEG K., BÓDI I.* and FRANK L.*

Central Laboratory, Chemistry, University of Pécs
Pécs, Pf. 99, H-7643, Hungary

*Alkaloida Chemical Factory
 H-4440 Tiszavasvári, Hungary

The first preparation of tetraaza macrocycles in a metal ion template synthesis has been introduced by Curtis et al. in 1960 [1]. The reaction only resulted in the formation of seven-membered cyclic compounds in the absence of metal ion [2,3].

The reaction between ethylenediamine mono salt (en.HClO$_4$) and aliphatic α,β-unsaturated ketones in alcohol led to macrocyclic dienes (e.g. Me$_6$[14]4,11-dieneN$_4$.2HClO$_4$)[4]. Polish chemists reported a detailed study for the reaction yielding (R^1,R^2)$_2$-[14]4,11-dieneN$_4$.2HClO$_4$ (R^1 and/or R^2 are H, Me, Et) [5]. This reaction seemed to be limited to β-arylated α,β-unsaturated ketones. By using the en.HClO$_4$ method with benzylidene acetone, Me$_2$Ph$_2$[14]4,11-dieneN$_4$.2HClO$_4$ could be obtained. The reaction was extended to other β-aryl-α,β-unsaturated ketones, as well [6-8].

The formation of crystalline dienes proceeds via complicated equilibria [8] (Scheme 1). The recrystallized dienes are trans-C-meso-diimines, in which the two carbon centers have different chiralities. The reduction of dienes with NaBH$_4$/EtOH gave only three tetramine diastereoisomers. The major diastereomer was proved to have a C-meso-C-meso-N-meso-N-meso configuration [9,10]. The stereoisomers can be separated by column chromatography.

Almost no significant data are available for biological activity of synthetic tetraaza macrocycles. This is even more striking if compared to the enormous chemical work which has been summarized recently in excellent text books [11,12].

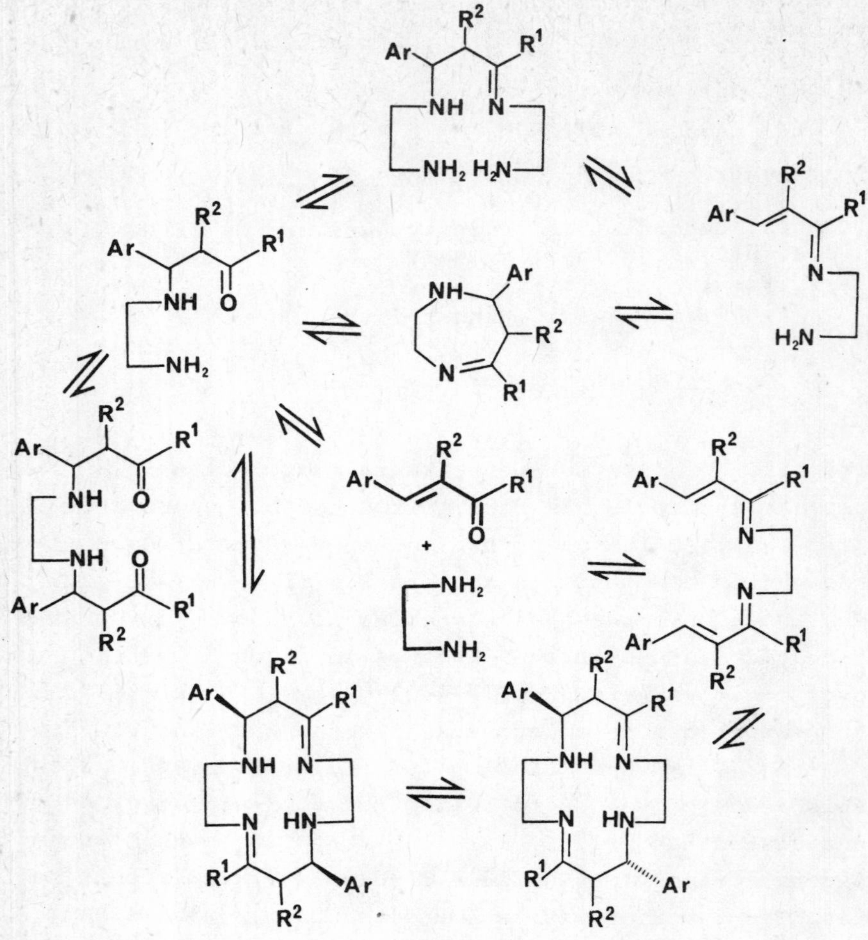

Scheme 1

For systematic investigations of the relationships between structure and biological activity, further series of diaryl[14]aneN$_4$ compounds were prepared. The syntheses have been carried out as shown in Scheme 2.

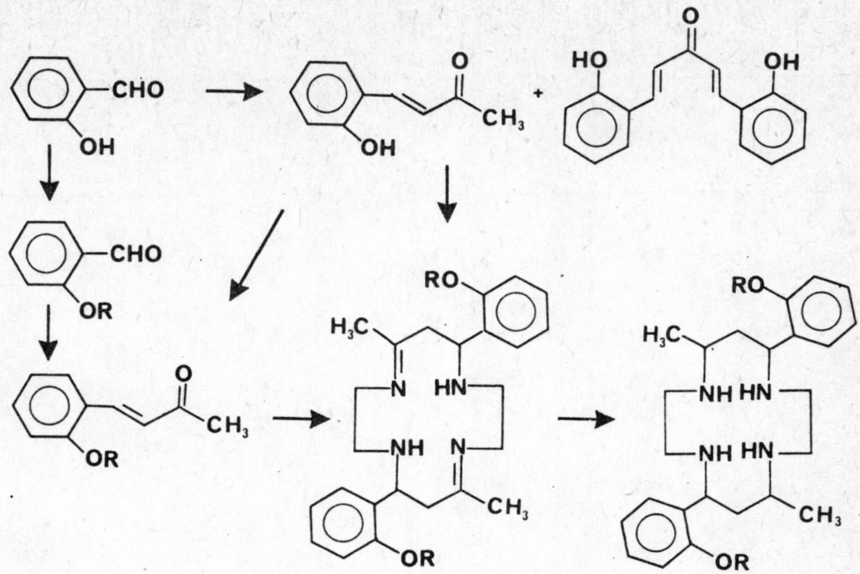

Scheme 2

Most of the diaryl[14]aneN$_4$.HCl salts showed significant antiarrhythmic activity on isolated preparations [13]. In vivo experiments (against aconitine arrhythmia) in anesthetized rats were performed as described earlier [14]. Some significant data of the experiments are summarized in Table 1. The data of relative activity, relative toxicity and, relative therapeutic index of some macrocyclic compounds are illustrated in Fig.1.

SUMMARY

The N$_4$-14-macrocycles may represent a new type of highly potent antiarrhythmic compounds in addition to the types already known [15].

Table 1. Antiarrhythmic Potencies and Toxicities of Diaryl-N_4-14-
membered Macrocycles

Compd	ED_{125}[a] mg/kg iv	ED_{150}[a] mg/kg iv	LD_{50}[b] mg/kg iv	Therapeutic index LD_{50}/ED_{150}	Relative activity (Quinidine = 1.0)
HA-68	1.10 (0.90-1.22)	1.50 (1.31-1.72)	15.0 (13.2-17.4)	10.0	7.9
H-2802	0.23 (0.20-0.26)	0.32 (0.28-0.36)	7.7 (6.4-9.1)	24.1	37.2
H-2807	0.47 (0.42-0.53)	0.64 (0.57-0.67)	4.7 (3.9-5.6)	7.3	18.6
HA-93	0.35 (0.31-0.39)	0.48 (0.43-0.55)	10.0 (9.2-10.2)	20.8	24.8
H-2817	0.16 (0.15-0.18)	0.20 (0.19-0.22)	5.34 (4.99-5.71)	26.7	59.5
H-2818	0.20 (0.19-0.22)	0.24 (0.22-0.25)	8.06 (7.07-9.20)	33.6	49.6
H-2815	1.20 (1.09-1.32)	1.55 (1.41-1.71)	13.2 (12.6-13.7)	8.5	7.7
H-2801	0.12 (0.11-0.13)	0.15 (0.14-0.17)	8.1 (6.4-10.3)	54.0	79.3
H-2803	0.06 (0.06-0.07)	0.08 (0.08-0.09)	3.4 (2.5-4.6)	42.5	148.8
H-2814	0.24 (0.22-0.26)	0.29 (0.27-0.32)	5.37 (4.48-6.44)	18.5	41.0
H-2800	1.01 (0.78-1.30)	1.97 (1.53-2.53).	21.8 (18.0-26.1)	11.1	6.0
Quinidine	6.8 (5.4-8.4)	11.9 (9.6-14.8)	50 (45-56)	4.2	1.0
Mexiletine	20.5 (18.8-22.3)	25.5 (24.1-28.1)	32 (28-36)		

a.) Effective doses (and 95%-confidence ranges) against the aconitine
arrhythmia in anesthetized rats. The duration of the aconitine infusion
(5ug/kg . min) was 6.1 min in the saline-pretreated controls. The ED_{125}
and ED_{150} correspond with an infusion time of 7.6 min and 9.2 min
respectively.

b.) Acut toxicity on mice.

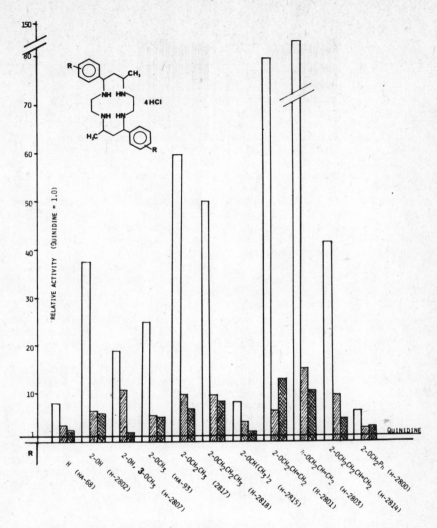

Fıg.1. Effect of some Diaryl-N_4-14-membered Macrocycles on Aconitine
Induced Arrhythmia

☐ relative antiarrhythmic activity
▨ relative toxicity
▦ relative therapeutic index

References

1. Curtis,N.F.: Transition-metal Complexes with Aliphatic Schiff Bases.
Part I. Nickel(II) Complexes with N-Isopropylidene-
ethylenediamine Schiff Bases
J. Chem.Soc., <u>1960</u>, 4409-4413; Curtis,N.F.: Macrocyclic coordination compounds formed by condensation of
metal-amine complexes with aliphatic carbonyl compounds
Coord. Chem. Rev., <u>3</u>, 3-47 (1968) and references therein.

2. Guareschi,I.: *Atti Accad. Sci.*, *Torino*, <u>29</u>, 694 (1894)
3.   Substituted 1,4-diaza-4-cycloheptenes,
                 Brit. Pat., 1,108,440 (1968); Ref.:*C.A.*,<u>69</u>, 52034h (1968)
4. Curtis,N.F., Hay,R.W.: A novel heterocycle synthesis. The formation
                 of 5,7,7,12,13,13-hexamethyl-1,4,8,11-tetraazacyclotetra-
                 deca-4,11-diene bis(hydroperchlorate) by reaction of di-
                 aminoethane monohydroperchlorate with mesityl oxide or
                 acetone
                 Chem. Commun., <u>1966</u>, 524-525
5. Kolinski,R.A., Korybut-Daszkiewicz,B.: Macrocyclic Ligands and Their
                 Metal Ion Complexes. Part VI. Preparation,Stereochemistry
                 and Conformational Analysis of Polyalkyl-1,4,8,11-tetra-
                 azacyclotetradeca-4,11-dienenickel(II) Diperchlorates
                 Inorg. Chim. Acta, <u>14</u>, 237-245 (1975)
6. Hankovszky,H.O., Hideg,K., Lloyd,D., McNab,H.: The Formation and Decom-
                 position of 1,4,8,11-tetraazacyclotetradeca-4,11-dienes
                 J. Chem. Soc., *Perkin I.* <u>1979</u>, 1345-1350
7. Hideg,K., Lloyd,D.: The Reaction between α,β-Unsaturated Ketones and
                 Ethylenediamine
                 Chem. Commun., <u>1970</u>, 929-930; Hideg,K., Lloyd,D.:
                 Reaction Products from α,β-Unsaturated Ketones and
                 Aliphatic Diamines or Dithiols
                 J. Chem. Soc., *(C)* <u>1971</u>, 3441-3445
8. Lloyd,D., Scheibelein,W., Hideg,K.: Further Studies of the Mixtures
                 obtained from Reactions between Conjugated Enones and
                 Ethylenediamine, and from Conjugated Enones and 1-Amino-
                 propane
                 J. Chem. Res., <u>1981</u>, (S) 62-63, (M) 0838-0858
9. Ferguson,F., Roberts,P.J., Lloyd,D., Hideg,K., Hay,R.W., Piplani,D.P.:
                 The Tetra-azacyclotetradecadiene formed from Benzylidene-
                 acetone and 1,2-Diaminoethane; the Crystal and Molecular
                 Structures of a Cooper(II) Complex of the Ligand,and a
                 Tetra-amine formed on Reduction
                 J. Chem. Res., <u>1978</u>, (S) 314-315, (M) 3734-3790
10. Hay,R.W., Jeragh,B., Piplani,P., Hideg,K., Hankovszky,H.O.: Cooper(II),
                 Nickel(II) and Cobalt(III) Complexes of the Macrocyclic
                 Ligand C-meso-7,14-Diphenyl-5,6-butano-12,13-butano-1,4,
                 8,11-tetraazacyclotetradeca-4,11-diene
                 Trans. Metal·Chem., <u>4</u>, 234-236 (1979)
11. Melson,G.A. (Ed.): *Coordination Chemistry of Macrocyclic Compounds*,
                 Plenum Press, New York, 1979
12. Gokel,G.W., Korzeniowski,S.H.: *Macrocyclic Polyether Syntheses*,
                 Springer-Verlag, Berlin-Heidelberg-New York, 1982
13. Daves,G.S.: Synthetic substitutes for Quinidine
                 Brit. J. Pharmacol., <u>1</u>, 90-111 (1946)
14. Zetler,G., Strubelt,O.: Antifibrillatory, Cardiovascular and Toxic
                 Effect of Sparteine, Butylsparteine and Pentylsparteine
                 Arzneim-Forsch/Drug Res., <u>30</u>, 1497-1502 (1980)
15. Thomas,R.E.: *Cardiac Drugs in "Burger's Medicinal Chemistry (Ed.:*
                 Wolff,M.E.) "Wiley-Interscience,New York,1981,Part 3,Ch.38

SYNTHESIS OF RAUNESCINONE ANALOGUES WITH HYPOTENSIVE AND ANTIHYPERTENSIVE ACTIVITY

TÓTH I., SZABÓ L., BOZSÁR G., SZÁNTAY Cs., SZEKERES L.*
and PAPP J.Gy.*

Central Research Institute for Chemistry,
The Hungarian Academy of Sciences,
Budapest, Pf.17, H-1525, Hungary

*Department of Pharmacology, University Medical School,
Szeged, Hungary

Abstract. Starting from the readily available ketoesters (2) the pharmacologically active methylenedioxy- and diethoxy-epialloberbane-ketoesters (1) can be synthesized. By appropriate reaction sequences both antipodes of ketoester 2a lead us to any enantiomer of the desired raunescinone analogue 1a. Hypotensive, antihypertensive and central depressant effects of 1a are described.

Among the berbane-derivatives, which can be regarded as depyrrolo analogues of the well-known and biologically potent yohimbine and reserpine alkaloids[1], many compounds show interesting pharmacological activity. According to our earlier investigations this type of compounds and certain intermediates of their synthesis show antiinflammatory[2], prostaglandin-like or -antagonist[3] as well as hypotensive and antihypertensive[4] effects.

Chemistry

Our linear synthetic method[2,3,5] developed earlier is flexible enough to build up either the normal-, allo- or epialloberbane skeleton; it furthermore permits wide variation of substituents at rings A and D.
As preliminary pharmacological examinations showed that la posseses outstanding hypotensive and antihypertensive effects, synthesis of both of its enantiomers seemed desirable in order to investigate whether one of the antipodes is responsible for the biological effect, or it is attributed to both of them.

For this purpose resolution at the stage of the ketoester 2a proved to be most suitable.

Now we illustrate our synthetic pathway starting with (+)-2a and leading to epiallo ketoester (+)-5.

One of the most interesting feature of our method was the synthesis of the same epiallo ketoester (+)-5a with (1R,12S,17S) absolute configuration from the other enantiomer of the starting key intermediate (-)-(3S,11bS)-2a.
(-)-6a ketoester can be reached via $\triangle^{12,\bar{1}3} \longrightarrow \triangle^{12,17}$ double bond migration and following hydrogenation.

Transformation of the allo ketoester (-)-6a into (+)-5a epimeric at C-1 was accomplished by the well known oxidation reduction method[6,7]. This transformation gives an excellent possibility to go over from the 1R series to the 1S series.

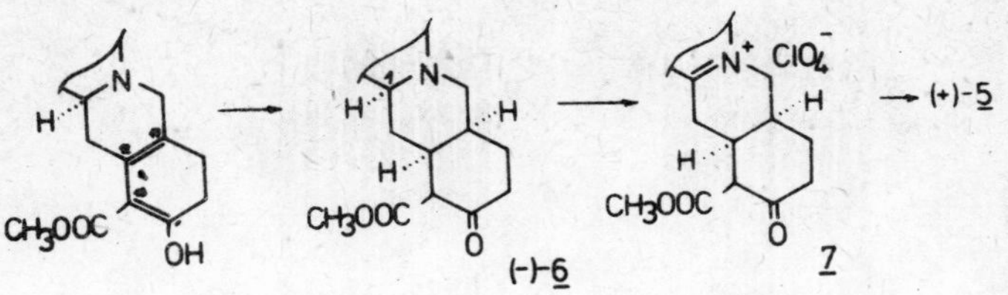

The reduction of the (12S,17S)-7 yields the (+)-(1R,12S,17S)-5a epiallo keto-ester, as the main product. Bromo-ketoester 9[3] with potassium salt of trimethoxy-benzoic-acid, ferulic-acid and siringoic-acid yields different 1 keto-ester.

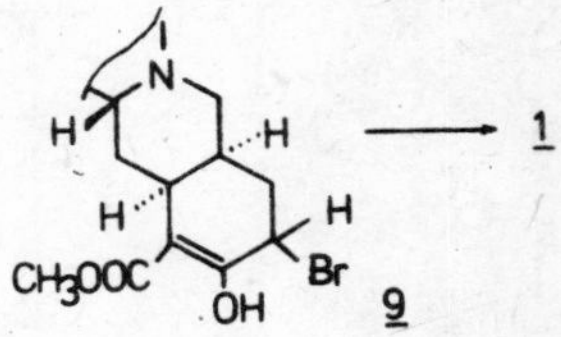

Pharmacological Investigations

The results summarized in Table 1 demonstrate that 1a possesses a significant and lasting hypotensive activity. Blood-pressure lowering of about the same intensity and duration can be achieved by administering 20 mg/kg of this compound and 5 mg/kg of reserpine, respectively. The fall in blood-pressure induced by 1a is associated with a biphasic change in heart rate: an initial tachycardia is followed by bradycardia. Similar results were obtained with derivatives containing two alkoxy groups in the place of the methylene-dioxy group.

The data given in Table 2 indicate that the (+)-(1R, 12S,17S) and (-)-(1S,12R,17R) enantiomers of 1a have anti-hypertensive potency. Furthermore, the degree of antihyper-tensive efficacy is the same with both enantiomers.

The results comprised in Table 3 show that at the time of the maximum blood-pressure lowering effect, 1a, unlike reserpine, does not induce depression of the central nervous system as measured by the forced coordinated motor activity.

15 Plas

Table 1. Effects of (±)-1a on Blood Pressure (BP)[a] and
Heart Rate (HR)[a] of Conscious Normotensive Rats

Compound mg/kg[b]	n	BP mmHg and HR beats/min prior to dosing		Change in BP (mmHg) or HR (beats/min) after dosing					
				3h	6h	12h	18h	24h	48h
(±)-**1a**									
10	5	BP	115± 2	-5±2	-16±2^c	-14±2^c	-11±2^c	-11±2^c	-11±4
		HR	364±10	+6±7	+40±1^c	+14±6	+4±3	-34±8^c	-78±5^c
20	5	BP	116±2	-9±2^c	-23±2^c	-20±2^c	-15±3^c	-11±2^c	-5±1^c
		HR	368±14	+8±6	+60±13^c	+42±8^c	+26±2^c	-2±1	-10±2^c
Reserpine									
5	5	BP	117±2	-18±5^c	-23±7^c	-16±1^c	-10±2^c	-12±3^c	-7±3
		HR	364±5	-34±5^c	-44±5^c	-40±6^c	-40±4^c	-32±5^c	-10±7

[a] All values are given as mean ± SEM. [b] Administered intraperitoneally as HCl salt [(±)-**1a**], or base (reserpine).
[c] $p < 0.05$ (Student's t test).

Table 2. Effects of the enantiomers of 1a on Blood Pressure (BP)[a] on Conscious Spontaneously Hypertensive Rats

Compound mg/kg [b]	n	BP mmHg prior to dosing	Change in BP mmHg after dosing		
			6 h	18 h	24 h
(+)- 1a	6	175 ± 2	-16 ± 3^{c}	-7 ± 2^{c}	-5 ± 3
(-)- 1a	6	176 ± 3	-15 ± 3^{c}	-8 ± 1^{c}	-2 ± 3

[a] All values are given as mean $\pm$ SEM. [b] Administered intraperitoneally as HCl salt.
[c] $p < 0.05$ (Student's t test.)

Table 3. Lack of Depressant Effect of 1a on the Central Nervous System as Measured by the Forced Coordinated Motor Activity on Mice

Compound (mg/kg)[a]	n [b]	Time of performance test after dosing (h)	Group performance time[c] (sec)	Activity ratio[d]
Control	10	-	98 ± 9	
(±)-1a				
20	10	6^{e}	102 ± 12	1.04
Control	10	-	117 ± 3	
Reserpine				
5	10	6^{e}	9 ± 3^{f}	< 0.08

[a] Administered intraperitoneally as HCl salt (1a) or base (reserpine). [b] Number of groups of ten mice used in test. [c] Mean$\pm$SEM. [d] Performance time under the effect of drug divided by the corresponding control performance time. [e] The time elapsed from dosing required to obtain maximum hypotensive or antihypertensive effect (see Table 1 and 2.). [f] $p < 0.05$ (Student's t test).

From these results it is concluded that the berbane derivative 1a possesses hypotensive and antihypertensive activity of particularly long duration. Contrary to reserpine, the compound exhibits its blood-pressure lowering effect without depressing the central nervous system.

REFERENCES

1. J. Jirkovsky, M.Protiva: Synthetische Versuche in der Gruppe Hypotensiv Wirsksamer Alkaloids XXVII. Synthese des Racemischen 10,11-Methylendioxydespyrrolo-deserpidins, Coll.Czech.Chem.Comm. 28, 2577 (1963)

2. L.Szabó, K.Nógrádi, I.Tóth, Cs.Szántay, L.Radics, S.Virág, E.Kanyó: Synthesis of Benzo[a]quinolizidine Derivatives showing Antiinflammatory Activity without Ulcerogen Side Effect, Acta Chim. Ac.Sci. Hung. 100, 1 (1979)

3. L.Szabó, I.Tóth, L.Tőke, P.Kolonits, Cs.Szántay: Über die Herstellung von halogenierten Despyrroloyohimbon-Derivaten mit prostaglandinartiger und prostaglandinantagonistischer Wirkung, Chem.Ber. 109, 3390 (1976)

4. Cs.Szántay, L.Szabó, I.Tóth, G.Bozsár, J.Gy.Papp, L.Szekeres: Investigations on the chemistry of Berbanes IX. Synthesis of Raunescinone Analogues with Hypotensive and Antihypertensive Activity, J.Med.Chem., in press

5. L.Szabó, I.Tóth, K.Honty, L.Tőke, J.Tamás, Cs.Szántay: Regiospezifische Synthese der Despyrroloyohimbinone durch Dieckmann-Kondensationen von ungesättigten Estern, Chem.Ber. 109, 1724 (1976)

6. J.Ernest, H.Protiva: Totalsynthese eines Stellungisomeren des Syrosingopins, Naturwiss 47, 156 (1960)

7. I.Tóth, L.Szabó, M.Kajtár-Peredy, E.Baitz-Gács, L.Radics, Cs.Szántay: Investigation on the Chemistry of Berbans-VII Synthesis of 10,11-Dimethoxy(despyrrolo)raunescine Stereoisomers, Tetrahedron 34, 2113 (1978)

CHALCONE AS STARTING MATERIAL FOR SYNTHESIS OF 1,2,4-TRIAZEPINES

RICHTER P. and STEINER K.

Sektion Pharmazie, Ernst-Moritz-Arndt-Universität
Greifswald, Friedrich-Ludwig-Jahn-Straße 17
2200 Greifswald, DDR

Chalcone 1 can add thiocyanic acid under formation of β-isothiocyanatoketone 2. By reaction of 2 with hydrazine or phenylhydrazine so far only 1-amino-4,6-diphenyltetrahydro-pyrimidine-2-thione 3 or the 1-phenylamino-derivative have been obtained [1]. It is also described that 2 in ethanolic solution adds hydrazine and forms 4, which cannot be cyclized to a 1,2,4-triazepine by influence of acids or bases.

We studied at first the reaction of 2 with equimolar amounts of methylhydrazine and obtained 6 (85 %). Heating of a solution of 6 in dioxane in presence of toluene-4-sulfonic acid gave a new compound (30 %). On the basis of the reaction and the structural data reported below, the product was assigned structure 7; 7 could also be prepared without isolation of 6 by heating of 2 with 10 mole methylhydrazine and catalytic amounts of hydrogen chloride in ethanol. Structure 7 is also confirmed by the reaction of 7 with methyl iodide yielding 9, which can also be produced via non-isolable 8. Considering these results, we heated a solution of 4 and catalytic amounts of toluene-4-sulfonic acid in propan-2-ol and obtained 5.

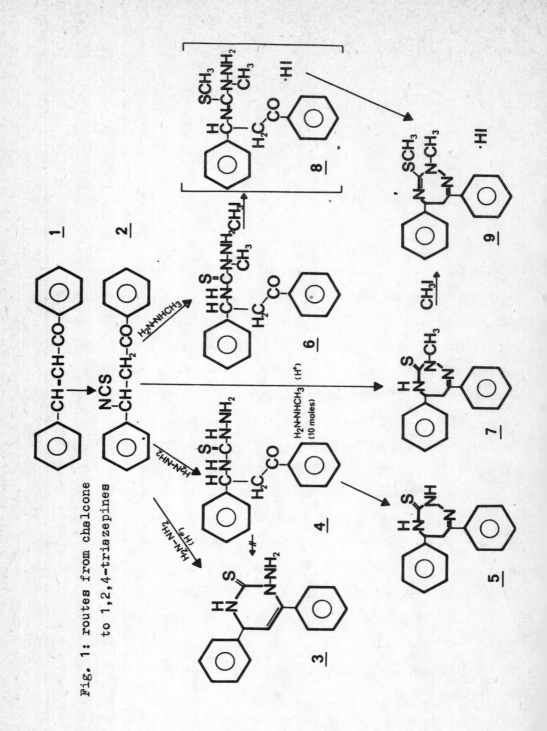

Fig. 1: routes from chalcone to 1,2,4-triazepines

218

The mass spectrum of $\underline{7}$ shows the mole peak at m/z 295(100 %) and that of $\underline{5}$ at m/z 281 (22 %).

In the ^{1}H-NMR spectrum of $\underline{7}$ (80 MHz, $CDCl_3$, TMS) signals were observed at 3,25 ppm (2H, m, CH_2), 3,80 ppm (3H, s, NCH_3), 4,85 ppm (1H, m, CH), 6,30 ppm (1H, s(b), NH) and 7,21-7,62 ppm (10H, m(b), aromatic). Owing to the presence of the signals for the protons at C-6, a pyrimidine or 1,3-thiazine structure could be excluded. The coupling of the nonequivalent C-6 protons with the proton at C-5 causes a splitting of the resonance signal into two quartets. Five of the expected peaks are found and represent the AB-part of an ABX-type spectrum. The other peaks are too weak to be observed.

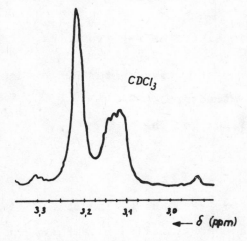

Fig. 2: ^{1}H-NMR signal of the C-6 protons of $\underline{7}$

The signal of the proton at C-5 shows seven peaks with intensity of 1:1:1:2:1:1:1. This is equivalent to two overlapping quartets (J = 5,7 Hz). After treatment with D_2O, the N-H signal (6,30 ppm) disappears and the signal for the proton at

C-5 splits only into two doublets (J = 5,7 Hz). These indicate the influence of the N-H proton and the methylene protons
on the proton at C -5.

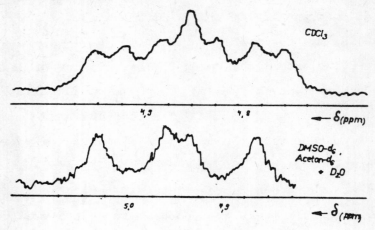

Fig. 3: ^{1}H-NMR signal of the C-5 proton of 7

The ^{1}H-NMR spectrum of 5 gave no further evidence for the
1,2,4-triazepine structure. Products 5 and 7 show similar
UV spectra differing from that of 3.

References

1) Weber, F. G., U. Pusch and B. Brauer,
   Thioharnstoffe und Pyrimidinderivate aus Chalkon
   Pharmazie 34, 443 - 444 (1979)

STEREOELECTRONIC CONTROL IN SOME REACTIONS OF SECOLOGANIN

PÁL Z., VARGA-BALÁZS M., SZABÓ L.F.,* TÉTÉNYI P.*

Semmelweis Medical University, Institute of
Organic Chemistry, Budapest, Hungary

*Research Institute for Medicinal Plants,
 Budakalász, Hungary

Secologanin 1 is the building stone of the large group of
indole as well as several other types of alkaloids [1, 2].
Battersby et al. have demonstrated that the first step of
the biosynthesis is a Mannich type reaction of secologanin
with biogenic amines (e.g. tryptamine or substituted phenyl-
ethyl amine)[3]. According to the generally accepted
mechanism of Mannich reactions, it starts with the formation of the appropriate Schiff base. Previously we reported
the reactions of secologanin with different nucleophiles
[4]. Now, we present some data on the transformation of the
Schiff base 2 derived from 1.

Secologanin with one mol of methylamine in isopropanol
under mild conditions gives the Schiff base 2 which can be
transformed by an excess of the same or a different amine
or another nucleophile into 5, probably through 3 and 4.
In most cases, 7α -H-5 is formed at first in a kinetically
controlled reaction, but the primary product can be
rearranged by acidic catalysis into the thermodynamically
more stable 7β -H-5. The reactions were checked by ^{1}H-nmr
spectroscopy, the compounds were isolated and their spectra
studied. ^{1}H-nmr chemical shifts and coupling constants are
shown in Table 1. In compounds 5A, 5C, and 5E, signal of
C(7)-H has a large doublet proving a diaxial interaction
with C(6)-β-H. In compounds 5B, 5D, and 5F, C(7)-H has a

$^\text{E}$Author for any correspondence.

glu=β-D-glucofuranosyl

CH₃NH₂

HX-Y

$\underline{1}$

$\underline{2}$

$\underline{3}$

$\underline{4}$

$\underline{5}$

X\Y	CH₃		H	
C/7/-H		C/7/-H		
	α	β	α	β
O	C	D	E	F
HN″	A	B		

narrow triplet demonstrating equatorial-axial and equatorial-equatorial interactions with the two H-atoms of C(6).

In the case of X = NH, interpretation of the chemical and spectroscopic data is as follows. The rate determining step to compounds A, C, and E is most probably the formation of the tetrahedral intermediate $\underline{4}$. According to the Bell-Evans-Polányi principle, stabilization of $\underline{4}$ requires less activation energy. Taken into consideration the possible alternative configurations at C(7), C(11), N' and N'' as well as the alternative ring closure at N' or N'', $2^5 = 32$ stereoisomers can be formed, in principle. 16 of them having an axial substituent at C(7) cannot be really expected because of the strong 1,3-diaxial interaction with the substituent at C(11). The principle of stereoelectronic control proposed by Deslongchamp [5] helps us to decrease drastically the number of potential stereoisomer intermediates $\underline{4}$. Supposing that facile elimination of the methoxy group at C(11) requires two primary stereoelectronic effects (dashed arrows in Figure 1), 8 structures having the C(11)-methoxy group

Table 1. ^{1}H-nmr signals of the compounds in δ ppm /J in H$_z$/[①]

5	A	B	C	D	E	F
C/1/	5,46 2,0/d/	5,45 2,0/d/	5,43 2,0/d/	5,40 2,0/d/	5,44 2,0/d/	5,45 2,0/d/
C/3/	7,37 1,5/d/	7,40 1,5/d/	7,35 1,5/d/	7,38 1,5/d/	7,37 1,5/d/	7,37 1,5/d/
C/6/ e		1,91 14,0/d/ 3,0/t/	2,10 14,0/d/ 4,5/t/	1,97 14,0/d/ 3,5/t/	2,09 14,0/d/ 4,5/t/	1,75
C/6/ a	1,55 14,0/t/ 3,0/d/	1,55 14,0/t/ 3,0/d/	1,43 14,0/t/ 10,5/d/	1,53 14,0/t/ 3,5/d/	1,46 14,0/t/ 10,5/d/	1,65 14,0/t/ 3,0/d/
C/7/	4,31 10,5/d/	4,17 3,0/t/	4,65 10,5/d/ 4,5/d/	4,62 3,5/t/	4,88 10,5/d/ 4,5/d/	5,02 3,0/t/
C/8/		2,55 8,0/d/ 6,5/d/ 2,0/d/	2,65 8,0/d/ 6,5/d/ 2,0/d/	2,55 8,0/d/ 6,5/d/	2,62 8,0/d/ 2,0/d/	2,58 8,0/d/ 6,5/d/ 2,0/d/
C/1'/	4,65 7,5/d/	4,63 7,5/d/	4,63 7,5/d/	4,63 7,5/d/	4,65 7,5/d/	4,65 7,5/d/
N'CH$_3$	2,90	3,05	2,90	3,00	2,95	2,99
C/7/XCH$_3$	2,20	2,41	3,32	3,38		

[①] d = doublet, t = triplet

in β equatorial position and further 4 structures having a β axial methyl group at N' can be disregarded. 2 of the remaining 4 isomers having a non-bonded electron pair at N'' in gauche position to the C(7)–N' bond are less stable because of a strong steric interference of N'-methyl and Y. The stability of the final 2 structures (or one, if Y = methyl) is increased by secondary stereoelectronic effects

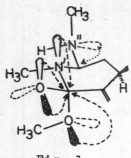

Fig. 1

(the dotted arrow in Fig. 1) between the non-bonded electron pair at N'' and the antiperiplanar C(7)-N' bond. The structure of the probably most stable stereoisomer (Y = CH_3) $\underline{4}$ is shown at Figure 1.

In lactams $\underline{5}$, no 1,3-diaxial interaction can be present between C(7)-α axial and C(11)-substituents. Therefore $\underline{5A}$, $\underline{5C}$ and $\underline{5E}$ can be transformed into $\underline{5B}$, $\underline{5D}$ and $\underline{5F}$ whose increased stability is due to one additional secondary stereoelectronic effect between the non-bonded electron pair of N' atom and the antiperiplanar C(7)-N'' bond. Similar considerations are effective in the case of X = O and Y = CH_3, H.

Financial support by the Scientific Research Council, Ministry of Health, Hungary, is gratefully acknowledged.

References

1./ D. S. Bhakuni, J. Scient. Ind. Res. $\underline{35}$. 449-460 /1976/

2./ R. T. Brown, Biomimetic Conversion of Secologanin into Alkaloids. In: Indole and Biogenetically Related Alkaloids. Edited by J. D. Phillipson and M. H. Zenk. Academic Press, London, 1980. Chapter 9. 171-184.

3./ A. R. Battersby, A. R. Burnett, P. G. Parsons, J. Chem. Soc. /C/ 1969, 1187-92; 1193-1200.

4./ L. Szabó, K. Böjthe-Horváth, F. Hetényi, Á. Kocsis, Z. Pál, M. Varga-Balázs, P. Tétényi, Proceedings of the First International Conference on Chemistry and Biotechnology of Biologically Active Natural Products. Varna, Bulgaria, 21-26. 9. 1981, Vol. 3_1, 87-90.

5./ P. Deslongchamps: Amides and Related Functions. In: Stereoelectronic Effects in Organic Chemistry. Pergamon Press, Oxford, 1983. Chapter 4. 101-162.

A SIMPLE SYNTHESIS OF SOME ANTIHISTAMINE GLYCINAMIDES

PIELICHOWSKI J. and POPIELARZ R.

Institute of Organic Chemistry and Technology
Technical University, Krakow, Poland

Some N-substituted derivatives of glycinamide including
N,N'-tetraethylglycinamide show biological activity [1]. Also
morpholine-substituted glycinamides, such as: N,N-dimethyl-
aminoacetic acid morpholide, N,N-diethylaminoacetic acid
morpholide and morpholineacetic acid morpholide are non-toxic,
antihistamine substances [2]. Hitherto known syntheses of
these compounds involved the reactions of corresponding
secondary amines with chloroacetyl chloride [2,3], with glyoxal
[4,5] or with dichlorodiacetamide [6]. Recently [7], a new,
simpler method, based on trichloroethylene was developed which
allows to obtain the N-substituted glycinamides in one step
directly from trichloroethylene and secondary amines, according
to the general equation:

$$2 \underset{R'}{\overset{R}{N}}H + \underset{Cl}{\overset{Cl}{C}}=\underset{H}{\overset{Cl}{C}} + 3\ NaOH \xrightarrow{TEBA^+Cl^-} \underset{R'}{\overset{R}{N}}-CH_2-\overset{O}{\overset{\|}{C}}-\underset{R}{\overset{R}{N}} + 3\ NaCl + 2\ H_2O \qquad (1)$$

The preparation of symmetrically substituted glycinamides:
$R_2NCH_2CONR_2$ (R = alkyl) do not present difficulties, such
as the synthesis of N,N'-tetraethylglycinamide and morpholineacetic acid morpholide, as we have described previously

---

$TEBA^+Cl^-$ — benzyltriethylammonium chloride.

[7]. Our further investigations were intended to apply this method to synthesize also the unsymmetrically substituted glycinamides: $R_2^1NCH_2CONR_2^2$ (where: R^1, R^2 = alkyl or cycloalkyl). In case of the trichloroethylene reaction with a 1:1 mixture of two secondary amines, the formation of four different glycinamides can be expected. Actually, if the basicity and also nucleophilicity of both amines are similar, four possible N,N-tetraalkylglycinamides are formed in the ratio of: 1:1:1:1. However, in case of morpholine (pK_b=5.3) and diethylamine (pK_b=3.07), which differ in their nucleophilicity, the formation of unsymmetrically substituted glycinamides is preferred.

Not giving all details of the reaction mechanism, six reactions can be considered as given by eqs. (2) - (7):

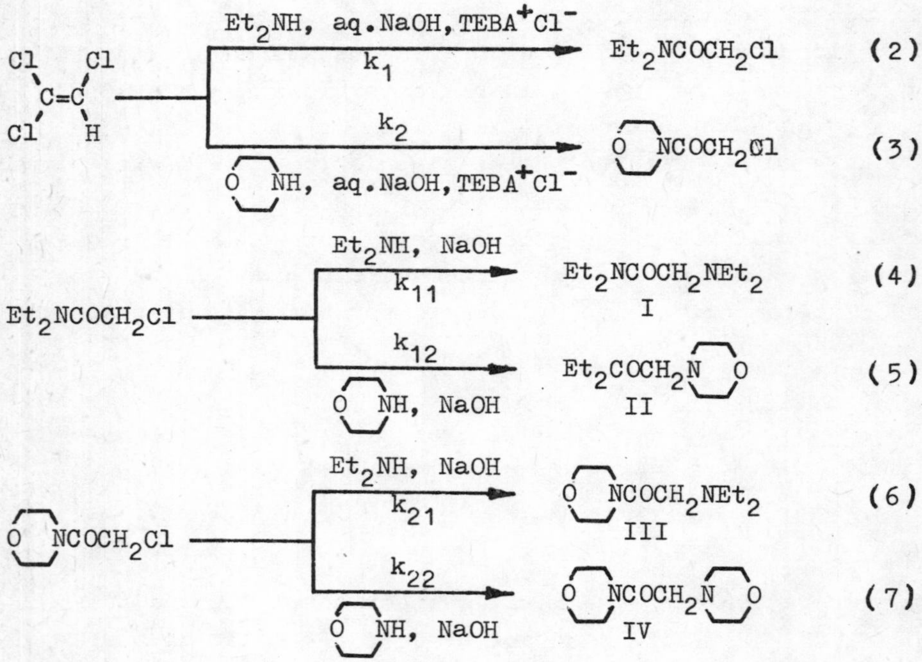

$$Cl_2C=CHCl \xrightarrow[k_1]{Et_2NH, aq.NaOH, TEBA^+Cl^-} Et_2NCOCH_2Cl \quad (2)$$

$$\xrightarrow[O\ NH,\ aq.NaOH, TEBA^+Cl^-]{k_2} O\ NCOCH_2Cl \quad (3)$$

$$Et_2NCOCH_2Cl \xrightarrow[k_{11}]{Et_2NH, NaOH} Et_2NCOCH_2NEt_2 \quad I \quad (4)$$

$$\xrightarrow[O\ NH,\ NaOH]{k_{12}} Et_2COCH_2N\ O \quad II \quad (5)$$

$$O\ NCOCH_2Cl \xrightarrow[k_{21}]{Et_2NH, NaOH} O\ NCOCH_2NEt_2 \quad III \quad (6)$$

$$\xrightarrow[O\ NH,\ NaOH]{k_{22}} O\ NCOCH_2N\ O \quad IV \quad (7)$$

The relative contents of individual compounds I - IV in the final product depends on the concentration of amines in the reaction medium and the rate constants: $k_1, k_2, k_{11}, k_{12}, k_{21}$ and k_{22}. As reported [7], the first stage comprising eqs. (2) and (3), proceeds via dichloroacetylene formation as active intermediate. Owing to the high reactivity of dichloroacetylene, reactions (2) and (3) are rather diffusion than kinetically controlled. In such case it may be assumed that proportions of individual chloroacetamides depend only on the mole fractions of corresponding amines and are independent of k_1 and k_2. Among the second-step reactions (4) - (7), the reactions (4) and (6) proceed faster than (5) and (7), because of higher nucleophilicity of diethylamine. Thus, if the concentration of diethylamine in the reaction medium is lower than that of morpholine, chloroacetylmorpholine is formed predominantly in the first step, which reacts faster with diethylamine and favours the formation of N,N-diethylamino-acetic acid morpholide, as a main product.

The above theoretical considerations were confirmed experimentally, indicating the applicability of the method for synthesizing unsymmetrically substituted glycinamides in some cases.

EXPERIMENTAL

21,8 ml /0.25 mole/ morpholine, 13 ml /0.125 mole/ diethylamine, 40 g /1 mole/ NaOH in 40 ml water and 0.5 g /0.0025 mole/ benzyltriethylammonium chloride mixture was warmed up to 70°C. Then a mixture of 22.5 ml /0.25 mole/ trichloroethylene and 13 ml /0.125 mole/ diethylamine was added for 1 hour. After all quantity of trichloroethylene had been added, stirring was continued further for 1 hour. The organic layer was decanted and the water phase was extracted with benzene. The extracts were mixed with the main part of product and the solvent was evaporated. Crude, oily product was distilled under vacuum to give 35 g colorless oil containing mainly N,N-diethyl-aminoacetic acid morpholide and several percent morpholine-acetic acid N,N-diethylamide, Yield of unsymmetrically substituted glycinamides: 70 %. Anal.:

Found: C:59.90, H:10.48, N:14.09, Calcd. for $C_{10}H_{20}N_2O_2$
/200.3/: C:59.97, H:10.06, N:14.0. From pot residue 4 g
/8% yield/ morpholineacetic acid morpholide was also isolated.

REFERENCES:

1. F.Bovet-Nitti., Rend.ist.super.sanita, 15, 989 /1952/.
2. S.Smolinski., Rocz.Chem., 30, 1111 /1956/.
3. P.Malatesta, G.Migliaccio., 11 Farmaco. /Pavia/, Ed.sci., 11, 113 /1956/.
4. P.Ferutti, A.Fere, L.Zetta, A.Betteli., J.Chem.Soc., C-1970, 2512.; ibid., C-1971, 2984.
5. J.M.Kliegman, R.K.Barnes., J.Heterocycl.Chem., 7, 1153 /1970/.
6. M.Backes., Compt.rend., 239, 1520 /1954/.
7. J.Pielichowski, R.Popielarz., Tetrahedron, in press.

STUDIES ON THE PROPERTIES AND STRUCTURE OF OPTICALLY ACTIVE 1-(3,4-DIMETHOXYPHENYL)-4-METHYL-5-ETHYL-7,8-DIMETHOXY-5H-2,3-BENZODIAZEPINE (TOFIZOPAM)

FOGASSY E., ÁCS M., SIMON K.,* LÁNG T.** and TÓTH G.***

Department of Organic Chemical Technology, Technical
University, Budapest, Hungary
  *CHINOIN Pharmaceutical and Chemical Works
   Budapest, Hungary
 **Institute of Drug Research, Budapest, Hungary
***Institute of General and Analytical Chemistry
   Technical University, Budapest, Hungary

The biological activity of mirror image molecules are different owing to the chiral structure of the living organism. As racemic tofizopam (I) is used as an anxiolytic agent, a number of investigations has been accomplished to study differences between the interactions of the antipodes of I with different tissues of the organism. The seven membered ring of the inherently chiral benzodiazepine exists in two stable conformations (Neszmélyi et al. 1974, Tóth et al. 1983), thus, a racemic mixture contains four species: A^R, A^S, B^R and B^S, where conformer A has the ethyl group in equatorial, while conformer B has it in axial position (Fig. 1).

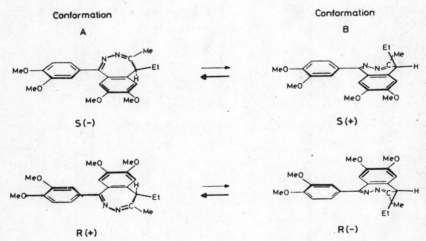

Fig. 1. Conformational equilibria for the enantiomers of
        compound I according to Simonyi and Fitos (1983)

Pharmacological tests indicated different biological activity for I stereoisomers (Petőcz et al. 1980), and differences between both enantiomers and conformers on their binding to human serum albumin were also found (Simonyi and Fitos, 1983). CNS investigations in mice (Petőcz et al. 1980) have shown that the activity of racemic substance does not correspond with the sum of enantiomer activities, and the activity differences of free bases, Cl^- and $CH_3SO_3^-$ salts cannot be explained by solubility differences.

For further biological studies and for explanation of the results we determined the crystal and molecular structure of compound I by X-ray analysis and investigated the effect of different conditions on the conformer ratio. The single crystal X-ray analysis of the Br^- salt of (-)-I has shown that in optically pure salt two crystallographically independent molecules are present (Fig. 2).

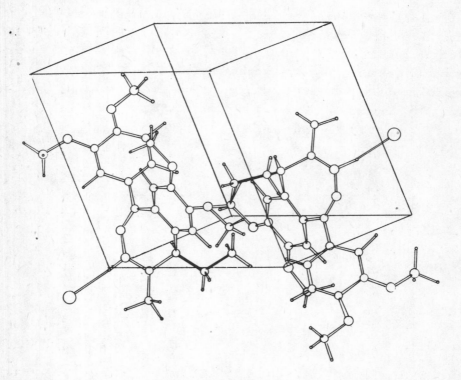

Fig. 2. Molecular and crystal structure of the Br^- salt of (-)-I

The two molecules are related by a *pseudo* center of symmetry. As both molecules are of the same chirality (S) at C-5, this *pseudo* symmetry is the consequence of the C-5 substituents found in different orientation (see Fig. 2). The asymmetric unit is, thus, a *quasi-racemic* pair $(A^S + B^S)$.

Dependence of conformer ratio on circumstances in case of configurationally pure samples was studied by ^{1}H-nmr spectro - scopy (Tóth et al. 1983 and Fogassy et al. 1984). The elevation of temperature decreases, whereas salt formation increases the amount of conformer B. Further investigations of enantiomer mixtures during recrystallization have shown that the conformer ratio depends on the optical purity. Here the formation of *quasi-racemic* pairs might play some role. In the course of a resolution procedure, the conformer ratio depends on chiral circumstances. Assuming a three-point attachment (Ogstone, 1948) we have succeded in predicting the absolute configuration by analyzing the possible interactions between I and the re- solving agent, dibenzoyl-R,R-tartaric acid (Fig. 3). According- ly, the stereoisomer in majority in the diastereoisomeric salt

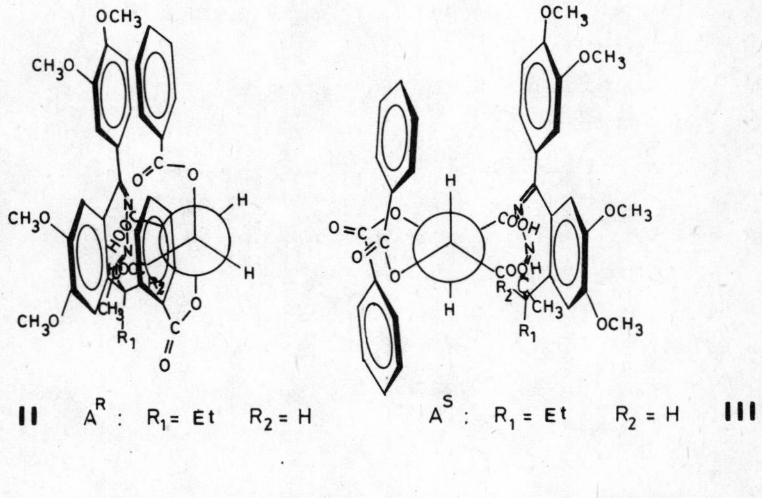

Fig. 3. Possible interactions between the stereoisomers and the resolving agent

mixture has R-configuration, as proved by X-ray analysis. As
Fig. 3 shows, in salt III the substituents can interact at only
two points, whereas in salt II a more favourable three-point
interaction is possible. This is in accord with X-ray powder
diagrams (Fig. 4) showing that salt III obtained by evaporating
the solvent, is amorphous.

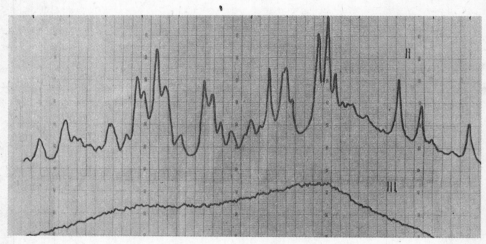

Fig. 4. X-ray powder diagrams of salts II and III

REFERENCES

Fogassy, E., Ács, M., Simon, K., Láng, T. and Tóth, G.: The
        Elucidation of Anomalous Chiroptical Behaviour and
        Determination of Configuration of 1-(3,4-Dimethoxy-
        -phenyl)-4-methyl-5-ethyl-7,8-dimethoxy-5H-2,3-
        -benzodiazepine Enantiomers, J.Chem.Res., submitted.

Neszmélyi, A., Gács-Baitz, E., Horváth, Gy., Láng, T. and
        Körösi, J.: Spektroskopische Konstitutions- und Kon-
        formationsbestimmung am 5-Äthyl-1-(3,4-dimethoxyphe-
        nyl)-7,8-dimethoxy-4-methyl-5H-2,3-benzodiazepin,
        Chem.Ber., 107, 3894-3903 (1974).

Ogstone, A. G.: Interpretation of Experiments on Metabolic
        Processes, using Isotopic Tracer Elements, Nature,
        162, 963 (1948).

Petőcz, L., Tóth, G., Fogassy, E., Ács, M., Láng, T., Grasser, K. and Kosotzky, I.: Resolution of Tofizopam Enantiomers and their Effects on Central Nervous System, 7th Int.Symp. on Med.Chem., Malaga Spain (1980), Chem.Abstr., $\underline{97}$, 38959 k.

Simonyi, M. and Fitos, I.: Stereoselective Binding of a 2,3-benzodiazepine to Human Serum Albumin. Effect of Conformation on Tofizopam Binding, Biochem.Pharmacol., $\underline{32}$, 1917-1920 (1983).

Tóth, G., Fogassy, E., Ács, M., Tőke, L. and Láng, T.: Racemspaltung von ($\pm$)-5-Athyl-1-(3,4-dimethoxyphenyl)-6,7-dimethoxy-4-methyl-5H-2,3-benzodiazepin und anomales chiroptisches Verhalten der Enantiomeren (1), J.Heterocyclic Chem., $\underline{20}$, 709-713 (1983).

INTERACTION OF BENZODIAZEPINES WITH NATURAL COMPOUNDS ON THE LEVEL OF GABA-BENZODIAZEPINE RECEPTOR COMPLEX

KORNEYEV A. Ya.

Laboratory of Biochemistry, Institute of Psychiatry, All-Union Research Centre for Mental Health, Acad. Med.Sci. USSR
113152, Zagorodnoe Shosse, 2, Moscow, USSR

An analysis of pharmacological and physiological proper-ties of heterocyclic compounds and their structural (conforma-tional) and functional relationship with naturally occurring physiologically active substances can be performed using receptor binding techniques. In many cases it is possible to establish the existence of binding sites for the heterocyclic compound on biological macromolecules or cell membranes. Otherwise, one can investigate the ability of a compound to inhibit receptor-binding of mediators and hormones.

The interaction of benzodiazepine heterocycles with the binding sites (benzodiazepine receptor) on the membranes of brain cells, widely investigated during the last decade, is inhibited by some recently characterized endogenous substances, e.g. by Prostaglandin A_1 (PGA_1). (Asano and Ogasawara, 1982). In the present work it is shown that specific binding of [^{3}H]flunitrazepam (FNZ) at 1 nM and also of the benzodiazepine antagonist [^{3}H]β-carboline-3-carboxylic acid ethyl ester (βCCE) at 1 nM is concurrently inhibited by PGA_1 (IC_{50}= 5 μM and 7.5 μM, respectively). Other prostaglandins (PGE_2, PGF_1) were ineffective up to 100 μM. These data point to the existence of some structural similiarity between PGA_1 and benzodiazepines and pose the question as to whether PGA_1 is agonist or antagonist of benzodiazepines?

This question also can be solved with the aid of receptor binding methods. In the case of the benzodiazepine receptor two criteria are available: 1. affinity of benzodiazepine ago-

nists to the binding site increases in the presence of GABA
and affinity of the antagonists decreases (Skolnick et al.,
1982); 2. benzodiazepine agonists increase the specific binding
of GABA and the GABA agonist muscimol to the postsynaptic
GABA receptor, and antagonists decrease the binding (Skerritt
et al., 1982, Korneyev, 1983). Investigation of PGA_1 according
to both criteria indicates that PGA_1 is a weak agonist of
benzodiazepines.

The presented data indicate an interaction of benzodiazepines with PGA_1, at the level of brain receptors, and
show the applicability of receptor binding studies for
investigations on pharmacological properties of heterocyclic
compounds and on their interaction with endogenous, physiologically active substances.

REFERENCES

Asano and Ogasawara, 1982, Eur. J. Pharmacol., 80, 271.
Skolnick et al., 1982, Eur. J. Pharmacol., 78, 133.
Skerritt et al., 1982, Neurosci. Lett., 33, 173.
Korneyev, 1983, Eur. J. Pharmacol., 90, 277.

ULTRASONIC BINDING OF 5-HYDROXYTRIPTAMINE TO CELL MEMBRANES

KORNEYEV A. Ya.

Laboratory of Biochemistry, Institute of Psychiatry,
All-Union Research Center for Mental Health, Acad.
Med. Sci USSR
113152, Zagorodnoe Shosse, 2, Moscow, USSR

5-Hydroxytryptamine (5-HT) can interact with cell membranes in many different manners. In the present report it is shown that ultrasonic treatment of the cell membrane suspension in the presence of [^{3}H]5-HT leads to a semi-irreversible binding (or incorporation) of this ligand to the membranes.

Membrane preparation. Fresh rat brain, bovine brain or rat kidney tissue was homogenized (Virtis-45, 30 sec, max. speed) in 10 volumes of 50 mM Tris-HCl buffer (pH 7.5 at 20°C) and centrifuged at 500 g for 10 min at 4°C. Supernatant was collected and centrifuged at 30000 g for 30 min and the pellet obtained was washed twice in the same manner (homogenization and centrifugation), then resuspended in 10 volumes of the same buffer and used for investigation.

Ultrasonic treatment. The sample (1 ml volume) containing the membrane suspension (0,1 ml) corresponding to 10 mg of initial wet tissue weight, [^{3}H]5-HT (15,5 Ci/mmol, Amersham) at a final concentration of 10 nM and various additives in 50 mM Tris-HCl buffer (pH 7.5) was sonicated in a water bath (10°C) with an "MSE" Ultrasonic Desintegrator (high power, amplitude 4, stainless steel probe 3 mm diameter) for 10 min. "Non-specific" binding (or incorporation) was determined with 1 mM of 5-HT. At the end of the ultrasonic treatment, the membrane suspension was filtered through GF/B filters (24 mm) under suction and washed 3 times with 5 ml of cold buffer and the radioactivity retained was determined by liquid scintillation counting.

In the above mentioned conditions, the "specific" binding (total binding minus "non-specific" binding) of $[^3H]$5-HT was proportional to the $[^3H]$5-HT concentration up to 150 nM (maximal concentration used), when 1,6 - 2,4 nmol/g initial tissue weight was determined for all tissues, and to the tissue concentration up to 15 mg of initial tissue weight per ml. The IC_{50} for 5-HT was in the range of 2 - 10 μM for all tissues used. The half-maximal incorporation was observed at about 2 - 4 min and the plateau was achieved at 5 - 7 min of the ultrasonic treatment.

Upon the termination of the sonication the binding of $[^3H]$5-HT was stable up to 30 min (maximal time studied) even after 10-fold dilution with the buffer.

"Non-specific" binding was less then 20% of the "specific" one at all concentrations of $[^3H]$5-HT used.

The binding of $[^3H]$tryptamine, $[^3H]$lysine, or $[^3H]$glycine to the membrane suspension was not stimulated by sonication.

The presence of ascorbic acid or of dithiothreitol in the incubation mixture prevents the ultrasonic binding of $[^3H]$5-HT to the cell membranes with IC_{50} of 5 - 10 μM and 20 - 40 μM, respectively. These data pose the possibility that the mechanism of ultrasonic binding may be connected with lipid peroxidation processes.

In conclusion, the presented data indicate that ultrasonic homogenization of tissues should be used with great care.

BINDING PROPERTIES OF 3-ALKYL-1,4-BENZODIAZEPINE-2-ONES TOWARDS SYNAPTIC MEMBRANES OF RAT BRAIN

KOVÁCS I., MAKSAY, G., TEGYEY Zs., VISY J., FITOS I., KAJTÁR M.,* SIMONYI M. and ÖTVÖS L.

Central Research Institute for Chemistry, The Hungarian Academy of Sciences, Budapest, Pf.17, H-1525, Hungary

*Institute of Organic Chemistry, Eötvös Lóránd University Budapest, Hungary

Benzodiazepines are known to bind with high-affinity and in a stereospecific manner to receptor sites present in the central nervous system [1,2]. We report here the investigation of two series of benzodiazepine derivatives (see formula) expected to be suitable for demonstrating the role of chirality, conformation and steric factors in receptor binding affinity.

R^1 = H or Me

R^2: H   Me   H   Me Pr^i H Bu^t H

R^3: H   H   Me   Me   H Pr^i H Bu^t

MATERIALS and METHODS

Enantiomers of 3-methyl- and 3-isopropyl-1,4-benzo-diazepine-2-ones were synthesized according to [3]. The 3-t.butyl- and 3-dimethyl derivatives were prepared by the analogy of [4]. The resolution of racemic 3-t.butyl-analogues was performed by albumin affinity chromatography as described [5,6]. The optical purity of all enantiomers was verified by CD spectroscopy (Fig. 1).

The incubation mixture for receptor binding assay contained P_2-fraciton (1,2 mg protein/ml) from whole rat brain [7] and 1,3 nM ^{3}H-diazepam in 50 mM Tris-HCl buffer (pH 7,1) at 4°C. Benzodiazepine derivatives used in the displacement experiments were dissolved in ethanol. The ethanol content in the incubation mixture amounted to 4,5 v/v%. The binding assay was performed by filtration. Non-specific binding determined in the presence of 1 μM diazepam represented about 10% of the total binding.

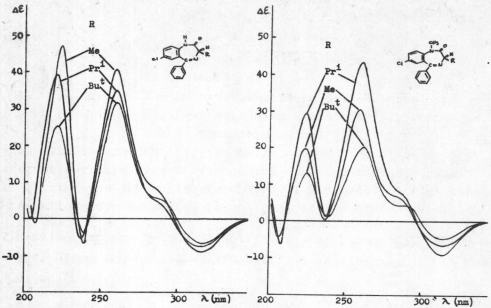

Fig.1. CD spectra of S(+)-3-alkyl-1,4-benzodiazepine-2-ones dissolved in ethanol

RESULTS and DISCUSSION

The two series represented by the formula have been investigated in respect to their potency (IC_{50}) to displace specifically bound ^{3}H-diazepam. The results are shown in Table 1.

The displacing potencies of the N(1)-Me series are slightly higher for compounds containing H or Me attached to position C(3) as compared to N(1) H series. In both series, the dimethyl substitution at C(3) position results in binding affinities about 1300 times weaker than those of the unsubstituted analogues. Stereoselectivity data from Table 1. expressed in terms of the ratio of IC_{50} values for R and S enantiomers show a sharp decrease with the growing bulk of C(3) substituents. The two series display somewhat contradictory stereoselectivities for larger substituents as a probable consequence of secondary (e.g. hydrophobic) binding components

240

being operative. In both series, the bulky substituents Pr^i, Bu^t reduce drastically the binding ability even of the more active enantiomer.

Table 1. Displacement of specifically bound ^{3}H-diazepam by 3-alkyl-1,4-benzodiazepine-2-ones

ABSOLUTE CONFIGURATION		R_2	R_3	IC_{50} (μM)	$\dfrac{IC_{50}\,(R)}{IC_{50}\,(S)}$
R_1 = H		H	H	0,010	
	S	CH_3	H	0,058	
	R	H	CH_3	66	1138
		CH_3	CH_3	13	
	S	$CH(CH_3)_2$	H	69	
	R	H	$CH(CH_3)_2$	500§	7,2
	S	$C(CH_3)_3$	H	40§	
	R	H	$C(CH_3)_3$	33	0,8
R_1 = CH_3		H	H	0,0042	
	S	CH_3	H	0,048	271
	R	H	CH_3	13	
		CH_3	CH_3	5,4	
	S	$CH(CH_3)_2$	H	107	1,1
	R	H	$CH(CH_3)_2$	115	
	S	$C(CH_3)_3$	H	29	20
	R	H	$C(CH_3)_3$	600§	

§ The limited solubility of these compounds allowed only the estimation of IC_{50} values.

It is proved by NMR [8] and by CD [9] measurements that in solution an equilibrium exists between two conformers of optically inactive 1,4-benzodiazepine-2-ones. Monosubstitution at C(3) stabilizes the quasi-boat-structured heterocyclic ring either in M or in P conformation [9]. The S enantiomer is found predominantly in M, while the R enantiomer in P conformation, both having the substituent in pseudoequatorial orientation [10,11]. Moreover, the S stereoisomer of conformationally restricted anthramycin-type benzodiazepines was preferentially recognized by benzodiazepine receptor sites [11].

Our binding data show that the high binding affinity of C(3) unsubstituted 1,4-benzodiazepine-2-ones decreases slightly on introduction of a 3-S-methyl substituent while the affinity is almost completely lost for 3-R-methyl or 3,3-dimethyl substitution. The slight preference of dimethyl derivatives over 3-R-methyl substituted compounds can be rationalized by the conformational preference of the receptor; while R-methyl derivatives have the wrong, P conformation, 50% of the dimethyl derivatives can be found in conformation M. The large decrease in binding affinity brought about by dimethyl substitution emphasizes, however, that the conformation itself is an insufficient criterion for receptor binding ability. In the case of 3-isopropyl and 3-t.butyl substituents, the conformational preference by the receptor may become even less important, as a probable consequence of the sterically hindered fit to the binding site. For these weakly binding derivatives, hydrophobic binding of the bulky 3-alkyl substituents may interfere with the effect of steric hindrance exhibited by the geometry of receptor binding sites.

REFERENCES

1. Squires, R.F. and Braestrup, C. Benzodiazepine receptors in rat brain. Nature 266, 732-734 (1977)
2. Möhler, H. and Okada, T. Benzodiazepine receptor: demonstration in central nervous system. Science 198, 849-851 (1977)
3. Sunjic, I., Kajfez, F., Stromar, I., Blazevic, N. and Kolbah, D. Chiral 1,4-benzodiazepines. Synthesis and properties of 1,4-benzodiazepin-2-ones containing α-amino acids as a part of the 1,4-diazepine ring. J. Heterocycl. Chem. 10, 591-599 (1973)

4. Bell, S.C., Sulkowski, T.S., Gochman, C. and Scott, J.C.
   1,3-Dihydro-2H-1,4-benzodiazepine-2-ones and their 4-oxides.
   J.Org.Chem. 27, 562-566 (1962)

5. Fitos, I., Simonyi, M., Tegyey, Zs., Ötvös, L., Kajtár, J.
   and Kajtár, M. Resolution by affinity chromatography:
   stereoselective binding of racemic oxazepam esters to human
   serum albumin. J. Chromatogr. 259, 494-498 (1983)

6. Simonyi, M. and Fitos, I. Stereoselective binding of a 2,3-
   -benzodiazepine to human serum albumin. Biochem. Pharmacol.
   32, 1917-1920 (1983)

7. Braestrup, C. and Squires, R.F. Specific benzodiazepine
   receptors in rat brain characterized by high-affinity
   ^{3}H-diazepam binding. Proc. Natl. Acad. Sci. USA 74,
   3805-3809 (1977)

8. Linscheid, P. and Lehn, J.-M. Études cinétiquies et
   conformationnelles par résonance magnétique nucléaire.
   Inversion de cycle dans des benzo-diazépinones. Bull. Soc.
   Chim. France, 992-997 (1967)

9. Snatzke, G. Experimental determination of the conformation
   of biomolecules by circular dichroism. In "Steric Effects
   in Biomolecules" Ed., G. Náray-Szabó, Akadémiai Kiadó,
   Budapest (1982) pp. 213-225.

10. Sikirica, M. and Vickovic, I.   7-chloro-5-phenyl-3(S)methyl-
    -2H-1,4-benzodiazepine-2-one. Cryst. Struct. Comm. 11,
    1293-1298 (1982)

11. Blunt, J.F., Fryer, R.I., Gilman, N.W. and Todaro, L.J.
    Conformational recognition of the receptor by
    1,4-benzodiazepines Molec. Pharmacol. 24, 425-428 (1983)

COMPRESSIONAL STERIC SUBSTITUENT EFFECTS IN DRUG-RECEPTOR INTERACTIONS

ÖTVÖS L.

Central Research Institute for Chemistry,
The Hungarian Academy of Sciences,
Budapest, Pf. 17, H-1525, Hungary

INTRODUCTION

Different forms of steric substituent constants (*i.e.* E_s, E_s^C, M_w, u) express quantitatively the influence of steric effects on the reactivity of congeneric series of compounds in organic, bio-organic, and biochemical reactions [1-5]. These constants describe the steric effects although generally do not explain them, especially in the case of stereospecific transformations [6]. A detailed study of the relationship between the steric requirement and the steric orientation of the substituents led to the concepts of *"orientational steric substituent effect"* and the *"compressional steric substituent effect"* [7]. These special steric effects has been applied for the explanation of several specific transformations of biopolymers [8-10] including receptor binding processes.

COMPRESSIONAL STERIC SUBSTITUENT EFFECTS IN DRUG-RECEPTOR INTERACTIONS

Receptors generally contain several functional groups. In *Fig. 1,* the plane of the paper represents the surface of the macromolecule or one of its cavities, generally termed in enzyme chemistry as "pocket". This part of the molecule contains A' and B' groups capable of binding the A and B

groups of the compound seen in the picture. In this way a
cross-linkage is formed. $A...A'$ and
$B...B'$ attachments may be covalent or
hydrogen bonds, ionic or hydrophobic
interactions, etc. Owing to this, R^1
and R^2 substituents of this tetrahedral carbon atom are necessarily
oriented. The R^2 substituent is
oriented towards the surface of the
macromolecule or the inside of its
cavity, R^1 is oriented towards the

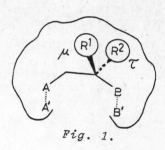

Fig. 1.

solution or biological media. The substituents are denoted
by τ and μ respectively. Since μ does not exert substantial
steric effect, reactivity mainly depends on the steric
hindrance of the τ substituent.

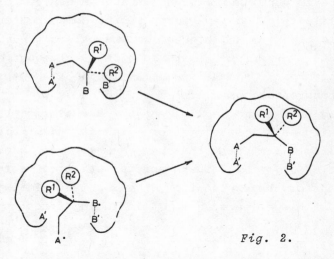

Fig. 2.

The compressional steric substituent effect implies that
after the formation of either of the $A-A'$ or $B-B'$ bond, the
R^2 atom or group hinders or prevents the other $B-B'$ or $A-A'$
bonding, due to steric compression between the macromolecule
and the R^2 substituent (Fig. 2).

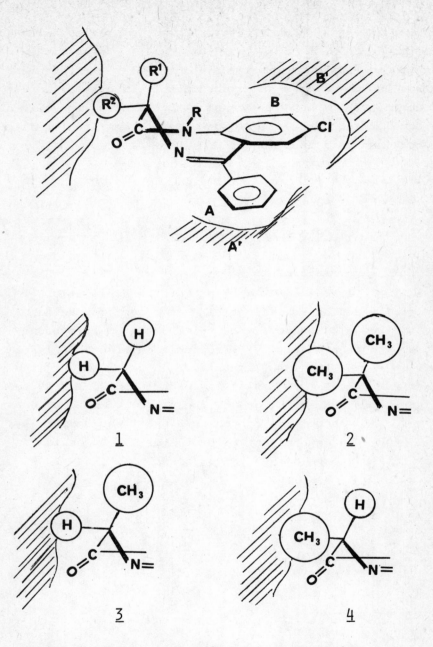

Figure 3. Illustration of steric compression

The probability of such a complex formation between drug and receptor depends on the steric requirement of substituent R^2 oriented towards the biopolymer and the steric requirement depends on the shape of the substituent. In case of congeneric series of compounds containing spherical R^2 substituents, the equilibrium constant of drug-receptor complex formation can be given as a function of Van der Waals radii.

$$D + R \xrightleftharpoons{K} DR \qquad [1]$$

$$\log K = \kappa^* r_{v,R}^2 + C \qquad [2]$$

$$\log \frac{K_x}{K_o} = \kappa^* (r_{v,R_x}^2 - r_{v,R_o}^2) + C' \qquad [3]$$

RECEPTOR BINDING OF 1,4-DIAZEPINES

The analysis discussed is applied for the explanation of specificity and enantioselectivity in receptor binding of 3-substituted-1,4-benzodiazepines. These drugs are bound to biopolymers probably by hydrophobic interaction of aromatic rings in $\underline{M}$ conformation [11] (*Fig. 3*).

Comparison of IC_{50} values of componds $\underline{1}$ and $\underline{2}$ clearly shows the steric effect of CH_3 group. The ratio of IC_{50} of enantiomeric $\underline{3}$ and $\underline{4}$ as well as compounds $\underline{3}$ and $\underline{1}$ furthermore $\underline{4}$ and $\underline{2}$ prove the determinant role of steric compression in specific receptor binding (*Fig. 4*).

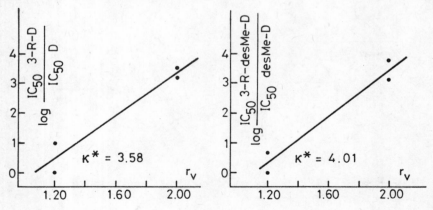

Fig.4. (D = diazepam)

248

REFERENCES

1. Taft, R.W.: Separation of polar, steric and resonace
   effects in Reactivity in <u>Steric Effects in Organic</u>
   <u>Chemistry</u> Ed.: M.S. Newman, Wiley- New York,
   pp. 556-675 (1977)

2. Hancock, C.K., Meyers, E.A., and Yanger, B.J.:
   Quantitative separation of hyperconjugation effects
   from steric substituent constants J.Am.Chem.Soc. <u>83</u>,
   4211-4213 (1961)

3. Verloop, A., and Tipker, J.Biological Activity and
   Chemical Structure, E.: J.A. Keverling Buisman, Elsevier,
   Amsterdam-Oxford-New York (1977) pp. 63-81.

4. Hansch, C., Unger, S.H. and Forsythe, A.B. Strategy in
   drug design. Cluster analysis as an aid in the selec-
   tion of substituents. J.Med.Chem. <u>16</u>, 1217-1222 (1973)

5. Steric effects. I. Esterification and acid-catalyzed
   hydrolysis of esters. Charton M.1. J.Am.Chem.Soc. <u>97</u>,
   1552-1556 (1975)

6. Ariens, E.J., Verloop, A., Hansch, C., Burgen, A.S.V.
   Round table discussion in <u>Biological Activity and</u>
   <u>Chemical Structure</u>. Ed.: J.A. Keverling Buisman,
   Elsevier, Amsterdam-Oxford-New York, pp. 282-284. (1977)

7. Ötvös, L., Elekes, I., Kraicsovits, F., and Moravcsik, E.
   in <u>Steric Effects in Biomolecules</u>. Ed.: Náray-Szabó, G.,
   Akadémiai Kiadó, Budapest, 1982. pp. 305-325.

8. Ötvös, L., and Elekes, I. Steric hindrance in the
   reactions of DNA with bifunctional alkylating agents.
   Definition of "τ" and "μ" substituents. Tetrahedron
   Letters, <u>1975</u>, 2477-2480.

9. Ötvös, L., and Elekes, I. Explanation of stereospecific
   and asymmetric reactivity of DNA in cross-linking
   alkylation. Tetrahedron Letters, <u>1975</u>, 2481-2484.

10. Ötvös, L., Moravcsik, E., and Kraicsovits, F. Steric effect of chiral substituents in enzyme-catalyzed reactions. Tetrahedron Letters, <u>1975</u>, 2485-2488.

11. Blount, J.F., Fryer, R.I., Gilman, N.W., and Toduro, L.J. Conformational recognition of the receptor by 1,4-benzodiazepines. Molec.Pharmacol. <u>24</u>, 425-428 (1983)

Bio-Organic Heterocycles
van der Plas H.C., Ötvös L., and Simonyi M. eds

PROTEIN-BINDING OF PIPECURONIUM BROMIDE (ARDUAN®) IN HUMAN PLASMA

SZELECZKY G., RÓNAI-LUKÁCS S., VERECZKEY L.

Department of Pharmacokinetics and Drug Metabolism, Chemical Works of Gedeon Richter Ltd. P.O.Box 27, H-1475 Budapest 10, Hungary

INTRODUCTION

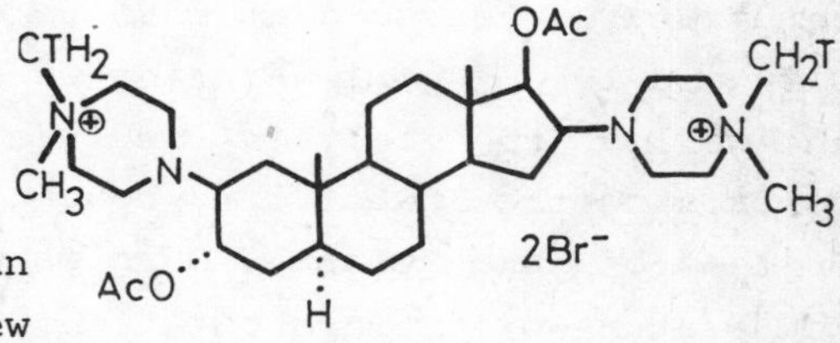

Among the bis-quaternary steroids inhibiting the nicotine-like effects of acetylcholine on neuromuscular junctions, Arduan (pipecuronium bromide) is a new non-depolarizing neuromuscular blocking agent with chemical and pharmacological properties resembling to those of Pavulon

Figure 1. Chemical structure of tritiated Arduan.

(pancuronium bromide) [1,2]. According to the results obtained on mice, chicken, cats and dogs, Arduan is 2-4 times as potent as Pavulon and the effectiveness of the drug exceeded that of pancuronium bromide by 20 % in the human studies, as well [3,4].

Considering the extreme low dose of Arduan required for complete relaxation as well as the disposition of the drug (accumulation in the liver) in experimental animals [5], it was a matter of interest whether or not Arduan binds to human plasma proteins.

MATERIALS and METHODS

All chemicals used were of analytical grade and were purchased from Reanal (Budapest), unless otherwise indicated.

Tritium-labelled Arduan was synthesized at the Central

Research Institute for Chemistry of the Hungarian Academy of Sciences. Labelling was made at the methyl groups of the two quaternary nitrogen atoms (Figure 1). Specific activity of the labelled compound was 73.26 GBq/mmol (1.98 Ci/mmol). The radiochemical purity was found to be 95 % by TLC.

Deep-frozen human plasma was obtained from the National Institute of Haematology.

Binding of Arduan to proteins was studied *in vitro* by using membrane filtration (Amicon Micropartition System) and equilibrium dialysis for the separation of free and protein-bound drug. Samples were prepared in the following way: the tritiated drug was added to 1 ml human plasma (at room temperature) to reach a final concentration ranging between 0.26 and 1290 ng/ml. Binding of Arduan to the membrane or to the wall of the dialysis sack was checked by using samples containing the lowest drug concentration and distilled water instead of human plasma. Samples were incubated at room temperature prior to separation.

Membrane filtration was performed by using Amicon Micropartition System filled with 0.5 ml aliquots of the samples. Filtration was achieved by centrifugation (1000 x g, 25 ^{O}C, 30 minutes; Janetzki K 24). Arduan concentration of initial plasma samples and that of the filtrates were determined by radioactivity measurements performed on a Packard Tri-Carb 2660 liquid scintillation spectrometer and by using Instagel (Packard) liquid scintillation solution.

For equilibrium dialysis, 0.5 ml of the samples was taken into dialysis sacks and immersed into 50 ml 0.1 M phosphate buffer (pH 7.4). Having allowed an overnight for reaching equilibrium, the concentration of Arduan both in buffer and in the sacks were determined as in the case of membrane filtration.

RESULTS and DISCUSSION

Results obtained by membrane filtration and expressed as
the percentage of free drug concentration are presented in
Table 1. As it is obvious, under the conditions applied and
in the concentration range studied, Arduan does not bind to
human plasma proteins; the mean ($\pm$ S.D.) of the percentage
of free drug concentration was found to be 98.2 $\pm$ 6.08.

Table 1. Ultrafiltration of Arduan from human plasma samples

Arduan concentration (ng/ml)	% of drug filtered	Arduan concentration (ng/ml)	% of drug filtered
0.26 (water)	96	55	101
0.26	104	63	102
0.5	92	122	96
1.6	105	253	92
13.0	104	703	86
48.0	100	1290	98

Equilibrium dialysis proved to be an inadequate method in
this case, since unlike to membranes, Arduan was bound in a
great extent (75 %) by dialysis sacks.

Control of the filtration method

After having observed that Arduan is quite unstable in
solution and having known that the drug undergoes rapid
metabolism in the organism [6], we checked whether our results
represented really the lack of   binding of Arduan or
indicated the same for some radioactive decomposition product.
In order to decide about the question, the filtrate of samples
was subjected to TLC analysis using Kieselgel 60 F_{254} plates
(Merck), pyridine-butanol-acetic acid-water (10:15:3:12)
chromatographic system and authentic radioactive Arduan as
reference compound. Location of the spots was determined by

radioactivity measurement. According to the results obtained, the radioactivity content of filtrates commigrated almost exclusively with authentic Arduan.

The reliability of the above results being confirmed, the following conclusion can be drawn: Arduan does not bind to human plasma proteins in the concentration range studied. Since the therapeutic plasma level of the drug falls within this range [6], it is very likely that the drug is present in the plasma only in free form, which may be an explanation for the extreme low dose required for complete muscle relaxation.

REFERENCES

(1) Tuba,Z., (1980) Synthesis of 2β,16β-Bis-/4'-dimethyl-1'-piperazino/-3α,17β-diacetoxy-5α-androstan dibromid and related compounds., Arzneim.-Forsch./Drug Res. 30(I) 342-346.
(2) Kárpáti,E., Biró,K.,(1980) Pharmacological study of a new competitive neuromuscular blocking steroid, pipecuronium bromide., ibid 30(I) 346-354.
(3) Alánt,O., Darvas,K., Pulay,I., Weltner,J., Bihari,I., (1980) First clinical experience with a new neuromuscular blocker, pipecuronium bromide., ibid 30(I) 374-379.
(4) Boros,M., Szenohradszky,J., Marosi,Gy., Tóth,I., (1980) Comparative clinical study of pipecuronium bromide and pancuronium bromide., ibid 30(I) 389-393.
(5) Vereczkey,L., Szporny,L.,(1980) Disposition of pipecuronium bromide in rats., ibid 30(I) 364-366
(6) Tassonyi,E., Szabó,G., Vereczkey,L.,(1981) Pharmacokinetics of pipecuronium bromide, a new non-depolarizing neuromuscular blocking agent in humans., ibid 31(II) 1754-1756.

ENANTIOSELECTIVE BINDING OF TIFLUADOM TO HUMAN SERUM ALBUMIN

SIMONYI M., VISY J., KOVÁCS I. and KAJTÁR M.*

Central Research Institute for Chemistry,
The Hungarian Academy of Sciences,
Budapest, Pf. 17, H-1525, Hungary

*Institute of Organic Chemistry, Eötvös Lóránd University,
 Budapest, Hungary

Tifluadom (1), the recently introduced compound of opioid activity [1] is structurally related to 1,4-benzodiazepines (2,3); instead of the oxo moiety, tifluadom contains a centre of asymmetry at position C-2.

(1)   (2)   (3)   (4)

Human serum albumin (HSA), the most abundant protein in human plasma binds enantioselectively 1,4-benzodiazepine-2-ones having the asymmetric centre either at C-3 (2)[2-5], or at C-5 (3)[6]. The binding site for benzodiazepines at HSA prefers the S(+) enantiomers, (2) and (3). Since the diazepine ring is a seven-membered non-planar heterocycle, S(+) enantiomers accomodate the boat conformation (4) assigned by the descriptor *M* [7]. Hence, the reason of enantioselective binding to HSA can be attributed to conformational preference exhibited by the binding site involved. Conformation *M* of benzodiazepines was recently indicated to be preferred by brain benzodiazepine receptors [8], too.

Tifluadom was found to show enantioselective binding affinity for the benzodiazepine receptor: the (+) enantiomer was 14-times more potent than its antipode [9]. The (+) sign refers to the optical rotation in toluol which is opposite to the sign in ethanol [9]. To the best of our knowledge, investigation on the binding of tifluadom to HSA has not yet been reported.

Affinity chromatography

Racemic tifluadom was a generous gift by Dr. D. Römer (Sandoz Ltd., Basle). It was resolved by affinity chromatography on HSA immobilized with cyanogen bromide-activated Sepharose 4B according to the method described previously [10]. Figure 1 demonstrates the separation of tifluadom enantiomers. Binding constants estimated from the maxima of elution peaks are, as follows:

$$K(I) = 6.3 \times 10^{-5} \ M^{-1}; \quad K(II) = 1.5 \times 10^{-6} \ M^{-1}.$$

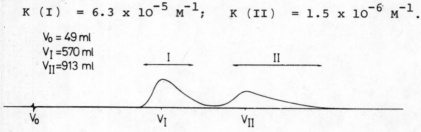

Fig.1. Elution of tifluadom enantiomers from HSA column.

Fractions I and II have been separately collected, extracted from aqueous buffer and purified to yield samples of tifluadom (I) and tifluadom (II) in quantities of about 0.1 mg.

CD analysis

Mirror image CD spectra were obtained from chromatographically separated tifluadom samples. Figure 2 shows the spectrum of tifluadom(I) in two solvents. The spectrum is subject to a net negative shift in cyclohexane as compared to that in ethanol. Hence, tifluadom(I) is identical with (−)tifluadom as defined by Kley et al. [9].

It can further be established from Figure 2 that tifluadom(I) and 3(R)-alkyl-1,4-benzodiazepine-2-ones exist in analogous conformations. This conformation of tifluadom(I) should be assigned $\underline{M}$ owing to the lacking carbonyl at C-2.

Consequently, tifluadom(II) which is more strongly bound to HSA (cf. Fig.1) exists in conformation $\underline{P}$.

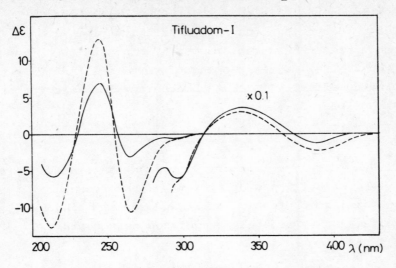

Fig.2. CD spectrum of tifluadom(I) in ethanol (——) and in cyclohexane (---).

Binding to brain benzodiazepine receptors

Figure 3 illustrates enantioselective displacement of ^{3}H-diazepam bound to receptor sites. The IC_{50} values are

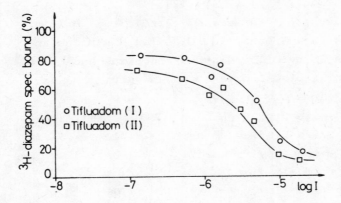

Fig.3. Displacement of ^{3}H-diazepam by chromatographically separated tifluadom enantiomers.

5.4 μM and 2.2 μM for tifluadom(I) and tifluadom(II),
respectively. The enantioselectivity was found much smaller
than reported on the displacement of ^{3}H-flunitrazepam [9].
The finding, however, on the higher affinity of tifluadom(II)
towards benzodiazepine receptors is in agreement with the
earlier report [9]. Of the two enantiomers, tifluadom(II) has
higher binding affinity for both benzodiazepine receptors and
HSA binding sites.

Conclusion

Tifluadom displays high affinity enantioselective binding
to HSA. Moreover, tifluadom represents a novel example
indicating similar preference of ligand conformation by HSA
binding sites and by central benzodiazepine receptors.

REFERENCES

1. Römer, D., Büscher, H.H., Hill, R.C., Maurer, R., Petcher,
   T.J., Zeugner, H., Benson, W., Finner, E., Milkowski, W.
   and Thies, P.W. An opioid benzodiazepine.
   Nature, 298, 759-760 (1982)
2. Müller, W.E. and Wollert, U. High stereospecificity of the
   benzodiazepine binding site on human serum albumin. Studies
   with d- and l-oxazepam hemisuccinate.
   Molec. Pharmacol. 11, 52-60 (1975)
3. Alebic-Kolbah, T., Kajfez, F., Rendic, S., Sunjic, V.,
   Konowal, A. and Snatzke, G. Circular dichroism and gel
   filtration study of binding of prochiral and chiral
   1,4-benzodiazepine-2-ones to human serum albumin.
   Biochem. Pharmacol. 28, 2457-2464 (1979)
4. Gratton, G., Rendic, S., Sunjic, V. and Kajfez, F.
   Stereoselective binding of chiral 7-chloro-1,3-dihydro-
   -3(S and R)-isopropyl-5-phenyl-2H-1,4-benzodiazepine-2-one
   to human serum albumin.
   Acta Pharm. Jugoslav. 29, 119-124 (1979)

5. Simonyi, M., Fitos, I., Kajtár J. and Kajtár, M.
   Application of ultrafiltration and CD spectroscopy for
   studying stereoselective binding of racemic ligands.
   Biochem.Biophys.Res.Commun. 109, 851-857 (1982)
6. Fitos, I. and Simonyi, M. Stereoselective binding of
   4,5-dihydrodiazepam to human serum albumin.
   Experientia 39, 591-592 (1983)
7. Konowal, A., Snatzke, G., Alebic-Kolbah, T., Kajfez, F.,
   Rendic, S. and Sunjic, V. General approach to chiroptical
   characterization of binding of prochiral and chiral
   1,4-benzodiazepine-2-ones to human serum albumin.
   Biochem. Pharmacol. 28, 3109-3113 (1979)
8. Blount, J.F., Fryer, R.I., Gilman, N.W. and Todaro, L.J.
   Conformational recognition of the receptor by 1,4-benzo-
   diazepines. Molec. Pharmacol. 24, 425-428 (1983)
9. Kley, H., Scheidemantel, K., Bering, B. and Müller, W.E.
   Reverse stereoselectivity of opiate and benzodiazepine
   receptors for the opioid benzodiazepine tifluadom.
   Eur.J.Pharmacol. 87, 503-504 (1983)
10. Fitos, I., Simonyi, M., Tegyey, Zs., Ötvös, L., Kajtár, J.
    and Kajtár, M. Resolution by affinity chromatography:
    stereoselective binding of racemic oxazepam esters to
    human serum albumin. J.Chromatogr. 259, 494-498 (1983)

ADDITIVE CONTRIBUTIONS OF GUVACINE- AND ISOGUVACINE-SENSITIVE COMPONENTS OF ^{3}H-GABA SPECIFICALLY BOUND TO SYNAPTIC MEMBRANES IN PHYSIOLOGICAL MEDIUM

KARDOS J. and SIMONYI M.

Central Research Institute for Chemistry,
The Hungarian Academy of Sciences,
Budapest, Pf. 17, H-1525, Hungary

Investigations on ^{3}H-γ-aminobutyric acid (GABA) binding to synaptic membrane receptor sites have usually been performed in the absence of Na$^+$ ions (in non-physiological buffers) in order to prevent interaction of ^{3}H-GABA with specific uptake sites, which may occur in the presence of Na$^+$ ions. The development of specific uptake inhibitors [1] (e.g. guvacine, nipecotic acid) afforded opportunities to detect separately the binding of ^{3}H-GABA to receptor sites even in the presence of Na$^+$ ion, i.e. under more physiological conditions.

The distribution of ^{3}H-GABA between receptor and uptake sites in physiological buffer was investigated in the presence of guvacine (1) and of the cyclic GABA analogue, isoguvacine (2), alternatively.

(1)
guvacine

(2)
isoguvacine

It is shown (Fig.1) that the isoguvacine-sensitive (receptor sites) and guvacine-sensitive (uptake sites) parts of specific ^{3}H-GABA binding are complementary and amount to the whole ^{3}H-GABA specifically bound.

The kinetics of specific ^{3}H-GABA binding in the presence of guvacine is presented (Fig.2). Analysis of data in terms of

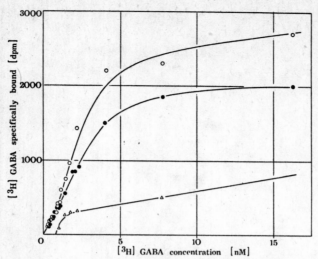

Fig.1. Saturation plots of specific ^{3}H-GABA binding to synaptic membranes (40 mg w.t.w./ml) in Tyrode buffer (pH 7.00) alone (o), with 10^{-4} M guvacine (●), with 10^{-4} M isoguvacine (△).

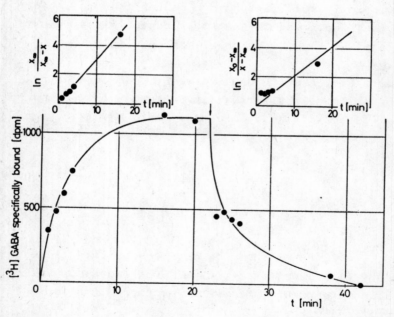

Fig.2. Time dependence of specific ^{3}H-GABA receptor binding to synaptic membranes in Tyrode buffer (pH 7.00).

reversible kinetics [2] gives the following parameters: dissociation equilibrium constant, K_d=3.3nM; association rate constant, k_2=7.5x10^{5}M^{-1}s^{-1}; dissociation rate constant, k_1=2.5x10^{-3}s^{-1}.

According to the slightly sigmoid character of the satura-tion plot (shown in Fig.1), the equilibrium binding of ^{3}H-GABA was interpreted by the model and equation (Lineweaver-Burk plot):

$$nF + R \xrightleftharpoons{K_d} B \qquad \frac{1}{B} = \frac{K_d^n}{B_{max}} \frac{1}{F^n} + \frac{1}{B_{max}}$$

where B is the number of receptors occupied by the ligand; R corresponds to free receptors; F is the amount of free ligand; n is the Hill coefficient, and B_{max} is the total number of receptors. Fig. 3 shows that GABA and the GABA antagonist, bicuculline methiodide are competing for the same binding sites.

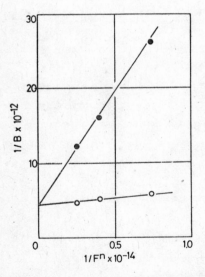

Fig.3. Lineweaver-Burk analysis of competition for receptor sites between GABA (o) and bicuculline methiodide (●) in Tyrode buffer (pH 7.00).

EXPERIMENTAL

Preparation of Synaptic Membranes. Rat brains (male, Wistar)
were homogenized in 15 vol. ice-cold 0.32M sucrose solution
with Potter-Elvehjem glass homogenizer. The homogenate was
centrifuged at 1000g for 10 min, the pellet discarded and the
supernatant centrifuged at 20000g for 10 min. The pellet was
resuspended in distilled water and after 2 hours centrifuged
at 8000g for 20 min. The supernatant together with the upper
layer of the pellet was resuspended and pelleted 3 times in
Tyrode at 45000g for 20 min. After the third centrifugation
the pellet was frozen for at least 18 hours at -20°C, then
thawed and centrifuged. This freezing-thawing-washing cycle (in
Tyrode solution) was repeated 3 times. Synaptic membranes were
freeze-dried [3]. Prior to binding experiments, the lyophilized
synaptic membranes were reconstituted in Tyrode buffer (pH=7.0).
Binding assays. Synaptic membranes (40 mg wet tissue weight/ml)
were incubated in the presence of either 10^{-4}M guvacine or
isoguvacine at 4°C with ^{3}H-GABA (65.4 Ci/mmole, Amersham),
or with ^{3}H-diazepam (64.0 Ci/mmole, NEN). After the binding
equilibrium had been settled (20 min), bound radioactivity was
separated by cold filtration (Whatman GF/C microporous glass
filters). Washing was performed with 8x2.5ml ice-cold Tyrode
buffer. Bound radioactivities were determined by liquid
scintillation spectrometry (Packard TRI-CARB spectrometer,
counting efficiency 37%). Non-specific binding determined in
the presence of 10^{-4}M GABA or 10^{-6}M diazepam was less than
20% of the total.

REFERENCES

1. Krogsgaard-Larsen, P. Inhibitors of the GABA uptake systems,
   Mol.Cell.Biochem., 31, 105-121 (1980).
2. Simonyi, M. and Mayer, I. Formal similarity between
   irreversible and reversible bimolecular kinetics, React.
   Kinet.Catal.Lett., 18, 431-432 (1981).
3. Kardos, J., Maksay, G. and Simonyi, M. German Patent
   DE 3247845 Hl, (1983).

Bio-Organic Heterocycles
van der Plas H.C., Ötvös L., and Simonyi M. eds

DISTINCTION OF THE HIGH AND LOW AFFINITY GABA RECEPTORS BY HETEROCYCLIC AGONISTS AND ANTAGONISTS

MAKSAY G. and TICKU M.K.*

Central Research Institute for Chemistry,
The Hungarian Academy of Sciences,
Budapest, Pf. 17, H-1525, Hungary

*University of Texas, Health Science Center
 San Antonio, TX 78284 USA

γ-Aminobutyric acid (GABA) is a major inhibitory neuro-transmitter. The aim of this study was to differentiate the molecular topography of the high and low affinity GABA receptors and the binding modes of GABA agonists and antagonists. **Methods**. A freeze-thawed membrane preparation of rat cerebral cortex was incubated with reagents and/or GABAergic ligands, then washed with Tris buffer [1]. The specific binding of [^{3}H]GABA and [^{3}H]muscimol was studied by a centrifugation method at 4°C [1].

Selective Protein Modification. Curvilinear Scatchard plots of [^{3}H]GABA binding revealed two populations of binding sites with affinities of 10$\pm$2 and 142$\pm$41 nM. Pretreatment of the membranes with tyrosine-(Tyr)-specific reagents selectively decreased the number of low affinity sites: a) 10 mM tetranit-romethane (TNM) decreased B$_{max,2}$ from 2900 to 1070 fmol/mg, the other parameters were unaffected [1]; b) 1 mM p-diazoben-zenesulfonic acid (DSA) completely eliminated the low affinity population [2]. The inhibitory effect of Tyr-selective reagents (TNM, DSA, N-Acetylimidazol) on GABA binding was selectively prevented by saturation of the receptors with direct GABAergic ligands (GABA and muscimol, Table 1), but allosteric ligands caused no or virtual protection (Table 1:(+)). The GABAergic selectivity of protection from inhibition by Tyr-specific reagents suggests the presence of an accessible Tyr residue at the low affinity GABA receptors.

Modification by the lysine-(Lys)-specific reagent pyridoxal-5-phosphate (PLP) also inhibited GABA binding [1] which was

265

Table 1. Protection of the GABA binding from the inhibitory
effect of protein-modifying reagents.

Protecting ligand	Reagents				
	DSA	TNM	N-AcIm	NBS	PLP
10^{-5}M GABA	+	++	++	+	+
10^{-6}M muscimol	+	++	++		+
3×10^{-6} M flurazepam	(+)	-	-	-	(+)
10^{-4}M etazolate	-	(+)	(+)	-	-

++: Complete protection; +: significant protection;
(+): virtual protection (due to the effect of the protecting
ligand itself); -: no protection. NBS: N-Bromosuccinimide.

selectively prevented by GABA agonists (Table 1). This suggests
that a Lys residue is situated in the GABA receptors [3].
However, the GABA antagonists bicuculline methiodide (BCM) and
R-5135 (Figure 1) did not protect from PLP [4]. In conclusion,
we suppose that a Lys residue takes part in the binding of
agonists by interacting with their anionic groups. Since the
antagonists lack such anionic groups, they cannot block the
Lys residue.

UV Affinity Labelling. Cerebral GABA receptors can be
UV-affinity labelled with 50 nM muscimol and irradiation, like
cerebellar receptors [5]. Affinity-labelling decreased the
B_{max} of muscimol binding by 27 % (Table 2). The displacing
potency of GABA decreased following affinity labelling but
the IC_{50} of the antagonist BCM did not change significantly
(Table 2). We assume that the UV radical of muscimol is
covalently incorporated into the polar core of GABA receptors.
This decreases the binding potency of the other agonist
(GABA), but not that of the antagonist BCM whose binding might
prefer other attachment points.

Effect of DSA and Thiocyanate. It has been demonstrated that
treatment with 1 mM DSA and 50 mM thiocyanate (SCN^-) abolish
the low and high affinity GABA sites, respectively, [6,2].
Accordingly, the two populations were studied separately.

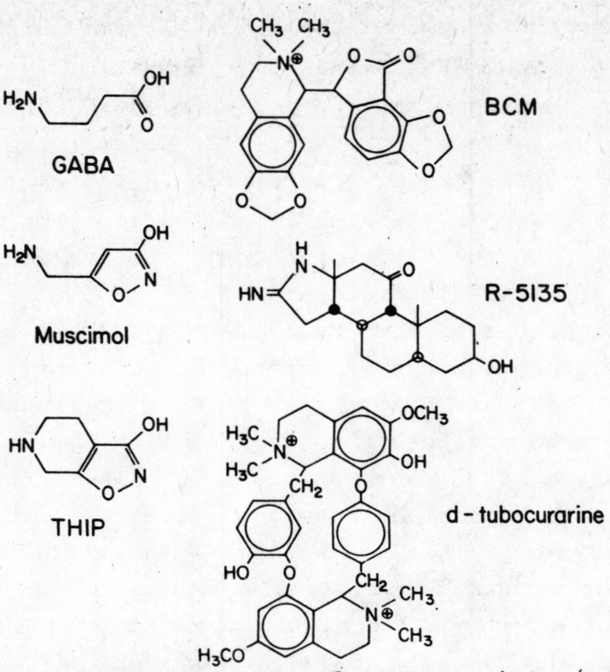

Figure 1. Structural comparison of GABA agonists (on the left) and antagonists (on the right).

Table 2. Scatchard analysis of [^{3}H]muscimol binding and displacing potencies of GABA and BCM.

	B_{max} (pmol/mg)	K_D (nM)	IC_{50} values	
			GABA (nM)	BCM (μM)
UV-Control	1.90±0.10	5.9±1.2	54.5±25.6	24.5±2.4
UV-affinity-labelled	1.38±0.08[a]	4.7±1.1	88.1±31.6	20.8±4.1
$\dfrac{IC_{50,labelled}[b]}{IC_{50,control}}$			1.72±0.34[c]	0.86±0.20

The membranes were UV-irradiated for 20 min in the presence of muscimol with (control) or without 10^{-4} M GABA. Data are means ±S.D. of 3-4 experiments. [a]p < 0.005: significantly less than in control; [b]calculated from simultaneous experiments; [c]p < 0.025: different from unity.

267

[^{3}H]GABA displacing potencies of agonists and antagonists are shown in Fig.2. In DSA-pretreated membranes, IC$_{50}$ values of the agonists decreased and that of the antagonists increased [4]. In the presence of SCN^{-}, the IC$_{50}$ values were shifted in an opposite manner. Figure 2 shows the shifts in IC$_{50}$ which results in separate regression lines for agonists and antagonists. Consequently, the antagonists preferentially bind to the low affinity sites.

   To interpret our data we have postulated that the low affinity GABA receptors contain hydrophobic accessory sites [4]. The effect of SCN^{-} is chaotropic; it further exposes the hydrophobic accessory sites, thus facilitating the binding of the ring system of the antagonists to the low affinity sites. In contrast, Tyr-specific reagents introduce ionized groups into the accessory sites and eliminate the low affinity receptors. In conclusion, DSA-pretreated membranes represent an abudance of the hydrophilic high affinity receptors, while SCN^{-} results in an excess of hydrophobic low affinity population of sites.

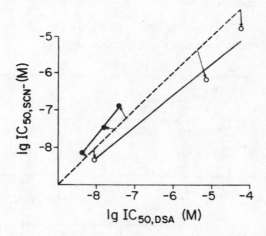

Figure 2. Correlation of IC$_{50}$ values for the GABA agonists (●) and antagonists (o) seen in Fig.1. The arrows show the shifts of IC$_{50}$ values in treated membranes.

REFERENCES

1. Maksay, G. and Ticku, M.K.: Characterization of GABA-benzodiazepine receptor-complexes by protection against inactivation by group-specific reagents.
   J.Neurochem., in press (1984a).
2. Burch, T. P., Thyagarajan, R. and Ticku, M.K.: Group-selective reagent modification of benzodiazepine-GABA receptor-ionophore complex reveals that low-affinity GABA receptors stimulate benzodiazepine binding.
   Mol. Pharmacol. 23, 52-59 (1983).
3. Tunicliff, G.: Inhibition by pyridoxal-5-phosphate of GABA receptor binding to synaptic membranes of cat cerebellum.
   Biochem. Biophys. Res. Commun. 87, 712-718 (1979).
4. Maksay, G. and Ticku, M.K.: Diazotization and thiocyanate differentiate agonists from antagonists for the high- and low-affinity receptors of GABA.
   J. Neurochem. in press (1984b).
5. Asano, T., Sakakibara, J. and Ogasawara, N.: Molecular sizes of photolabeled GABA and benzodiazepine receptor proteins are identical.
   FEBS Lett. 151, 277-280 (1983).
6. Browner, M., Ferkany, J.W. and Enna, S.J.: Biochemical identification of pharmacologically and functionally distinct GABA receptors in rat brain.
   J. Neurosci. 1, 514-518 (1981).

(±)-SALUTARIDINE: A PARTIAL AGONIST
AT THE GABA/BENZODIAZEPINE RECEPTOR COMPLEX

KARDOS J., BLASKÓ G., SIMONYI M. and SZÁNTAY CS.

Central Research Institute for Chemistry
The Hungarian Academy of Sciences,
Budapest, Pf. 17, H-1525, Hungary

Salutaridine (1) is a principal morphinandienone alkaloid and occurs in nature both in optically active and racemic forms [1]. Since salutaridine contains an atomic sequence isosteric with γ-aminobutyric acid (GABA), the possibility of its GABAergic activity might be suggested. In contrast with GABA, salutaridine can be considered as a conformationally restricted GABA analogue lacking the negative charge.

CHEMISTRY

The preparation of (±)-salutaridine (1) for investigation is shown in the Scheme. Phenolic oxidative coupling of (±)-N-formylnorreticulin (4)   with lead tetraacetate in the presence of trichloroacetic acid resulted regioselectively [2] in (±)-N-formylnorsalutaridine (2) which was deformylated in aqueous hydrogen chloride to provide (±)-N-norsalutaridine (3) and the latter was immediately N-methylated by Eschweiler-Clarke method to yield (±)-salutaridine (1)[3].

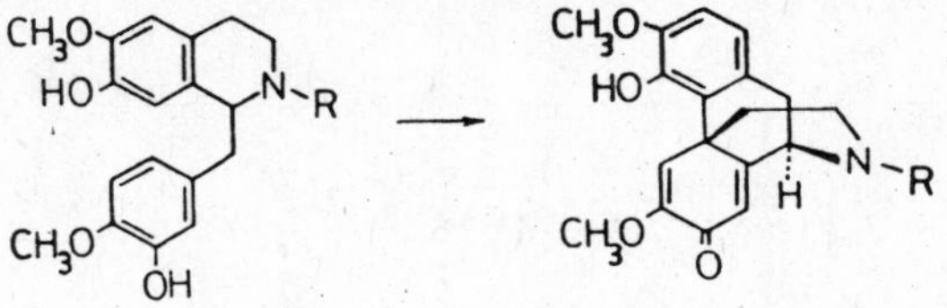

4; R=CHO

1; R = CH$_3$
2; R = CHO
3; R = H

PHARMACODYNAMIC INVESTIGATION

(±)-Salutaridine displaces ^{3}H-GABA from rat brain synaptic membranes in Tris-HCl buffer solution (IC_{50} = 9,2 ± 0.8 μM). The ^{3}H-GABA displacing activity of (±)-salutaridine was tested in bicarbonate-buffered physiological solution (Tyrode), too. This alteration leads to increased potency of (±)-salutaridine (IC_{50} = 0.66 ± 0.04 μM, Figure 1). Similarly, increased potency in Tyrode was established for the antagonist, bicuculline [4].

The enhancement of specific ^{3}H-diazepam binding by GABA agonists is inhibited by the antagonist, bicuculline [5]. The coupling of benzodiazepine binding sites to the GABA receptor (GABA/Benzodiazepine receptor complex), thus, offers an in vitro test system to differentiate agonists from antagonists. The influence of (±)-salutaridine on specific ^{3}H-diazepam binding was investigated. Increasing concentration of (±)-salutaridine significantly enhances specifically bound ^{3}H-diazepam; concentrations higher than μM, however, abolish the enhancement (Figure 2).

In conclusion, our in vitro data indicate (±)-salutaridine to be a partial agonist at the GABA/Benzodiazepine receptor complex.

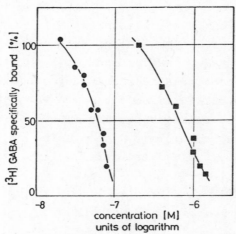

Figure 1. Displacement of ^{3}H-GABA specifically bound from rat brain synaptic membranes by GABA (●) and (±)-salutaridine (■) in Tyrode.

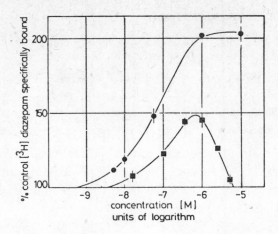

Figure 2. Influence of increasing concentration of GABA (●) and (±)-salutaridine (■) on ³H-diazepam binding to rat brain synaptic membranes in Tyrode buffer.

EXPERIMENTAL SECTION

1-Methyl-^{3}H-diazepam and 2,3-^{3}H-GABA were purchased from NEN, with radioactivities 64 and 29 Ci/mmol, respectively. Guvacine was a gift from Prof. P. Krogsgaard-Larsen. <u>Preparation of (±)-salutaridine</u>. (±)-N-Formylnorreticuline (<u>4</u>)(343 mg, 1.0 mmol) was dissolved in a mixture of dry dichloromethane (700 ml) and trichloroacetic acid (490 mg, 3.0 mmol) at -78 °C. Lead tetraacetate (443 mg, 1.0 mmol) was added to the mixture in two portion. The solution was stirred for additional 4 h at -78 °C and then kept overnight in refrigerator at -15 °C. The reaction mixture was treated with water (80 ml), the organic layer was separated and dried, finally the solvent was removed under reduced pressure. The remaining material was purified by column chromatography (Brockmann II-III aluminium oxide, dichloromethane-methanol 100:1 v/v solvent system) to supply (±)-<u>2</u> (71.6 mg, 21 %). (±)-N-Formylnorsalutaridine (<u>2</u>) (60 mg, 0.17 mmol) was dissolved in methanol (10 ml) and 18 % aqueous hydrogen chloride (2 ml) and kept at room temperature for 3 days. Methanol was removed under reduced pressure. The residue was basified with ammonium hydroxide then extracted with dichloro-

methane (4 x 10 ml). Removal of the solvent gave amorphous
(±)-N-norsalutaridine (3) (42 mg) which was dissolved in 98 %
formic acid (3 ml) and 38 % aqueous formaldehyde solution
(3 ml). The mixture was refluxed for 1 h, cooled, basified
with ammonium hydroxide, then extracted with dichloromethane
(5 x 10 ml). The organic layer was dried and evaporated. The
remaining material was purified by preparative thin layer
chromatography (Silica gel PF$_{254}$ coated plates, dichloromethane-methanol 150:12 v/v solvent system) to give
(±)-salutaridine (22 mg, 38 %, for spectral data of (±)-1 see
ref. [3]).

Binding assays. A three-times frozen-thawed and washed
synaptic membrane preparation was used [6,7]. Washing procedures were performed both with 50 mmol/l Tris-HCl (pH=7.1) and
bicarbonate-buffered physiological salt solution, Tyrode
(pH=7.1). Synaptic membrane preparations were freeze-dried.
Before use lyophilized membranes were reconstituted in the
appropriate buffer solutions. 2.74 nmol/l ^{3}H-GABA and
1.32 nmol/l ^{3}H-diazepam were incubated at 4 °C for 20 min.
In order to inhibit the binding of ^{3}H-GABA to uptake sites in
the presence of Na$^+$ ions [8](when Tyrode buffer was applied),
the specific uptake inhibitor, guvacine was used in 0.1 mmol/l
concentration [9]. The incubation was terminated by filtration
of 1000 μl samples through Wahtman GF/C glass fiber filters
and washed with (6 x 2.5 ml) ice-cold buffer [7]. Non-specific
binding was measured in the presence of 0.1 mmol/l GABA or
1 μmol/l diazepam. Radioactivity was determined in a liquid
scintillation spectrometer (Packard, model 2650; counting
efficiency 37 %). IC$_{50}$ values were determined as the
concentration which decreases the specific binding of ^{3}H-GABA
by 50 %.

REFERENCES

1. Stuart, K.L. Morphinandienone alkaloids, Chem.Rev., 71,
   47-72 (1971)
2. Szántay, Cs., Blaskó, G., Bárczai-Beke, M., Péchy, P. and
   Dörnyei, G. Studies aiming at the synthesis of morphine
   II. Studies on phenolic coupling of N-norreticuline
   derivatives, Tetrahedron Lett., 21, 3509-3512 (1980)
3. Szántay, Cs., Bárczai-Beke, M., Péchy, P., Blaskó, G. and
   Dörnyei, G. Studies aimed at the synthesis of morphine
   III. Synthesis of ($\pm$)-salutaridine via phenolic oxidative
   coupling of ($\pm$)-reticuline, J.Org.Chem., 47, 594-596
   (1982)
4. DeFeudis, F.V., Somoza, E. Density of GABA-receptors in
   rat cerebral cortex measured with bicuculline-methiodide
   in the presence of Na^+ and other inorganic ions,
   Gen.Pharmacol., 8, 181-187 (1977)
5. Braestrup, C., Nielsen, M., Krogsgaard-Larsen, P. and
   Falch, E. Partial agonists for brain GABA/benzodiazepine
   receptor complex, Nature /London/, 280, 331-333 (1979)
6. Chiu, T.H., Rosenberg, H.C. GABA receptor-mediated modula-
   tion of ^{3}H-diazepam binding in rat cortex, Eur.J.Pharmacol.
   58, 337-345 (1979)
7. Maksay, G., Kardos, J., Simonyi, M., Tegyey, Zs. and
   Ötvös, L. Specific binding of racemic oxazepam esters
   to rat brain synaptosomes and the influence of bioactivation by esterases, Arzneim.-Forsch./Drug Res., 31,
   979-981 (1981)
8. Churchill. L., Redburn, D.A. Effects of chaotropic anions
   on bicuculline inhibition of high affinity γ-aminobutyric
   acid binding in bovine retina and rat brain membranes,
   Neurochem.Internatl., 5, 221-226 (1983)
9. Krogsgaard-Larsen, P., Inhibitors of the GABA uptake systems, Mol.Cell.Biochem., 31, 105-121 (1980)

275

Bio-Organic Heterocycles
van der Plas H.C., Ötvös L., and Simonyi M. eds

ENZYMATIC *IN VITRO* REDUCTION OF KETONES. PART 13. HLAD-CATALYZED REDUCTION OF 3-PIPERIDONE DERIVATIVES TO PIPERIDINOLS WITH HIGH ENANTIOMERIC PURITY

VAN LUPPEN J.J., LEPOIVRE J.A., LEMIERE G.L. and
ALDERWEIRELDT F.C.

Laboratory of Organic Chemistry, University of Antwerp,
Groenenborgerlaan 171, B-2020 Antwerp, Belgium

In a previous paper (Van Luppen et al. 1984) the first reduction of a nitrogen heterocyclic ketone with a HLAD-NAD$^+$-ethanol system has been described. The reduction of 1-n-butyl-4-piperidone was shown to follow a reaction mechanism kinetically similar to the reduction of cyclohexanone. However, slow reduction rates were found, partly ascribed to the strong solvation of these substrates. It was proven that only the free amine form is reacting. Hence pH-values of the reaction medium were maintained around or above the pK$_a$-value of the piperidone. Previous findings are applied now in the preparative reduction of 1-substituted-3-piperidones.

THE ENZYMATIC REDUCTION.

The reaction conditions are shown in Scheme 1.

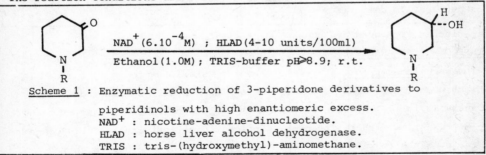

$$NAD^+(6.10^{-4}M) ; HLAD(4-10 \text{ units}/100ml)$$
$$Ethanol(1.0M) ; TRIS\text{-buffer pH}\geqslant 8.9; r.t.$$

Scheme 1 : Enzymatic reduction of 3-piperidone derivatives to

piperidinols with high enantiomeric excess.
NAD$^+$: nicotine-adenine-dinucleotide.
HLAD : horse liver alcohol dehydrogenase.
TRIS : tris-(hydroxymethyl)-aminomethane.

The results are summarized in Table 1.

Compared to cyclohexanones, 3-piperidones are reduced very slowly. The major cause can be the hydrophobic nature of the active site on the enzyme (Van Luppen et al. 1984). The different behaviour of substrates [7] and [8] is comparable to the reducibility of the corresponding bicycloalkanones. The large difference in reactivity between the latter compounds is fully explained by sterical arguments shown in the reaction model of the HLAD-reductions (Lemière et al 1982). Substrate [9] and also chlorohydrates of

Table 1 : Enzymatic reduction of 1-substituted-3-piperidones.

1-substituent	substrate concentration (M)	time of reaction	yield of purified alcohol	% enantiomeric excess	k* (rel)
[1] $-CH_3$	0.1	54 hours	10	93 ± 3	1.9
[2] $-CH_2CH_3$	0.1	4 weeks	45	~100	1.4
[3] $-CH_2CH_2CH_3$	0.18	5 weeks	57	~100	1.7
[4] $-COOC_2H_5$	0.02	8 weeks	22	~100	0.05
[5] $-CO(CH_2)_3CH_3$	0.02		analyt. exp.		1.6
[6] $-CO\emptyset$	0.02		analyt. exp.		0.3
[7] CH_3-N ... $N\equiv C-$ (±)	0.04	5 days	49	89 ± 3 (2S)-2-exo	10.3
			7	17 ± 3 (2R)-2-exo	2.1

[8] CH_3-N (structure)   no reduction

[9] (structure) N^+ CH_3 CH_3 no reduction

* Relative rate values are obtained from reaction times needed for a 30% reduction. In comparison : $k_{(cyclohexanone)}$ is about 10^3.

piperidones are found to deactivate the enzyme.

STEREOCHEMICAL ANALYSIS.

- The enantiomeric purity is determined by esterification of the alcohols with Mosher's reagent to the corresponding (R)-MTPA esters and further analysis in ^{19}F-NMR spectroscopy with the use of lanthanide shift reagent (Merckx et al. 1983). The high enantiomeric selectivity is fully comparable to the results obtained in the reduction of the corresponding cycloalkanones.

- Absolute configuration determinations were possible, mainly by correlation with an already described product (3S)-3-hydroxypiperidine (Ringdall et al. 1981). The alkylated (3S)-compounds are compared with the enzymatically obtained products via their MTPA-esters and the above mentioned ^{19}F-NMR technique. For 1-ethoxycarbonyl-3-hydroxypiperidine no differentiation between (3S)- and (3R)-forms could be made in this way. Its absolute configuration is determined by chemical correlations as

shown below

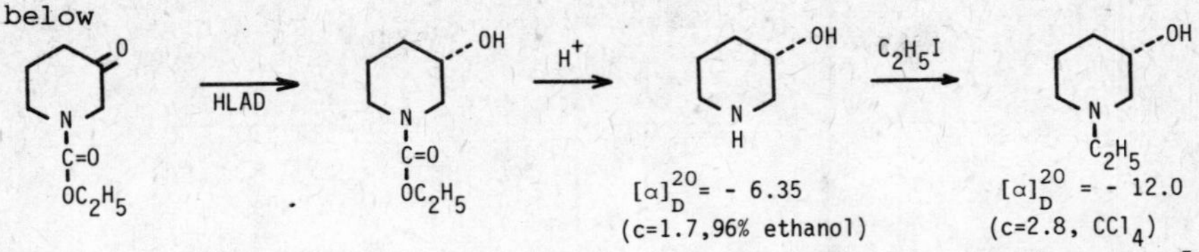

$[\alpha]_D^{20} = -6.35$
$(c=1.7, 96\% \text{ ethanol})$

$[\alpha]_D^{20} = -12.0$
$(c=2.8, CCl_4)$

For the absolute configuration of the alcohols derived from substrate [7]
no direct proof is given. However, according to the reaction model of the
HLAD-catalyzed reduction (Lemière et al. 1982), a (2S)-exo configuration
is expected to be the major compound while also a (2R)-endo compound can
be formed at a smaller rate.

- The conformational analysis of 1-_n_-propyl-3-hydroxypiperidine is already
described (Van Luppen et al. 1982). It is shown that in _n_-heptane
solution the conformation with an axial hydroxyl group is present in 69%.
In aqueous solution, the same conformation predominates in the protonated
form (56% at pH=5) while only 43% is present at pH=11.

SIDE REACTIONS.

In aqueous solution, as shown, acid-base and keto-diol equilibria are found
(Scheme 2). Measurements obtained from ^{1}H- and ^{13}C-NMR spectroscopy are
given in Table 2.

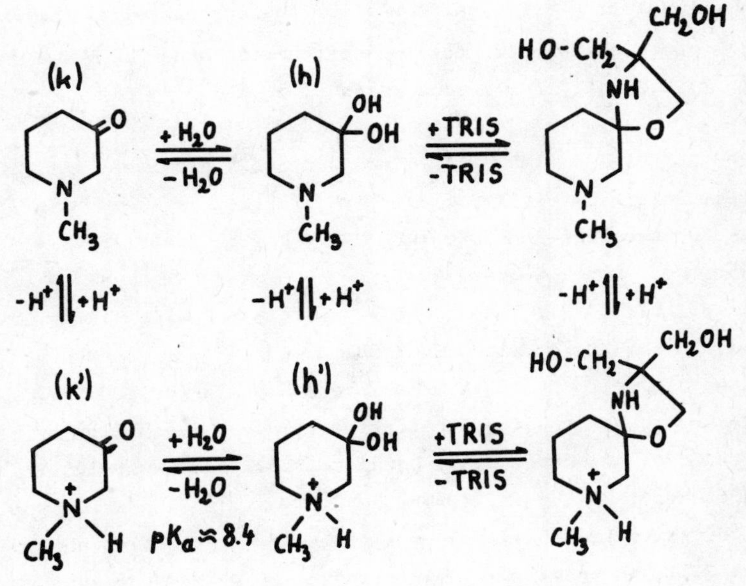

Scheme 2 : Side reactions on the substrate during the enzymatic reduction.

19*

Table 2 : Acid-base equilibria and keto-diol equilibria in aqueous
solution, studied with ^{1}H- and ^{13}C-NMR spectroscopy.

pH [a]	10.94	9.68	9.11	8.60	8.10	7.42	4.50
%k	70	70	54	38	19	7	-
%k'	-	-	7	19	13	7	4
%(h+h') [b]	30	30	39	43	68	86	96

[a] A 2.5 M solution of 3-piperidone is titrated with HCl at 36°C
The symbols k, k', h and h' correspond to the structures
shown in Scheme 2.

[b] Signals of (h) and (h') forms are always coinciding.

At pH above 8.6 the unprotonated ketone is the major form. Similar results
are found for 4-piperidone derivatives (Van Luppen et al. 1979). Therefore
TRIS-buffer (pH=8.9) is suitable for the enzymatic reduction, especially
because it is known that the enzyme is more stable in this type of buffer
than in a phosphate-type. However, we have recently found that TRIS-
molecules can form adducts with activated carbonyl compounds; structure is
proved for adducts with 3-cyanocyclohexanone (Willaert et al. 1984) and
1-n-butyl-4-piperidone (Van Luppen - unpublished results). Similar oxazolidine derivatives, shown in Scheme 2, are probably formed with 3-piperi-
dones but no attempt was made yet to isolate these compounds.

ACKNOWLEDGMENTS.

The authors thank the "Nationaal Fonds voor Wetenschappelijk Onderzoek" for
financial support of this project. One of the authors (J.J.Van Luppen)
thanks the "I.W.O.N.L." for a scolarship.

REFERENCES.

- Lemière G.L., Van Osselaer T.A., Lepoivre J.A., Alderweireldt F.C.,
  J.Chem.Soc.Perkin Trans. II, 1123-1128 (1982).

- Merckx E.M., Lepoivre J.A., Lemière G.L., Alderweireldt F.C.,
  Org. Magn. Res. 21, 380-387 (1983).

- Ringdall B., Ulrich F., Ohnsorge F.W., Cymerman C.J.,
  J.Chem.Soc.Perkin Trans. II, 697 (1981).

- Van Luppen J.J., Lepoivre J.A., Lemière G.L., Alderweireldt F.C.,
  Heterocycles ; accepted for publication (1984).

- Van Luppen J.J., Lepoivre J.A., Dommisse R.A., Alderweireldt F.C.,
  Org. Magn. Res. 12, 399-405 (1979).
  ibidem 18, 199-206 (1982).

- Willaert J.J., Lemière G.L., Dommisse R.A., Lepoivre J.A.,
  Alderweireldt F.C., Bull.Soc.Chim.Belg. 93, 139-149 (1984).

Bio-Organic Heterocycles
van der Plas H.C., Ötvös L., and Simonyi M. eds

OXIDATION OF N-ALKYL AND N-ARYL AZAHETEROCYCLES BY FREE AND IMMOBILIZED RABBIT LIVER ALDEHYDE OXIDASE

ANGELINO S.A.G.F., BUURMAN D.J., van der PLAS H.C. and MÜLLER F.

Laboratory of Organic Chemistry, Agricultural University, De Dreijen 5, 6703 BC Wageningen, The Netherlands

SUMMARY

Immobilized rabbit liver aldehyde oxidase can be profitably used for small scale oxidations of N-substituted azahetero-cycles. The site of oxidation is determined by the size and the conformation of the N-substituent, whereas the rate of oxidation is strongly influenced by electronic effects.

Objectives

The oxidation of azaheterocycles is usually difficult because the aromatic rings are π-deficient, making electron-donation to an oxidizing agent less feasible or impossible. Aldehyde oxidase (AO) isolated from rabbit liver efficiently catalyzes the oxidation of many of these electron-deficient heteroarenes. The enzyme preferentially oxidizes azaheterocyclic compounds, substituted at the ring nitrogen with an alkyl or aryl group. Since it was reported for AO that oxidation can occur at two different sites with some substrates, we investigated whether regiospecific synthesis is feasible applying immobilized enzyme. In addition we tried to elucidate important oxidation site-determining factors for AO.

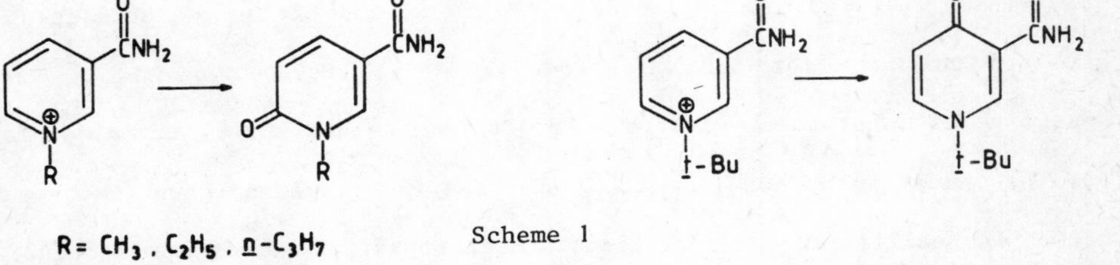

$R = CH_3 , C_2H_5 , \underline{n}-C_3H_7$

Scheme 1

Results

The oxidation of 1-alkyl-3-aminocarbonylpyridinium chlorides (Scheme 1)
with free or immobilized AO showed that highly regiospecific reaction occurred (1). The 1-methyl, 1-ethyl and 1-n-propyl derivatives gave exclusively formation of the 6-oxo compound, while the 1-t-butyl analogue
yielded the 4-oxo compound as sole product.

The results indicate that the enzyme-mediated reaction is susceptible to
steric effects of the substituent at N-1. Determination of the maximum
rates of oxidation and of the Michaelis constants K_M presented evidence
for the existence of a hydrophobic site in the vicinity of the active
centre of the enzyme. Moreover, the bulkiness of the t-butyl group at N-1
influences the binding orientation in such a way that attack at C-6 by the
enzymic nucleophilic species is sterically more hindered than attack at
C-4. This hypothesis is supported by inhibition experiments with the
1-ethyl, 1-n-propyl and 1-t-butyl compounds (1).

The oxidation of 3-aminocarbonyl-1-arylpyridinium salts (Scheme 2) by AO
is somewhat less regiospecific than that of the analogous 1-alkyl derivatives (2). The 6-oxo products are obtained predominantly, but also 4-oxo

X = H , CH₃ , F , Cl, OH   Scheme 2

compounds are formed as minor products. On the other hand, 3-aminocar-
bonyl-1-mesitylpyridinium chloride yielded exclusively 4-oxo product. It
appeared that the rate of oxidation at C-6 of these compounds is very
sensitive to substituent effects. A positive ρ-value (= 3.6) is calculated
for the oxidation by AO, indicating that the rate-limiting step in the reaction is facilitated by a low electron-density at the oxidation site (2).

1-Alkyl(aryl)quinolinium chlorides (Scheme 3) are also oxidized by AO essentially at two positions, C-2 and C-4 (3). Regiospecificity is

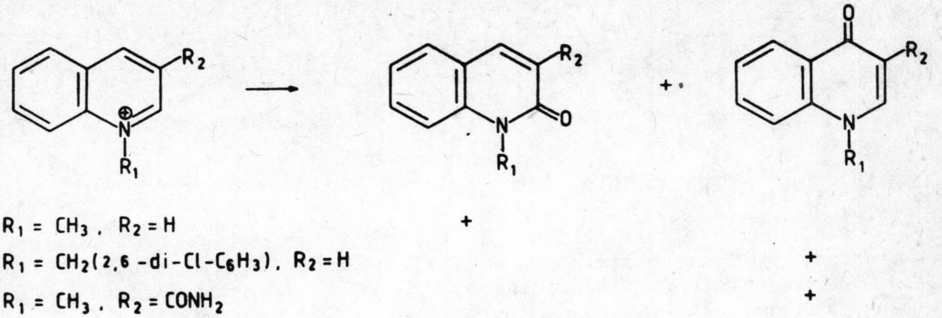

$R_1 = CH_3$, $R_2 = H$

$R_1 = CH_2(2,6-di-Cl-C_6H_3)$, $R_2 = H$

$R_1 = CH_3$, $R_2 = CONH_2$

Scheme 3

complete for 3-aminocarbonyl-1-methylquinolinium chloride and 1-(2,6-dichlorobenzyl)quinolinium chloride, which are both only converted to their corresponding 4-oxo product. The 1-methylquinolinium salt yielded predominantly 2-oxo product and some of the 4-oxo compound. It is of interest to note that the presence of an aminocarbonyl group at C-3 directs the oxidation for R_1 = methyl from the preferred site C-2 to C-4. Oxidation of N-methyl- and N-benzylpyrimidin-2- and -4-ones (Scheme 4) by AO resulted exclusively in the formation of the corresponding uracil derivatives (4).

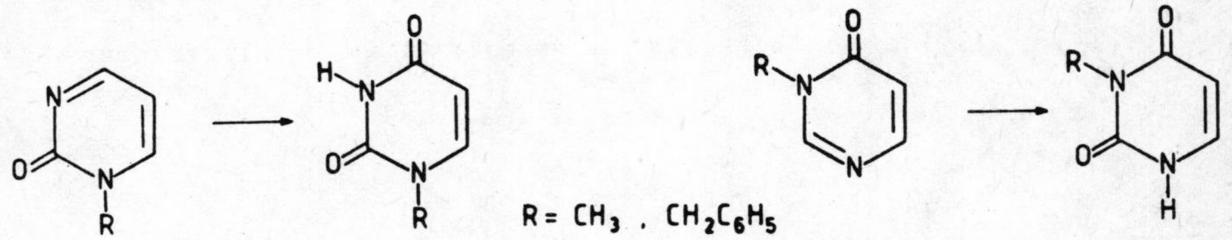

$R = CH_3$, $CH_2C_6H_5$

Scheme 4

No formation of isomeric products or change in the site of oxidation due to the size or conformation of the N-substituent, was established.

A very simplified schematic representation of the orientation of two different azinium substrates in the catalytic centre of AO is illustrated in Fig. 1, based on the three important features of the active site suggested in this study (1-3).

283

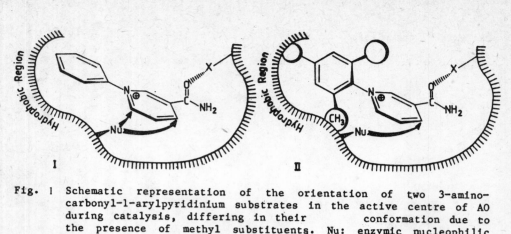

Fig. 1 Schematic representation of the orientation of two 3-amino-
carbonyl-1-arylpyridinium substrates in the active centre of AO
during catalysis, differing in their   conformation due to
the presence of methyl substituents. Nu: enzymic nucleophilic
species; X: interactive species.

The immobilization of AO to several supports by various methods was
investigated in order to make continuous operation feasible and to improve
the enzyme stability (5). This study showed that of the various matrices
and coupling methods tested, the activity of AO is best retained upon
adsorption to n-alkylamine-substituted Sepharose 4B and diethylaminoethyl
(DEAE) Sepharose CL 6B. The operational stability of immobilized AO
increased substantially compared to that of free enzyme. Inactivation of
AO during turnover is dependent on the rate of oxidation. A positive
effect on the operational stability and the corresponding productivity is
found for AO/n-alkylamine Sepharose 4B preparations by increasing the
amount of enzyme per unit weight of support (5).

References

1) S.A.G.F.Angelino, D.J.Buurman, H.C.van der Plas and F.Müller, Recl.
Trav.Chim.Pays-Bas, 101, 342-346 (1982).

2) S.A.G.F.Angelino, D.J.Buurman, H.C.van der Plas and F.Müller, Recl.
Trav.Chim.Pays-Bas, 102, 331-336 (1983).

3) S.A.G.F.Angelino, B.H.van Valkengoed, D.J.Buurman, H.C.van der Plas and
F.Müller, J. Heterocyclic Chem., 21, 107-112 (1984).

4) S.A.G.F.Angelino, D.J.Buurman, H.C.van der Plas and F.Müller, J.
Heterocyclic Chem., in press (1984).

5) S.A.G.F.Angelino, F.Müller and H.C.van der Plas, Biotechnol.Bioeng.,
submitted.

Bio-Organic Heterocycles
van der Plas H.C., Ötvös L., and Simonyi M. eds

OXIDATION OF N-ALKYL(ARYL)PYRIDINIUM SALTS BY FREE AND IMMOBILIZED RABBIT LIVER ALDEHYDE OXIDASE

ANGELINO S.A.G.F. and van der PLAS H.C.

Laboratory of Organic Chemistry, Agricultural University, De Dreijen 5, 6703 BC Wageningen, The Netherlands

1 INTRODUCTION

The use of enzymes in organic synthesis has been a research subject in the Department of Organic Chemistry at the Agricultural University of Wageningen since 1975. A group of hydroxylases, belonging to the class of oxido-reductases were chosen as model enzymes because of their broad substrate specificities towards azaheterocycles. The study of the chemistry of these compounds is a major topic in this laboratory (1-5). Representatives of this group of enzymes, xanthine oxidase and xanthine dehydrogenase, have been employed succesfully, both in free and immobilized form, in laboratory scale oxidations of azaheterocyclic compounds (6-11). In general, the hydroxylation of azaheterocycles by aldehyde oxidase, another hydroxylase within this group of enzymes, has been investigated far less extensively. It has been reported that aldehyde oxidase, isolated from rabbit liver, catalyzes the hydroxylation of purines, pteridines, pyrimidines, other heterocyclic nitrogenous compounds and aliphatic, aromatic or heteroaromatic aldehydes (12-16). Quaternization of an azine usually leads to an enhanced reactivity of the heterocycle

285

towards aldehyde oxidase, as can be exemplified by the higher reactivity of 3-aminocarbonyl-1-methylpyridinium chloride compared to 3-pyridinecarboxamide (13). There exists, however, little information on the factors determining the site of oxidation in N-alkyl(aryl)azinium salts. In view of our general interest to study the relation between structural requirements of organic compounds and their reactivity towards enzymes, we wanted to elucidate the effect of steric and electronic factors in N-alkyl- and N-arylpyridinium salts with regard to the site of oxidation. In context with this study we also investigated whether immobilized aldehyde oxidase could be profitably applied in synthetic organic chemistry.

2 OXIDATION OF N-ALKYL(ARYL)PYRIDINIUM COMPOUNDS BY ALDEHYDE OXIDASE

2.1 1-Alkyl-3-aminocarbonylpyridinium chlorides

The products of the oxidation of 1-alkyl-3-aminocarbonyl-pyridinium chlorides 1 (Scheme 2.1) by aldehyde oxidase were analyzed with HPLC, utilizing conditions in which separation of the possible products 2 - 4 could be achieved (17).

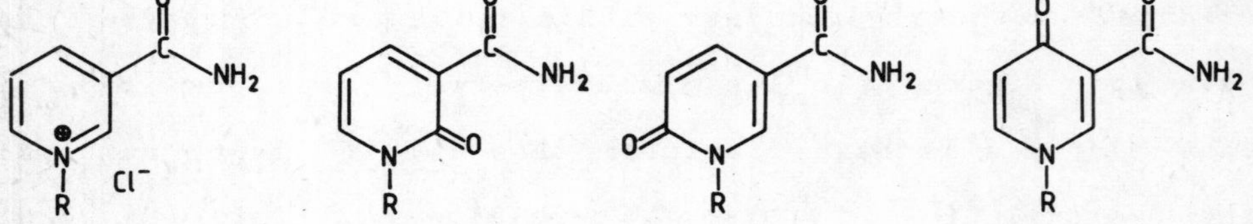

R= a: CH_3 , b: C_2H_5 , c: $\underline{n}-C_3H_7$, d: $\underline{i}-C_3H_7$, e: $\underline{t}-C_4H_9$

Scheme 2.1

It was found that the 1-methyl, 1-ethyl and 1-n-propyl salts
1a-c gave the corresponding 1-alkyl-1,6-dihydro-6-oxo-3-
pyridinecarboxamides (2a-c). The enzymic oxidation of 3-
aminocarbonyl-1-i-propylpyridinium chloride (1d) yielded two
products; from 3-aminocarbonyl-1-t-butylpyridinium chloride
(1e) only a single product was acquired. Since no reference
compounds of any of the 1-i-propyl- or 1-t-butyldihydro-oxo-3-
pyridinecarboxamides are known, we prepared these products on a
small scale using immobilized aldehyde oxidase in order to have
enough material to determine their structures.

It was established by using UV-spectrometry at different pH,
^{1}H NMR and mass spectrometry that the oxo compounds obtained
from 1d are the isomeric 6-oxo compound 3d and 4-oxo compound
4d. The compound being exclusively formed from 1e, is the
4-oxo product 4e (17).

From the peak heights in the HPLC pattern of the reaction
mixture obtained from 1d a product ratio 3d/4d of approximately
3.5:1 was calculated.

The results of our experiments indicate that increasing the
size of the 1-alkyl group enhances the steric hindrance to
oxidation at position 6 of the ring, making oxidation at
position 4 more favourable or the only alternative. In case of
the 1-t-butyl group, oxidation at C-4 seems the only possible
reaction.

Experiments using 3-aminocarbonyl-1-methylpyridinium salts 5 as
substrates (Scheme 2.2) showed that methyl substituents at
position 6 and/or 4 prevent the formation of dihydro-oxo-3-
pyridinecarboxamides: 5a and 5c gave no product and 5b
yielded, in a very slow reaction, a trace of a single product

287

(as observed by HPLC) which is probably the corresponding 6-oxo compound (17).

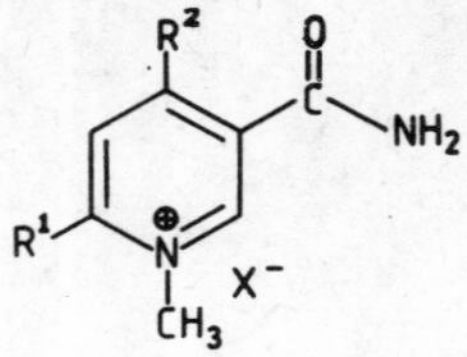

a: $R^1 = CH_3$, $R^2 = H$, $X = Cl^-$
b: $R^1 = H$, $R^2 = CH_3$, $X = I^-$
c: $R^1 = CH_3$, $R^2 = CH_3$, $X = I^-$

5

Scheme 2.2

It has been suggested (18) that the formation of the 4-oxo derivatives in the enzymic reaction could be due to a different binding orientation of the substrate. We examined this hypothesis more closely by determining the kinetic constants V and K_M for some of these substrates. Assuming simple Michaelis-Menten kinetics (19), we obtained the results summarized in Table 2.1. As one can see, the maximum rate V drastically drops to about 6% when the 1-alkyl substituent is changed from methyl to ethyl to n-propyl. The very low rates for 1d and 1e (the oxidation rate for 1d is about 7% of that of 1c at a substrate concentration of 0.8 mM) prevent accurate kinetic assays with these compounds.

The considerable drop in oxidation rate observed when comparing 1a, 1b and 1c cannot be explained simply by a decreased reactivity towards nucleophilic attack by an enzymic species (see section 2.2) resulting from the somewhat more electron-donating character of the ethyl and n-propyl substituent; the effective size of these groups must also play a role.

The Michaelis constant K_M was found to decrease with increasing size of the alkyl substituent suggesting that the association of the enzyme-substrate complex for compounds 1b and 1c

288

increases with larger hydrophobic substituents. These results point to the existence of a hydrophobic site in the vicinity of the active centre of rabbit liver aldehyde oxidase; this proposal parallels that of Baker et al. (20) for bovine milk xanthine oxidase. To obtain further support for this proposal, we performed inhibition experiments utilizing 1b-e as inhibitors in the oxidation reaction of 1a. The inhibition constants K_i (19) are given in Table 2.1. We established competitive inhibition which increased with the size of the alkyl substituent at position 1, pointing to a higher affinity

Table 2.1 Kinetic data for the oxidation of compounds
1a-e by free aldehyde oxidase at pH 9.0.

Compound	K_M^a	V^b	$K_i^{a,c}$
1a	310 ± 14	0.330 ± 0.020	
1b	100 ± 6	0.020 ± 0.002	124 ± 6
1c	14 ± 1	0.020 ± 0.003	21 ± 2
1d	–	–	100 ± 6
1e	–	–	39 ± 3

a. In μmol/l.
b. In μmol/min.mg.
c. The small contributions of the rate of the oxidation of the inhibitors to the oxidation rate of 1a were neglected.

However, the competitive inhibition decreased on branching (K_i (1d) and K_i (1e) versus K_i (1c)), although 1e gave rise to greater inhibition in this reaction as did 1d. This result seems to support the hypothesis that, due to affinity of the hydrophobic site close to the active centre of the enzyme and

to the size of the t-butyl substituent, the binding orientation
has changed in such a way that attack of the enzymic nucleophilic species is only possible at the C-4 position.

2.2   3-Aminocarbonyl-1-arylpyridinium chlorides

Oxidation of 3-aminocarbonyl-1-arylpyridiniumchlorides (6a-f)
(Scheme 2.3) with aldehyde oxidase gave as major products the
corresponding   1-aryl-1,6-dihydro-6-oxo-3-pyridinecarboxamides
(7a-f), together with the 4-oxo isomers 8a-f as minor products.

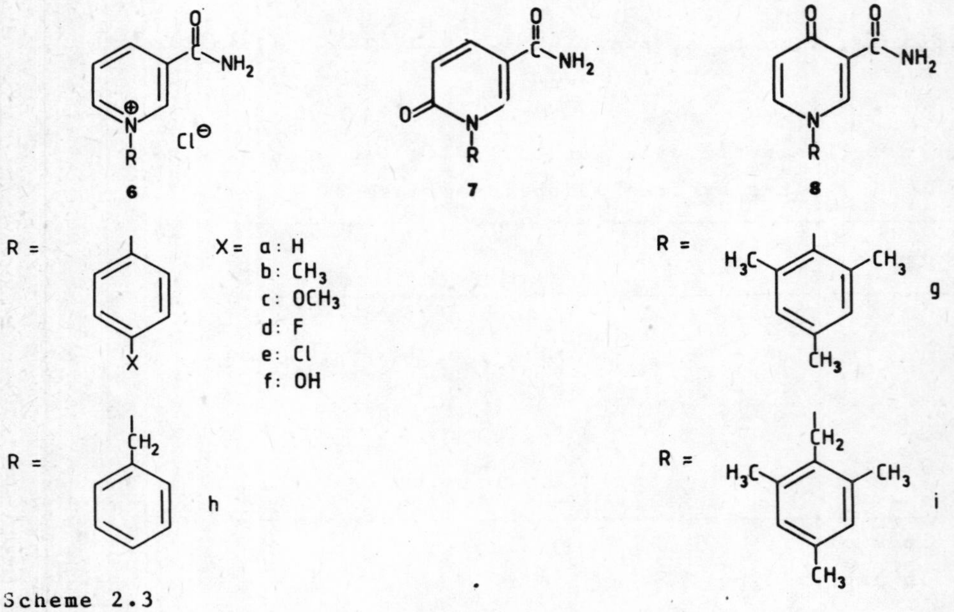

Scheme 2.3

In contrast the substrates 6g-i gave only one product: the 4-
oxo product 8g from compound 6g, whereas substrates 6h-i both
yielded 6-oxo products (7h-i). The structure assignment of the
oxidation products was based upon ^{1}H NMR data and upon UV
spectroscopy (21).

Kinetic data obtained with free aldehyde oxidase at pH 9.8 are
collected in Table 2.2. The logarithmic plot of the relative

maximal oxidation rates at the C-6 position by aldehyde oxidase versus the substituent constant σ of X is shown in Figure 2.1 for substrates 6a-e.

Table 2.2   Maximum rates for the oxidation of 3-aminocarbonyl-
1-arylpyridinium   chlorides by free aldehyde
oxidase at pH 9.8

Substrate	Aldehyde oxidase	
	V_{6-oxo}[a]	V_{4-oxo}[a,b]
6a	0.43 ± 0.02	–
6b	0.151 ± 0.009	–
6c	0.058 ± 0.008	0.023 ± 0.003
6d	1.46 ± 0.23	–
6e	2.92 ± 0.53	–
6g		0.56 ± 0.03
6h	0.050 ± 0.003	
6i	0.133 ± 0.007	
1a	0.48 ± 0.02	

a. In μmol/min mg.
b. No accurate data could be determined for substrates 6a, 6b, 6d and 6e.

A reaction constant ρ of approximately 3.6 can be calculated from the slope of the line. This large positive reaction constant indicates that the rate-limiting step in the reaction is facilitated by a low electron density at the reaction site and is very sensible to substituent effects. For the addition of methoxide ions to the analogous 1-(3- or 4-substituted phenyl)pyridinium salts at C-6 (or C-2) at ρ-value of 3.56 was found (22), suggesting that a similar attack by an oxygen-containing nucleophile occurs in the oxidation mechanism of aldehyde oxidase.

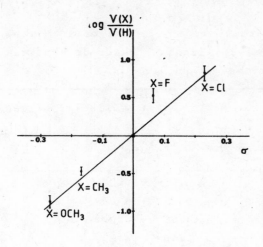

Figure 2.1   Hammet plot for the oxidation at the C-6 position
             of 3-aminocarbonyl-1-arylpyridinium chlorides.
             Oxidation by free aldehyde oxidase at pH 9.8.

Since no kinetic and chemical model for the catalysis by rabbit liver aldehyde oxidase has yet appeared in the literature, we adapted the model of Olson et al. (23) for bovine milk xanthine oxidase, which seems to be a correct transaction because of the close structural relationship (24-28).

In our model a mechanism is proposed in which the substrate is attacked by an oxygen-containing nucleophile (Nu) in the Michaelis-complex I (Figure 2.2). Subsequent rehybridization of the sp^3-carbon to a sp^2-carbon by proton abstraction and electron transfer takes place (II), and finally the O-linkage to the reaction intermediate is hydrolyzed by a water molecule with release of the product (III). The nature of the enzymic nucleophile was originally suggested to be a persulfide, but more recent investigations indicated that the nucleophile could possibly be an oxo ligand of molybdenum (29, 30). It is still not clear whether the electron transfer actually occurs as a coupled proton/electron transfer (as indicated in II) or as a hydride transfer (12). Also the exact nature of the proton acceptor is not yet known (12, 30, 31).

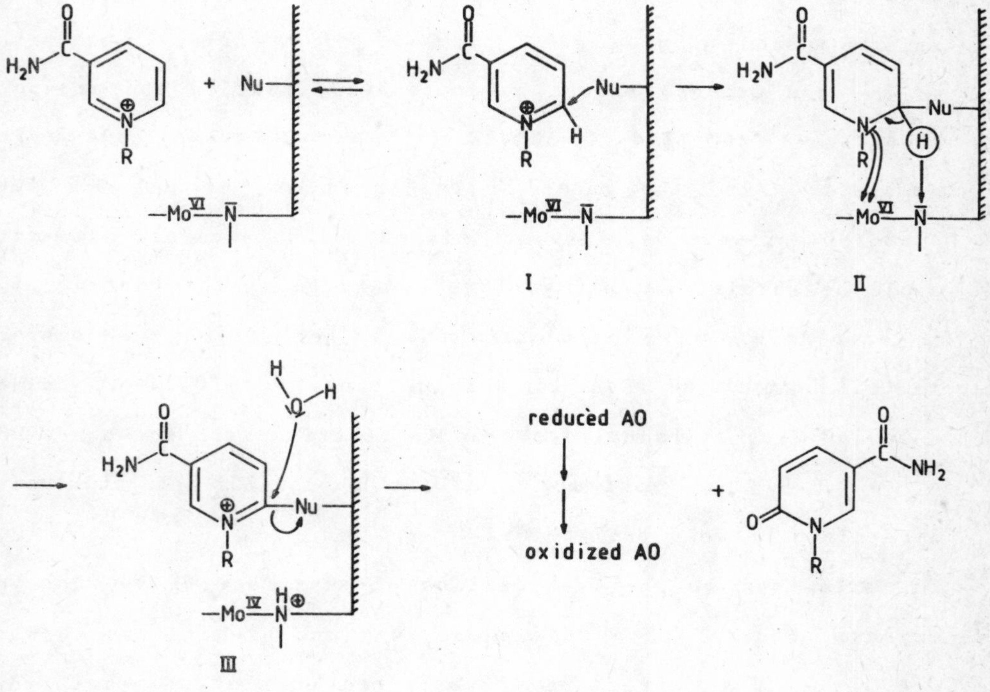

Figure 2.2. Representation of the oxidation of 1-alkyl(aryl)-3-aminocarbonylpyridinium chloride at C-6 by aldehyde oxidates (AO) in a model adapted from Olson et al. (23). Complex II was introduced as additional intermediate complex to make the model chemically more feasible.

Considering the data collected in Table 2.2, it is shown that for substrate 6c having a p-methoxy substituent, the maximum rate of formation of the 4-oxo product is about 2.5 times lower than that for oxidation at C-6.

It is very unlikely that this increase in the rate of oxidation at C-4 arises from the electron-donating character of the para substituent in the phenyl ring. This is based on the argument that incubation of substrate 6f, containing the stong electron-donating hydroxy substituent (at pH 9.8 the hydroxy group in this compound, having a pK_a = 9.25 ± 0.05 (21), is ionized to a large extent) showed no increased formation of the 4-oxo product as compared to that of the substrates 6a-e (HPLC analysis). Also the rates of oxidation at C-4 and C-6 with substrate 6f were very low at this pH and no accurate rate data could be obtained. Testing of substrate 6a in the pH-range 6.4 to 10.2 gave, after correction for buffer effects, a constant maximal oxidation rate in the pH range 7.5-10.2. At values below pH 7.5, a gradual decrease in reaction rate was observed. At pH 6.4, the maximal rate was about 75% of the value determined between pH 7.5-10.2.

The oxidation at the C-4 position of substrate 6g by aldehyde oxidase occurred at a rate about 24 times higher than that at C-4 in 6c. This effect may be ascribed to a strong preference of 6g for a conformation in which both aromatic rings are out of plane, assuming that this conformation will be partially retained in the enzyme-substrate complex since it is energetic-ally most favourable. As a consequence, the orientation of the mesityl group in the hydrophobic pocket of the enzyme may force

294

the enzymic nucleophilic species in such a position that it can only attack at C-4.

The maximal oxidation rate at the C-6 position of substrate $6h$ is rather low in comparison to the rate of oxidation for compound $6a$; compared to the oxidation of 3-aminocarbonyl-1-methylpyridinium chloride $1a$, the rate is only 10%. As previously noted for a number of alkyl substituents, this effect is probably mainly due to steric hindrance, caused by the N-1 substituent, towards nucleophilic attack. The effect is not present or masked for compounds $6a-f$ by the overriding electronic effects on the oxidation site. In this way the fact that the oxidation rates for $6h$ and $6i$ differ by a factor of 2.5 can be explained by the disparity in accessibility of the C-6 position in both substrates.

This phenomenon is illustrated by the ^{1}H NMR data (Table 2.3) of compounds $6h$ and $6i$. All the protons in the pyridinium ring of compound $6i$ show upfield shifts compared to those in $6h$. Calculation of the net upfield shifts of H-2 and H-6 in relation to H-4 or H-5 result in significant values for both protons. These net upfield shifts may be rationalized by assuming a shielding effect of the 2,4,6-trimethylbenzyl group. Two extreme conformations, in which both ring planes A and B face each other (I) or are at right angles (II), are shown in Figure 2.3 (32). These conformations are transformed into one another by a rotation of ring B 90° about the C-N bond.

As a consequence of steric hindrance it is to be expected that, especially for R = CH_3, conformation II will be more favourable than I, leading to the shielding effect of H-2 or H-6 of the

Table 2.3 Chemical shifts of the pyridinium ring protons of compounds 6h and 6i in D$_2$O

Hydrogen	Compounds		
	6h	6i	$\Delta \delta^a$
H$_2$	9.52	9.09	
H$_4$	9.07	8.93	
H$_5$	8.39	8.19	
H$_6$	9.29	8.80	
H$_2$ - H$_4$	0.45	0.16	0.29
H$_2$ - H$_5$	1.13	0.90	0.23
H$_6$ - H$_4$	0.22	-0.13	0.35
H$_6$ - H$_5$	0.90	0.61	0.29

a. Upfield shifts relative to 6h.

pyridinium ring by the 2,4,6-trimethylbenzylgroup. The upfield shifts are in agreement with data which have been found by Verhoeven et al. (32) for 1-benzylpyridinium derivatives ($\Delta \delta$ = 0.25 ppm). It is concluded that the steric conformation of substrate 6i tends to approximate structure II, which we presume to be partially retained in the enzyme-substrate complex. Thus, in 6i, C-6 is more accessible towards nucleophilic attack than it is in compound 6h. Compared to 6g, the interaction of the aryl substituents of 6h and 6i with the hydrophobic site of the enzyme (17) results in oxidation at C-6 because of the different stereochemical distortion induced by the methylene group. These results support our assumption that steric factors control the position of nucleophilic attack in the reaction pathway of the oxidation of these compounds by aldehyde oxidase.

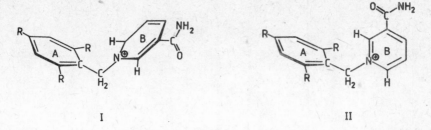

I   II

Figure 2.3   Two conformations of compounds 6h and 6i resulting from rotation about the C-N bond.

2.3 Discussion

In the previous sections 2.1 and 2.2 results were presented on the catalytic activity of rabbit liver aldehyde oxidase towards 1-alkyl(aryl)-3-aminocarbonylpyridinium salts, in which the alkyl(aryl) substituents differ in size and/or electronic effects. Comparison of substrate specificities of aldehyde oxidase and bovine milk xanthine oxidase towards 1-alkyl(aryl)-pyridinium salts (Scheme 2.4) shows quite distinct differences for both enzymes (21, 33).

It is obvious that aldehyde oxidase possesses a broader specificity towards these compounds than xanthine oxidase (Table 2.4). Moreover, xanthine oxidase only significantly converts N-alkyl(aryl)azinium salts at pH > 9.0 (34), whereas aldehyde oxidase can operate in a wider pH-range, viz. 6.5-11.0 (21, 35). The variation in size and thus steric conformation of the N-substituent restricts the oxidation of these compounds by xanthine oxidase more than by aldehyde oxidase. Substitution of the pyridinium ring at other positions than at the ring nitrogen atom affects the substrate specificity of xanthine oxidase to a much lesser degree (34,36).

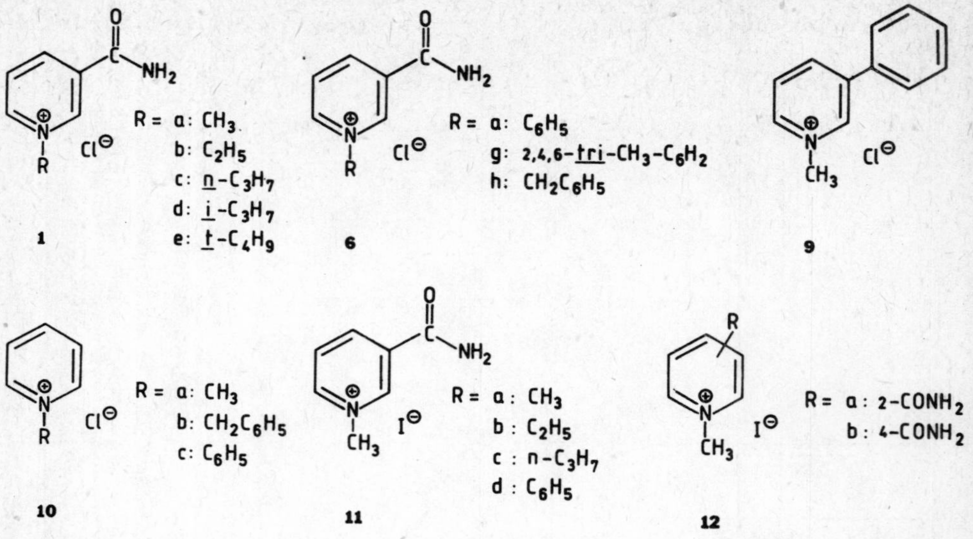

Scheme 2.4.

1 R = a: CH_3
b: C_2H_5
c: n-C_3H_7
d: i-C_3H_7
e: t-C_4H_9

6 R = a: C_6H_5
g: $2,4,6$-tri-CH_3-C_6H_2
h: $CH_2C_6H_5$

9

10 R = a: CH_3
b: $CH_2C_6H_5$
c: C_6H_5

11 R = a: CH_3
b: C_2H_5
c: n-C_3H_7
d: C_6H_5

12 R = a: 2-$CONH_2$
b: 4-$CONH_2$

Table 2.4   Comparison of the substrate specificities of aldehyde oxidase (AO) and xanthine oxidase (XO) for various N-alkyl(aryl)pyridinium salts

Compound	Oxidation		Site of oxidation[a]	
	AO	XO	AO	XO
1a	+	+[b]	6	6[b]
1b	+	+	6	6
1c	+	−	6	−
1d	+	−	6(4)	−
1e	+	−	4	−
6a	+	+	6(4)	4(6)
6g	+	−	4	−
6h	+	−	6	−
9	+	−[c]	6	−[c]
10c	+	+[b]	4	4[b]

a. Site of oxidation of minor product in parentheses.
b. In agreement with reference (34).
c. Reference (36).

298

As an extension of this work on the 1-alkyl(aryl)-3-amino-
carbonylpyridinium salts we also investigated the oxidation of
pyridinium derivatives by aldehyde oxidase in dependence of the
presence, the position and the nature of the C-substituent.
Semi-quantitative results of these reactions are collected in
Table 2.5.

Table 2.5 Oxidation of various pyridinium compounds by
aldehyde oxidase at pH 7.8

Compound	Oxidation rate[a]	Site of oxidation
1a	+++	6
2	+	6
10a	-	-
10b	-	-
10c	+	4
11a	++++	6
11b	++++	6
11c	++++	6
11d	++++	6
12a	-	-
12b	+	b

a. Rate of oxidation: ++++ = fast, +++ = moderate, ++ = slow,
   + = very slow, - = no oxidation.
b. Not determined.

1-Methylpyridinium derivatives which contain a 3-alkanoyl-
(benzoyl) substituent (11a-d) are found to be oxidized at a
higher rate than 3-aminocarbonyl-1-methylpyridinium chloride
1a; the oxidation of these compounds occurs exclusively at C-6.
These results show that the enzyme has a great steric
'tolerance' at the C-3 site. The presence of a phenyl group

instead of a benzoyl group at C-3 (i.e. in compound 2) results in a low rate of oxidation by aldehyde oxidase, indicating the importance of the carbonyl moiety. This moiety most probably orientates the substrate molecule in the catalytic centre of the enzyme through interaction with a (proton-donating) active site species (35). Another indication, showing the importance of a substituent at C-3 is obtained by studying the reaction of pyridinium salts 10, which contain no substituent at C-3. Compounds 10a and 10b are not converted at all and with the 1-phenyl analogue 10c oxidation takes place at a very low rate exclusively at C-4, which is comparable to the oxidation of 1-(4'-pyridyl)pyridinium chloride by this enzyme (37).

Oxidation of 1-methylpyridinium compounds with an aminocarbonyl substituent at C-2 or C-4 (12a,b) is not feasible or occurs very slowly (Table 2.5), indicating that these compounds are not oriented in the right fashion in the catalytic centre due to interaction of the carbonyl moiety with an active site species.

A very simplified schematic representation of the orientation of two different azinium substrates in the catalytic centre of aldehyde oxidase is illustrated in Figure 2.4, based on the three important features of the active site suggested in this study, viz. the presence of a hydrophobic region (17), a rate-determining nucleophilic attack (21) and the interaction with a (proton-donating) active site species (35). The depicted pyridinium substrates differ in their steric conformation due to the presence of methyl substituents. In the oxidation of 3-aminocarbonyl-1-phenylpyridinium chloride the pyridinium ring

is rotated in a position which brings about a favourable
interaction of the carbonyl moiety of the 3-aminocarbonyl group
with the active site species (I). The phenyl substituent
interacts with the hydrophobic region in the vicinity of the
catalytic site. In this orientation C-6 is much better
accessible for nucleophilic attack than C-4, as has been
established from the respective maximum rates of oxidation
(21). The conformation of the aminocarbonyl group in the
substrate and in the intermediate formed after nucleophilic
attack at C-4 can have an effect on the overall orientation
during catalysis as well (38-40).

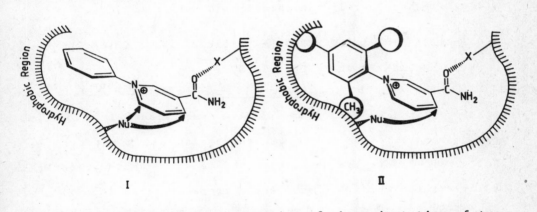

I   II

Figure 2.4   Schematic representation of the orientation of two
             3-aminocarbonyl-1-arylpyridinium substrates in the
             active centre of aldehyde oxidase during catalysis.
             Nu: enzymic nucleophilic species; X: interactive
             species.

The substrate 3-aminocarbonyl-1-mesitylpyridinium chloride has
a steric conformation in which both ring planes are nearly
perpendicular, because of the interaction of the ortho-methyl
groups. This conformation will be at least partly retained in

301

the active centre during catalysis. The approach of the nucleophilic species at C-6 (or C-2) is now severely hindered by the ortho-methyl groups and therefore, the oxidation appears to occur exclusively at C-4 (II). Because of the difference in hydrophobicity of the N-substituent it can be expected that the interaction of this substituent with the hydrophobic region of the enzyme will be slightly different, resulting in another orientation of the substrate molecule. This may lead to a greater accessibility of the C-4 site for the enzymic nucleophilic species and correspondingly the maximum rate of oxidation is much higher than that for oxidation of 3-amino-carbonyl-1-phenylpyridinium chloride at this site (21).

3 OXIDATION OF 3-AMINOCARBONYL-1-ARYLPYRIDINIUM CHLORIDES WITH IMMOBILIZED ALDEHYDE OXIDASE

To test the suitability of immobilized aldehyde oxidase for organic synthesis, aldehyde oxidase was adsorbed onto DEAE Sepharose CL 6B (25 mg enzyme preparation per g matrix) (41) and applied for the oxidation of some 3-aminocarbonyl-1-arylpyridinium chlorides on a small preparative scale (21). The immobilized enzyme preparation was packed in a column and washed with 10 mM potassium phosphate buffer, pH 7.0 (0.1 mM EDTA) at 4°C. It was established that the yield of products was maximal when the reaction was performed at neutral or slightly acidic pH, i.e. pH 6.4-7.0. The amount of aldehyde oxidase used per column for each conversion was 12 units (21).

300 ml of a 0.5 mM substrate solution (in the same buffer) was slowly (0.25 ml/min) passed through the column at 4°C and the conversion of substrate detected by registration of the formation of product(s) at 254 nm (HPLC). In the case of a

302

Table 3.1 Product yields of 1-aryl-1,4-dihydro-4-oxo- and
          1-aryl-1,6-dihydro-6-oxo-3-pyridinecarboxamides
          obtained by oxidation using immobilized aldehyde
          oxidase.

Substrate	Product	Yield (%)
6a	7a	81
	8a	< 5
6b	7b	65
	8b	10
6c	7c	52
	8c	34
6d	7d	75
	8d	< 5
6e	7e	42
	8e	< 5
6f	7f	76
	8f	11
6g	8g	88

rather slow reaction, the solution was recirculated using a
pump. The collected effluent was evaporated to dryness and the
residue purified by column chromatography (chloroform/methanol
19:1) to separate both oxidation products and coloured by-
products. After evaporation of the solvent, the crude
product(s) was (were) weighed and recrystallized two or three
times from distilled water. The yields of the crude products
and the analytical data of the recrystallized products are
given in Table 3.1. All product yields are up to 75% or higher,
except for the oxidation of 6e. This is very probably due to
the formation of by-products during the conversion, caused by
ring opening reactions (22).

The oxidation of these compounds has not yet been achieved

chemically, thus showing the potentiality of the use of
immobilized aldehyde oxidase in organic chemistry.

4 COVALENT AMINATION OF 1-ALKYL(ARYL)-3-AMINOCARBONYL-
   PYRIDINIUM SALTS AS 'MODEL' FOR THE ENZYMIC ACTIVITY OF
   ALDEHYDE OXIDASE

In the mechanism for the oxidation by aldehyde oxidase an
initial nucleophilic attack by an enzymic species is assumed
to occur, leading to an intermediary covalent σ-adduct. Based
on the proposed mechanism (see section 2.2), it is clear that
the carbon atom at which this addition takes place, is the
same as where the oxo group is introduced into the substrate.
Since there is in our laboratory an ongoing interest in σ-
adduct formation between N-alkylpyridinium salts and liquid
ammonia, we observed that there exists a striking similarity
between the site of σ-adduct formation with liquid ammonia
(i.e. C-6) and the site of oxidation with aldehyde oxidase in
the 1-methyl-3-aminocarbonylpyridinium salt (17).
The addition of nucleophiles such as nitromethane (42),
nitromethide ion (43), sulfite (42, 44), methanethiolate (45),
ethanethiolate (43), ethoxide (46), cyanide (42, 46-48), or
hydroxide ions (42, 45, 46, 49) usually takes a somewhat
different course, resulting in the formation of σ-adducts in
which the nucleophile is attached to C-4 and/or to C-2 and C-6.
These data induced us to study the covalent amination of 1-
alkyl(aryl)-3-aminocarbonylpyridinium salts in more detail, in
order to establish whether the σ-adduct formation with liquid
ammonia could possibly serve as an in-vitro model for the
initial σ-adduct formation in the aldehyde oxidase-mediated
oxidation.

304

4.1   1-Alkyl-3-aminocarbonylpyridinium chlorides

The reaction of compounds 1a-c (Scheme 4.1) with liquid ammonia gives rise to exclusive formation of the 6-amino-1,6-dihydro compounds 13a-c, as evidenced by ^{1}H NMR spectroscopy (50). All proton signals are shifted upfield compared to the corresponding signals of 1a-c in D_2O (Table 4.1).

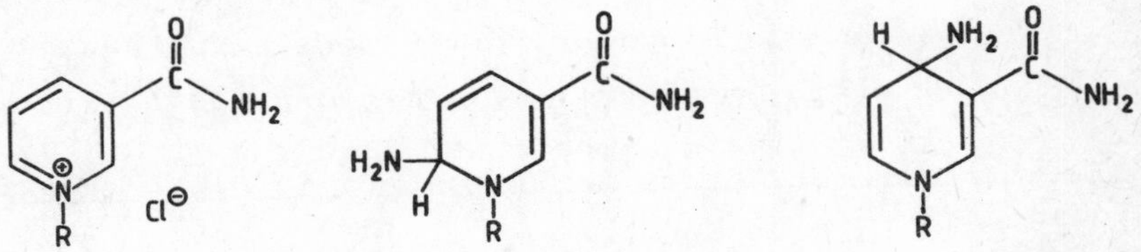

R= a: CH_3 , b: C_2H_5 , c: n-C_3H_7 , d: i-C_3H_7 , e: t-C_4H_9

Scheme 4.1

The shifts are most pronounced for the hydrogens attached to C-6 ($\Delta\delta$ = 4.32-4.51 ppm) due to the newly formed tetrahedral centre at C-6. The correct signal assignment is based on the chemical shift values and the coupling patterns and confirmed by the data, obtained by measurement of the ^{1}H NMR spectrum of 3-aminocarbonyl-4-deuterio-1-ethylpyridinium chloride in liquid ammonia. These data are in agreement with values published (51). Variation over a wide temperature range (-70° to 0°C) does not change the addition pattern.

The ^{1}H NMR spectra of compounds 1d-e in liquid ammonia are rather complex. From the spectra (Figure 4.1) it is concluded that two aminodihydro compounds are obtained from 1d-e, viz.

the C-6 adducts 13d-e and in addition the C-4 adducts 14d-e
(Table 4.1). This conclusion is based on the chemical shifts
of the C-6 adducts 13a-c, the upfield shift values ($\Phi\phi$), the
coupling constants and especially comparison with a simplified
spectrum obtained from the reaction of 3-aminocarbonyl-1-t-
butyl-4-deuterio-pyridinium salt with liquid ammonia.

Table 4.1 ^{1}H NMR data of the ring protons for 1-alkyl-6-
amino-1,6-dihydro-3-pyridinecarboxamides 13 and
1-alkyl-4-amino-1,4-dihydro-3-pyridinecarboxamides
14 in liquid ammonia at -45°C[a].

Compound	H-2	$\Delta\delta$[b]	H-4	$\Delta\delta$[b]	H-5	$\Delta\delta$[b]	H-6	$\Delta\delta$[b]
13a	7.29	1.97	6.55	2.35	5.06	3.14	4.66	4.32
13b	7.33	2.05	6.49	2.44	5.04	3.22	4.72	4.39
13c	7.32	2.15	6.54	2.51	5.07	3.31	4.71	4.51
13d	7.40	1.99	6.49	2.44	5.02	3.24	4.73	4.45
14d	7.16	2.23	4.15	4.78	c	-	6.20	2.98
13e	7.64	1.82	6.66	2.29	5.26	3.07	5.01	4.35
14e	7.47	1.99	4.26	4.69	c	-	6.52	2.84

a. Adducts 13: $J_{2,4}$ = 1.5-1.8 Hz; $J_{2,6}$ = 1.1 Hz; $J_{4,5}$ =
9.1-9.8 Hz; $J_{5,6}$ = 4.8-5.4 Hz;
adducts 14: $J_{2,6}$ = 2.0 Hz; $J_{4,5}$ = 4,2 Hz; $J_{5,6}$ could not be
determined due to overlap of signals.
b. Upfield shifts relative to the corresponding compounds 1 in
D_2O.
c. Difficult to interpret due to overlap by the H-6 signal of
the corresponding 6-adduct and to the low intensity for
compound 14d.

The ratio of 13d/14d is 9:1, the ratio of 13e/14e 6:4. These
ratios are independent of the temperature in the range from

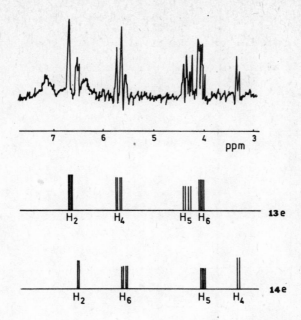

Figure 4.1 ^{1}H NMR spectrum of 3-aminocarbonyl-1-t-butylpyridinium chloride 1e in liquid ammonia showing the signals and their assignments due to C-6 adduct 13e and C-4 adduct 14e

-70° to 0°C. It is obvious from these results that the position of nucleophilic addition is not only dependent on the substituent at position 3 (51,52), but certainly also on the nature of the substituent at position 1 as well. Moreover, it is evident that with an increasing size of the alkyl group at position 1, the addition at C-4 is promoted at the cost of addition at the C-6 position. Covalent amination in liquid ammonia is apparently rather susceptible to steric effects.

In Table 4.2 the qualitative data of the covalent amination patterns with 1a-e are compared with the oxidation patterns found for these compounds in reaction with aldehyde oxidase

(17). A good agreement exists between the position of addition by liquid ammonia and the nucleophilic species active in the aldehyde oxidase-mediated reaction. Both reactions are susceptible to steric interference of the substituent at position 1, the enzymic oxidation to a greater extent than the amination reaction. This is understandable because the nucleophilic species in the enzymic reaction is fixed in the catalytic site of the enzyme molecule and therefore has a greater steric interaction with the N-1 substituent.

Table 4.2   Comparison of the site of amination with liquid
             ammonia at -45°C and the site of oxidation by
             aldehyde oxidase of compounds 1a-e, 6a, 6c and 6g.

Compound	Oxidation	Amination
1a	C-6	C-6
1b	C-6	C-6
1c	C-6	C-6
1d	C-6/C-4	C-6/C-4
1e	C-4	C-6/C-4
6a(j)	C-6/C-4	C-6/C-2
6c	C-6/C-4	C-6/C-2
6g	C-4	C-6/C-2/C-4

In addition, the orientation of the substrate in the active site is of course sterically governed, leading to exclusive oxidation into a 4-oxo product in the case of the 1-t-butyl derivative 1e.

308

4.2 3-Aminocarbonyl-1-arylpyridinium chlorides

The chemical shifts and coupling constants derived from the ^{1}H NMR spectra of 1-aryl compounds **6c** and **6j** in liquid ammonia (Scheme 4.2) show that two σ-adducts are obtained from both compounds, viz. the C-6 adducts **15c** and **15j** and the Cl-2 adducts **16c** and **16j** (Table 4.3).

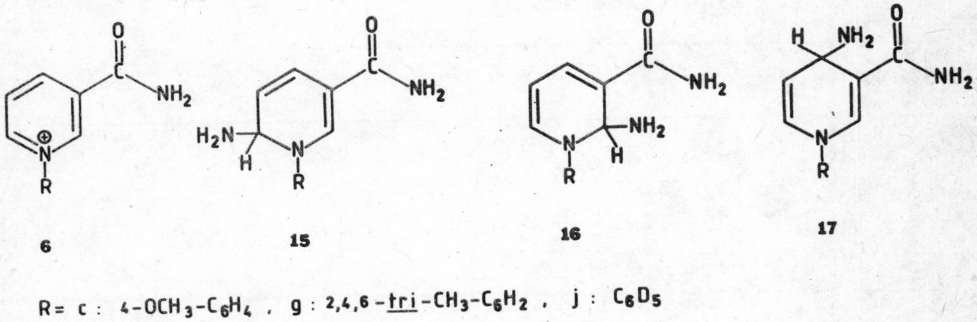

R= c : 4-OCH$_3$-C$_6$H$_4$, g : 2,4,6-<u>tri</u>-CH$_3$-C$_6$H$_2$, j : C$_6$D$_5$

Scheme 4.2.

The adduct structures have been assigned based on our knowledge of the ^{1}H NMR data of C-6 adducts (**13a-e**), the upfield shifts, the magnitude of the coupling constants and especially the ^{1}H NMR spectra of the <u>4-deuterio</u> derivatives of **6c** and **6j** in liquid ammonia. The main difference between the covalent amino adducts obtained from the 1-alkyl and 1-aryl compounds is that the upfield shift values (Δδ) of the hydrogens attached to the tetrahedral centres in the adducts are significantly smaller for the 1-aryl compounds.

Compound **6g** yields a more complicated ^{1}H NMR spectrum which arises from the presence of three σ-adducts (**15g**, **16g** and **17g**); the C-6 adduct is formed in excess. Complete assignment

21 Plas

of signals could not be made because of the complexity of the spectrum. Additional proof for the formation of the C-6, C-4 and C-2 adducts with compound 6g has been obtained from ^{13}C NMR spectroscopy (50).

Table 4.3 ^{1}H NMR data of the ring protons for 6-amino-1-aryl-1,6-dihydro-3-pyridinecarboxamides 15 and 2-amino-1-aryl-1,2-dihydro-3-pyridinecarboxamides 16 in liquid ammonia at -45°C[a].

Compound	H-2	$\Delta\delta$[b]	H-4	$\Delta\delta$[b]	H-5	$\Delta\delta$[b]	H-6	$\Delta\delta$[b]
15j	7.65	1.94	6.71	2.48	5.52	3.00	5.19	4.19
16j	5.73	3.86	7.05	2.14	c	–	6.93	2.45
15c	7.55	1.97	6.70	2.46	5.45	3.03	5.14	4.19
16c	5.65	3.87	d	–	c	–	d	–

a. Adducts 15: $J_{2,4}$ = 1.5 Hz; $J_{4,5}$ = 9.0 Hz; $J_{4,6}$ = 1.8 Hz; $J_{2,6}$ < 1.0 Hz;
    adducts 16: $J_{2,6}$ = 1.8 Hz; $J_{4,5}$ = 6.0 Hz; $J_{5,6}$ = 7.4 Hz; $J_{2,4}$ < 1.0 Hz; $J_{4,6}$ < 1.0 Hz.
b. Upfield shifts relative to the corresponding compounds 6 D_2O.
c. Difficult to interpret due to overlap by the H-5 signal of the corresponding 6-adduct.
d. These signals lie under the phenyl multiplet.

Variation of the temperature from -70° to -20°C shows that the ratios C-2 adduct/C-6 adduct obtained from compounds 6c and 6j alter. In Table 4.4 this is illustrated for the σ-adducts obtained from 6j. The amount of C-2 adduct decreases in favour of the C-6 adduct at higher temperature. It is interesting to note that the ratio of the three σ-adducts obtained from 6g remains unaffected by temperature variation over this range.

Table 4.4   Isomer distribution of σ-adducts from 3-aminocar-
             bonyl-1-(pentadeuteriophenyl)pyridinium chloride 6j
             at various temperatures.

T (°C)	C-6 adducts 15j (%)	C-2 adducts 16j (%)
-70	60	40
-45	65	35
-20	80	20

Comparing these results of covalent amination of 6c, 6g and 6j
with those obtained for oxidation by aldehyde oxidase (21), it
is evident that the similarity between these two reactions is
very small (Table 4.2) and certainly less convincing as an 'in
vitro-model' than the corresponding reactions of the 1-alkyl
derivatives (1a-e). This leads to the conclusion that in the
1-aryl compounds besides the steric influence of the aryl
group, an electronic effect is operative which strongly
influences the site of amination. The combined effect of steric
and electronic effects makes comparison between σ-adduct
formation and oxidation of limited value, since they probably
operate in a different manner in both reactions.

21*

5 REFERENCES

1. M. Woźniak and H.C. van der Plas, Advances in the chemistry of 1,7-naphthyridine. Heterocycles, 19, 363-405 (1982).
2. H.C. van der Plas, Reactivity of pteridines and some deazapteridines towards nucleophiles in enzymatic and non-enzymatic reactions. Lectures in Heterocyclic Chemistry, (Eds. R.N. Castle and T. Kappe), Hetero Corporation, Tampa, Florida (1982), Vol. 6, S-1 - S-23.
3. H.C. van der Plas, M. Woźniak and H.J.W. van den Haak, Reactivity of naphthyridines toward nitrogen nucleophiles. Advances in Heterocyclic Chemistry, (Ed. A.R. Katritzky), Academic Press, New York (1983), Vol. 33, 95-146.
4. H.C. van der Plas and F. Roeterdink, Six-membered didehydroheteroarenes. The Chemistry of Functional Groups, Suppl. C, Part I, (Eds. S. Patai and Z. Rappoport), Wiley, New York (1983), 421 - 511.
5. H.C. van der Plas, Ring degenerate transformations of azines. Tetrahedron Reports, in press.
6. J. Tramper, F. Müller and H.C. van der Plas, Immobilized xanthine oxidase: kinetics, (in)stability and stabilization by coimmobilization with superoxide dismutase and catalase. Biotechnol. Bioeng., 20, 1507 - 1522 (1978).
7. J. Tramper, A. Nagel, H.C. van der Plas and F. Müller, The oxidation of 7-(p-X-phenyl)pteridin-4-ones (X = OMe, H, Br, CN, NO_2) with free and immobilized xanthine oxidase. Recl. Trav. Chim. Pays-Bas, 98, 224-231 (1979).
8. J. Tramper, S.A.G.F. Angelino, F. Müller and H.C. van der Plas, Kinetics and stability of immobilized chicken liver xanthine dehydrogenase. Biotechnol. Bioeng., 21, 1767-1786 (1979).
9. J. Tramper, H.C. van der Plas and F. Müller, Direct immobilization of milk xanthine oxidase in milk protein with and without addition of gelatin. Biotechnol. Lett., 1, 133-138 (1979).

10. J. Tramper, A. van der Kaaden, H.C. van der Plas, F. Müller and W.J. Middelhoven, Xanthine oxidase activity of Arthrobacter X-4 cells immobilized in glutaraldehyde-crosslinked gelatin. Biotechnol. Lett., 1, 397-402 (1979).

11. J. Tramper, W.E. Hennink and H.C. van der Plas, Oxidation of 7-alkylpteridin-4-ones to 7-alkyllumazines by free and immobilized xanthine oxidase. J. Appl. Biochem., 4, 263-270 (1982).

12. M.P. Coughlan, Aldehyde oxidase, xanthine oxidase and xanthine dehydrogenase: hydroxylases containing molybdenum, iron-sulphur and flavin. Molybdenum and Molybdenum-containing Enzymes, (Ed. M.P. Coughlan), Pergamon Press, Oxford (1980), 119 - 185.

13. T.A. Krenitsky, S.M. Neil, G.B. Elion and G.H. Hitchings, A comparison of the specificities of xanthine oxidase and aldehyde oxidase. Arch. Biochem. Biophys., 150, 585 - 599 (1972).

14. C.N. Hodnett, J.J. McCormack and J.A. Sabean, Oxidation of selected pteridine derivatives by mammalian liver xanthine oxidase and aldehyde oxidase. J. Pharm. Sci., 65, 1150-1154 (1976).

15. J.J. McCormack, B.A. Allen and C.N. Hodnett, Oxidation of quinazoline and quinoxaline by xanthine oxidase and aldehyde oxidase. J. Heterocyclic Chem., 15, 1249-1254 (1978).

16. C. Stubley, J.G.P. Stell and D.W. Mathieson, The oxidation of azaheterocycles with mammalian liver aldehyde oxidase. Xenobiotica, 9, 475-484 (1979).

17. S.A.G.F. Angelino, D.J. Buurman, H.C. van der Plas and F. Müller, The use of immobilized enzymes in organic synthesis. Part 6. Oxidation of 1-alkyl-3-carbamoyl-pyridinium chlorides by rabbit liver aldehyde oxidase. Recl. Trav. Chim. Pays-Bas, 101, 342-346 (1982).

18. R.L. Felsted, A.E.-Y.Chu and S. Chaykin, Purification and properties of the aldehyde oxidases from hog and rabbit livers. J. Biol. Chem., 248, 2580-2587 (1973).

19. I.H. Segel, Kinetics of unireactant enzymes. Enzyme Kinetics: Behavior and Analysis of Rapid Equilibrium and Steady-State Enzyme Systems, Wiley-Interscience, New York (1975), 18 - 99.

20. B.R. Baker, W.F. Wood and J.A. Kozma, Irreversible enzyme inhibitors. CXXVI. Hydrocarbon interaction with xanthine oxidase by phenyl substituents on purines and pyrazolo-[3,4-d]pyrimidines. J. Med. Chem., 11, 661-666 (1968).

21. S.A.G.F. Angelino, D.J. Buurman, H.C. van der Plas and F. Müller, The use of immobilized enzymes in organic synthesis. Part 8. Oxidation of 1-aryl-3-carbamoyl-pyridinium chlorides by rabbit liver aldehyde oxidase and bovine milk xanthine oxidase. Recl. Trav. Chim. Pays-Bas, 102, 331-336 (1983).

22. J.W. Bunting, Heterocyclic pseudobases. Advances in Heterocyclic Chemistry, (Eds. A.R. Katritzky and A.J. Boulton), Academic Press, New York (1979), Vol. 25, 1-82.

23. J.S. Olson, D.P. Ballou, G. Palmer and V. Massey, The mechanism of action of xanthine oxidase. J. Biol. Chem., 249, 4363-4382 (1974).

24. M.P. Coughlan and I. Ní Fhaoláin, On the sites of interaction of oxidizing substrates with molybdenum iron/sulphur flavin hydroxylases. Proc. R.I.A., 79B, 169-175 (1979).

25. M.J. Barber, M.P. Coughlan, K.V. Rajagopalan and L.M. Siegel, Properties of the prosthetic groups of rabbit liver aldehyde oxidase: a comparison of molybdenum hydroxylase enzymes. Biochemistry, 21, 3561-3568 (1982).

26. M.J. Barber, M.P. Coughlan, K.V. Rajagopalan and L.M. Siegel, Rabbit liver aldehyde oxidase: reduction potentials and spectroscopic properties. Dev. Biochem., 21, 805-809 (1982).

27. R.C. Bray, G.N. George, S. Gutteridge, L. Norlander, J.G.P. Stell and C. Stubley, Studies by electron-paramagnetic-resonance spectroscopy of the molybdenum centre of aldehyde oxidase. Biochem. J., 203, 263-267 (1982).

28. M.J. Barber and L.M. Siegel, Electron paramagnetic resonance and potentiometric studies of arsenite interaction with the molybdenum centers of xanthine oxidase, xanthine dehydrogenase and aldehyde oxidase: a specific stabilization of the molybdenum (V) oxidation state. Biochemistry, 22, 618-624 (1983).

29. S. Gutteridge and R.C. Bray, Oxygen-17 splitting of the very rapid molybdenum (V) e.p.r. signal from xanthine oxidase. Biochem. J., 189, 615-623 (1980).

30. R.C. Bray, The flavin and the other catalytic and redoxcenters of xanthine oxidase and related enzymes. Dev. Biochem., 21, 775-785 (1982).

31. R.C. Bray, S. Gutteridge, D.A. Stotter and S.J. Tanner, The mechanism of action of xanthine oxidase. Biochem. J., 177, 357-360 (1979).

32. J.W. Verhoeven, I.P. Dirkx and Th.J. de Boer, Studies of inter- and intramolecular donor-acceptor interactions. Conformational population study of some substituted N-aralkyl 4-cyanopyridinium ions by NMR spectroscopy. J. Mol. Spectrosc., 36, 284-294 (1970).

33. S.A.G.F. Angelino, D.J. Buurman, H.C. van der Plas and F. Müller, unpublished results.

34. J.W. Bunting, K.R. Laderoute and D.J. Norris, Specificity of xanthine oxidase for nitrogen heteroaromatic cation substrates. Can. J. Biochem., 58, 49-57 (1980).

35. S.A.G.F. Angelino, B.H. van Valkengoed, D.J. Buurman, H.C. van der Plas and F. Müller, The oxidation of 1-alkyl-(aryl)quinolinium chlorides with rabbit liver aldehyde oxidase. J. Heterocyclic Chem., 21, 107-112 (1984)

36. J.W. Bunting and A. Gunasekara, An important enzyme-substrate binding interaction for xanthine oxidase. Biochem. Biophys. Acta, 704, 444-449 (1982).

37. K.V. Rajagopalan and P. Handler, Hepatic aldehyde oxidase III. The substrate-binding site. J. Biol. Chem., 239, 2027-2035 (1964).

38. V. Skála and J. Kuthan, Contribution to EHT conformation analysis of 1-methyl-3-carbamoylpyridinium cation. Coll. Czech. Chem. Commun., 43, 3049-3055 (1978).

39. H.-J. Hofman and J. Kuthan, On the conformational structure of nicotinamide and 1-methyl-1,4-dihydronicotinamide. Coll. Czech. Chem. Commun., 44, 2633-2638 (1979).

40. P.M. van Lier, M.C.A. Donkersloot, A.S. Koster, H.J.G. van Hooff and H.M. Buck, A model study of the role of the $CONH_2$-group in the redox co-enzyme $NAD(P)^+$-$NAD(P)H$. Recl. Trav. Chim. Pays-Bas, 101, 119-120 (1982).

41. S.A.G.F. Angelino, F. Müller and H.C. van der Plas, The stability of free and immobilized rabbit liver aldehyde oxidase. Biotechnol. Bioeng., submitted.

42. K. Wallenfels and H. Schüly, Über den Mechanismus der Wasserstoffübertragung mit Pyridinnucleotiden. VIII. Anionen-addition bei DPN^+-Modellen. Justus Liebigs Ann. Chem., 621, 86-105 (1959).

43. J.A. Zoltewicz, L.S. Helmick and J.K. O'Halloran, Competitive addition of carbon, sulfur and nitrogen nucleophiles to quaternized heteroaromatic compounds in liquid ammonia. J. Org. Chem., 41, 1308-1313 (1976).

44. G. Pfleiderer, E. Sann and A. Stock, Das Reaktionsverhalten von Pyridinnucleotiden (PN) und PN-Modellen mit Sulfit als nucleophilem Agens. Chem. Ber., 93, 3083-3099 (1960).

45. H. Minato, E. Yamazaki and M. Kobayashi, Reactions of several nucleophiles with quaternary salts of nicotinamide. Chem. Lett., 525-530 (1976).

46. A.G. Anderson and G. Berkelhammer, Action of base on certain pyridinium salts. J. Org. Chem., 23, 1109-1112 (1958).

47. R.N. Lindquist and E.H. Cordes, Secondary valence force catalysis. IV. Rate and equilibrium constants for addition of cyanide ion to N-substituted 3-carbamoylpyridinium ions. J. Am. Chem. Soc., 90, 1269-1274 (1968).

48. V. Skála and J. Kuthan, EHT study of interaction of 1-methyl-3-carbamoylpyridinium with cyanide. Coll. Czech. Chem. Commun., 43, 3064-3070 (1978).

49. D.C. Dittmer and J.M. Kolyer, Action of base on quaternary salts of nicotinamide. J. Org. Chem., 28, 2288-2294 (1963).

50. S.A.G.F. Angelino, A. van Veldhuizen, D.J. Buurman and H.C. van der Plas, Covalent amination of 1-alkyl- and 1-aryl-3-carbamoylpyridinium chlorides as "model" for the enzymic activity of rabbit liver aldehyde oxidase. Tetrahedron, 40, 433-439 (1984).

51. J.A. Zoltewicz, L.S. Helmick and J.K. O'Halloran, Covalent amination, substituent effects on the site of addition of ammonia to quaternized pyridines and pyrazines. J. Org. Chem., 41, 1303-1308 (1976).

52. J.A. Zoltewicz, T.M. Oestreich, J.K. O'Halloran and L.S. Helmick, Covalent amination of heteroaromatic compounds. J. Org. Chem., 38, 1949-1952 (1973).

ANTIBACTERIAL OXOLINYL-AMINO ACIDS

FRANK J., PONGRÁCZ K., KULCSÁR G. and MEDZIHRADSZKY K.*

CHINOIN Pharmaceutical and Chemical Works Ltd.,
Budapest, Pf. 110, H-1325, Hungary

*Central Research Institute for Chemistry, The Hungarian
Academy of Sciences, Budapest, Pf. 17, H-1525, Hungary

Oxolinic acid (1) is an efficient antibacterial agent [1] against gram-negative bacteria. It is used for the treatment of urinary tract infections under the trade names of UTIBIDE, URITRATE, PRODOXOL, GRAMURIN etc.

In search of compounds with a broader spectrum of activity new oxolinyl-amino-acids (4) were prepared by reacting the active ester of oxolinic acid (2) with an amino-acid derivative (3) of D- or L-configuration at the N-terminal. The C-terminals were acids, esters, or substituted amides.

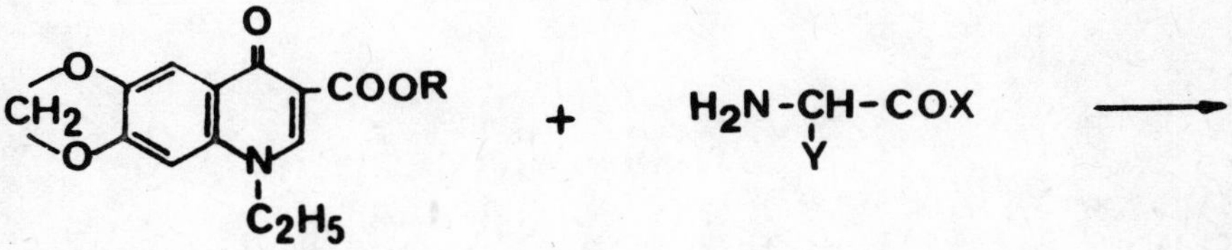

$$CH_2 \quad \text{quinolinone-COOR} \quad + \quad H_2N-CH-COX \longrightarrow$$
$$C_2H_5 \qquad\qquad\qquad Y$$

(1) R = H   (3)
(2) R = pentachlorophenyl

$$CH_2 \quad \text{quinolinone-CO-NH-CH-COX}$$
$$C_2H_5 \qquad\qquad\qquad Y$$

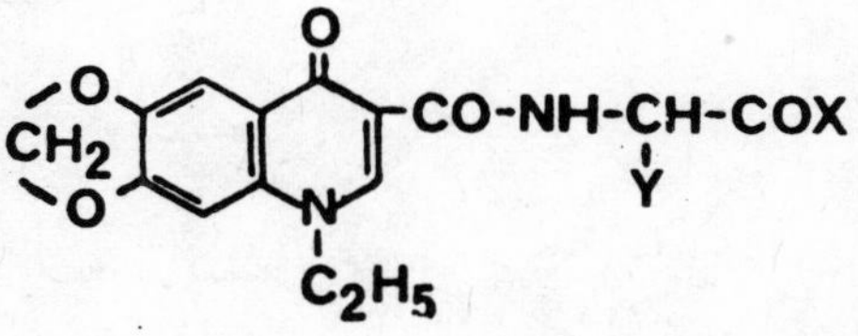

(4)

X = OH, OMe, NH-octyl,
     NH-cetyl, HN-Et-N/Et/$_2$

Y = H, alkyl, aralkyl,
     substituted alkyl

319

METHODS

Amino-acids of L- and D-configuration were commercially available, their ester and amide derivatives were prepared by known methods. Protecting groups were removed by general procedures: Boc (t-butyloxycarbonyl) by hydrogen chloride in ethylacetate, Z(benzyloxycarbonyl) by catalytic reduction.

The antibacterial activity of oxolinyl-amino-acids was tested with *multiresistant* bacteria by agar dilution technique. The minimal inhibitory concentration (MIC) was defined as the lowest concentration of the test agent which inhibited the formation of visible colonies. MICs were determined with Difco-Bouillon broth after 24h of incubation at $37^{\circ}C$ using double control.

RESULTS AND DISCUSSION

Oxolinyl-glycine derivatives (6-11) were chosen to study the relationship between different C-terminals and antibacterial activity. The MIC values are summarized in Table 1.

Table 1. MIC (μg/ml) values of oxolinyl-glycine derivatives

No	Compound	Pseud.aerug. CCM 1960	Prot.vulg. CCM 1799	Esch.coli DSM 30038	Vibrio par. CCM 5938	Staph.aur. CCM 885	Strept.faec. CCM 1875	Strept.agal. CCM 5534	Bac.cereus CCM 2010
5	Gly-NH-octyl	150	150	200	150	>200	>200	5	>200
6	Ox-Gly-NH-octyl	10	10	25	10	50	50	50	25
7	Ox-Gly-NH-cetyl	150	100	100	100	150	150	150	150
8	Ox-Gly-NHEtNEt$_2$	50	50	75	50	75	75	100	
9	Ox-Gly-NHEtNEt$_2$Me I$^{\ominus}$	>200	>200	>200	200	200	200	>200	150
10	Ox-Gly-OMe	150	150	200	150	200	200	200	150
11	Ox-Gly-OH	200	200	>200	200	>200	>200	200	>200

The results in Table 1. indicate that C-terminal amides (6-8) are more active than the ester (10) or acid (11) and for the substitution of amide the lipophilic shorter aliphatic non-polar chain (6) is favourable. For comparison MIC values of 5 glycine-octyl-amide are also shown. Quaternarization of the amine side chain of 8 results in a water-soluble compound 9 but the activity is lost.

Based on the good antibacterial activity of 6 a great number of different oxolinyl-amino-acid-octylamides were prepared and tested against several microorganisms. In Table 2. a few examples are shown with MIC values against four gram-negative and four gram-positive bacteria. Oxolinic acid is included for comparison.

Table 2. MIC (μg/ml) values of oxolinyl-amino-acid-octylamides

No	Compound	Pseud.aerug. CCM 1960	Prot.vulg. CCM 1799	Esch.coli DSM 30038	Vibrio par. CCM 5938	Staph.aur. CCM 885	Strept.faec. CCM 1875	Strept.agal. CCM 5534	Bac.cereus CCM 2010
1	Oxolinic acid	150	10	25	1	50	200		50
12	Ox-L-Leu-NH-octyl	25	50	50	25	25	10	50	50
13	Ox-ILeu-NH-octyl	25	25	50	2,5	50	50	25	25
14	Ox-L-Lys-NH-octyl. .2HCl	50	150	50	25	10	50		10
15	Ox-L-Val-NH-octyl	25	150	75	25	50	75	50	50
16	Ox-L-Phe-NH-octyl	10	25	50	10	50	75	50	50
17	Ox-D-Phe-NH-octyl	>200	>200	>200	>200	>200	>200	>200	>200

The most active compounds in the series were 16 L-phenyl-alanine and 6 glycine containing derivatives, with significant activity against Pseudomonas aeruginosa compared with that of oxolinic acid.

The effect of configuration is represented by the phenyl-alanine octylamides: while 16 of L-configuration is active, 17 of D-configuration is inactive.

Although some amino-acid derivatives improved the activity of oxolinic acid (1) against gram-negative bacteria, the search was continued for a broader spectrum of activity by preparing di-, tri- and tetra-peptide derivatives of 1. The results of this work will be published elsewhere.

REFERENCE

1. D. Kaminsky and R. I. Meltzer: Quinolone Antibacterial Agents. Oxolinic acid and related Compounds. J. Med. Chem. 11 160-163 (1968)

CHEMICAL REACTIVITY OF ANTICONSULSIVE 1,2,4-OXADIAZINE DERIVATIVES

ÜRÖGDI L., PATTHY Á., VEZÉR Cs. and KISFALUDY L.

Chemical Works of Gedeon Richter Ltd., Hungary
H-1475 Budapest 10 P.O.Box 27

INTRODUCTION

Anticonvulsive tetrahydro-1,2,4-oxadiazin-5-ones were first prepared in our laboratory [1] . Their chemical reactivity has been investigated to obtain a wide variety of derivatives and preliminary information on their expectable in vivo chemical reactions (metabolism) [2] . The stability of the 1,2,4-oxadiazine ring has also been investigated under various conditions.

RESULTS

Chemical transformations

2-Acyl-tetrahyro-1,2,4-oxadiazin-5-ones (III) were prepared from α-aminooxyacid derivatives (I) and aldehydes (II) in acid-catalyzed reactions (Scheme 1) :

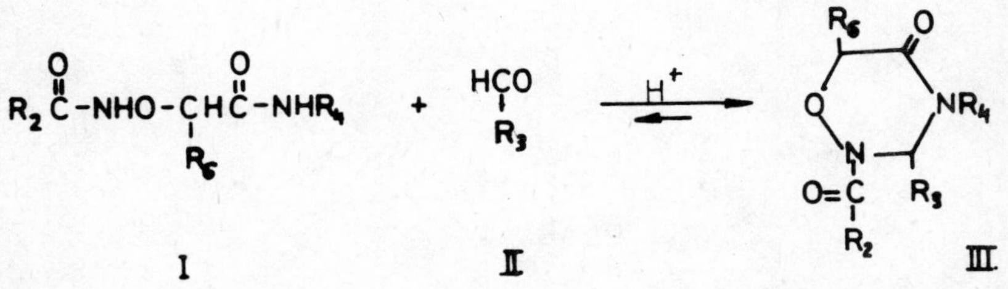

$$\tag{1}$$

In further transformations of III several known lactam reactions - such as O- and N-alkylation, N-acylation, silyla-
tion and hydroxymethylation - were carried out (Scheme 2):

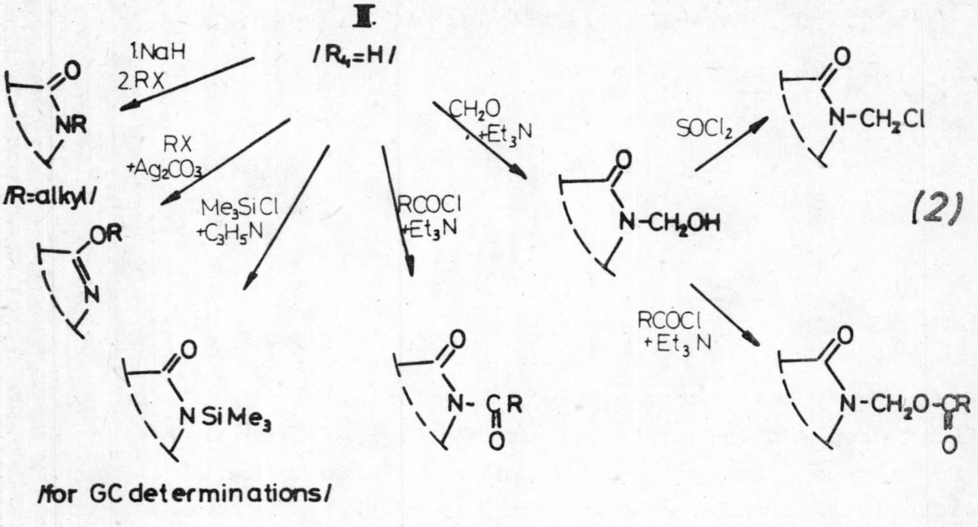

III.

/for GC determinations/

Removal of the 2-acyl moiety from III by hydrolysis or
acidolytic splitting of the 2-benzyloxycarbonyl group led
to ring opening (Scheme 5). However, carefully controlled
catalytic hydrogenolysis of the 2-benzyloxycarbonyl derivatives resulted in the expected deacylated compounds IV
(Scheme 3):

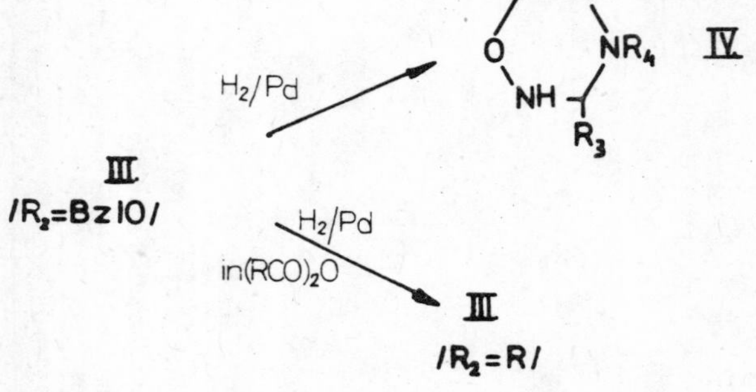

(3)

Various acylation reactions of IV gave a wide variety of 2-acyl derivatives (Scheme 4). Higher yields of these derivatives were obtained by _in situ_ acylation of IV, when the hydrogenolysis of the benzyloxycarbonyl group of III was affected in the appropriate anhydride as a solvent (Scheme 3).

Hydroxymethylation of IV could be carried out stepwise, selectively. Oxidation _via_ bromination followed by HBr elimination resulted in the known [3] 5,6-dihyro-4H-1,2,4-oxadiazin-5-one derivatives (V). This conversion also proves the structure of the terahydro ring (Scheme 4):

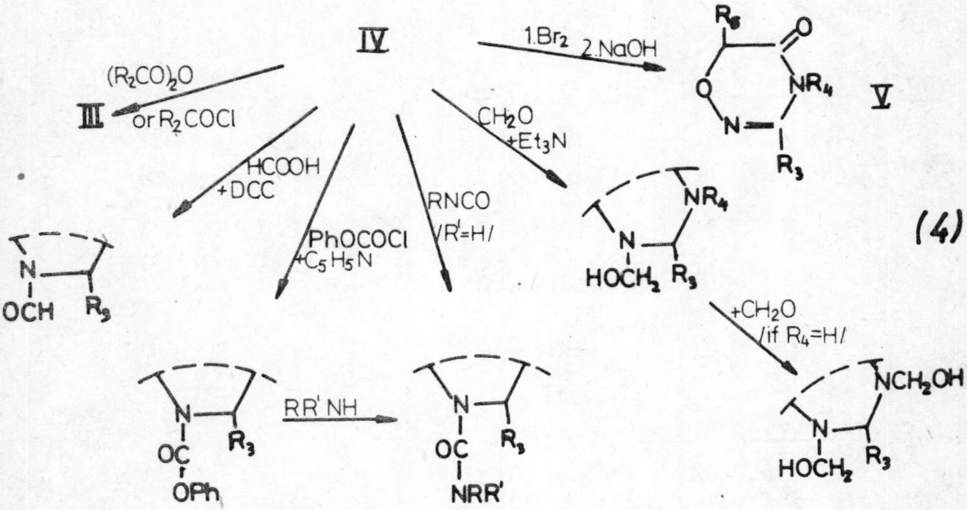

Stability

Characteristic difference was observed between the stabilities of the 2-acyl (type III) and 2-unsubstituted (type IV) derivatives. Catalytic hydrogenation of IV led to ring opening (Scheme 5); subsequent debenzylation and splitting of the N-O bond gave finally hydroxy-acetamide (VI). Compounds III proved to be more resistant to hydrogenation than IV. They are also very stable under oxidative conditions.

A similar difference was observed in the hydrolytic stability of III and IV. The 2-acyl derivatives are quite stable in acidic or alkaline media, and only more vigorous conditions led - obviously <u>via</u> deacylation - to ring opening to give VII. In contrast, the 2-unsubstituted derivatives (IV) are extremly unstable, especially in acidic media. Two different ring cleavage reactions were observed, depending on the concentration of the acid (Scheme 5):

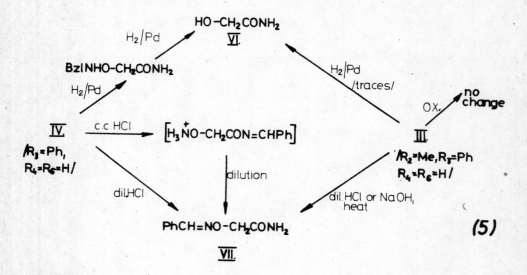

$$(5)$$

Anticonvulsive properties

Anticonvulsive effects of the synthesized compounds were tested by the maximal electroshock and pentetrazole methods. Results obtained with the more than 2oo derivatives showed strong substituent-dependence. The most effective compound (RGH-4615 = III, when R_2 = Me, R_3 = Ph, R_4 = R_6 = H) has an anticonvulsive effect equal to that of Phenytoin.

REFERENCES

[1] Kisfaludy,L. et al.: Tetrahydro-1,2,4-oxadiazin-5-one
    derivatives and medicines containing these compounds
    Eur.Pat.Appl.: EP 55484 A1; and EP 5549o A1 /1982

[2] Pálfi-Ledniczky,M. et al.: Biotransformation of 2-acetyl-
    -3-phenyl-tetrahydro-1,2,4-oxadiazon-5-one (RGH-4615)
    in rats. /in this volume/

[3] McKee,R.L., The Chemistry of Heterocyclic Compounds
    /A.Weissberged ed./ Vol.17. /R.H.Wiley ed./ p.447
    /Interscience Publ., N.Y. 1962/

Bio-Organic Heterocycles
van der Plas H.C., Ötvös L., and Simonyi M. eds

BIOTRANSFORMATION OF 2-ACETYL-3-PHENYL-TETRAHYDRO-1,2,4-OXADIAZIN-5-ONE (RGH 4615) IN RATS

PÁLFI-LEDNICZKY M., SZINAI I., UJSZÁSZY K., HOLLY S., WEISZ I., KEMÉNY V., TÜDŐS H., HORVÁTH M., and ÖTVÖS L.

Central Research Institute for Chemistry, The Hungarian Academy of Sciences, Budapest, Pf. 17, H-1525, Hungary

2-Acetyl-3-phenyl-tetrahydro-1,2,4-oxadiazin-5-one (Formula, RGH 4615), possesses significant anticonvulsive effect [1]. In order to elucidate the biological transformation of RGH 4615 in rats, distribution, metabolism and elimination were studied *in vivo*. The biotransformation of RGH 4615 and model compounds was studied under *in vitro* conditions, too.

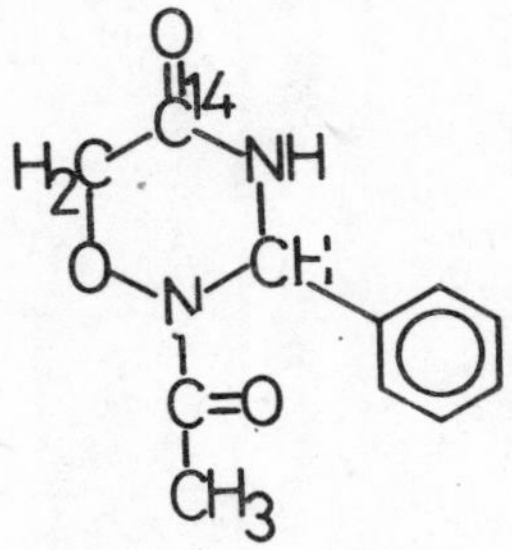

METHODS

5-^{14}C-labelled racemic RGH 4615 synthesized in our laboratories was administered p.o. in a single dose of 40 mg/kg to male RG-Wistar rats. The elimination of radioactivity was measured in urine and faeces for a period of 0-24 hours and in bile for 6 hours. Distribution of radioactivity was observed by using S. Ulberg's whole body autoradiography technique. *In vitro* experiments were carried out with homogenized brain, liver and by a liver perfusion system by incubation of different model compounds at 37 ^{O}C for 60 min. Biological samples were separated by stepwise extraction with chloroform and ethanol;

the metabolites were separated by Florisil column chromatography. Radioactive metabolites weré purified by TLC/silica
G/ and HPLC /on silica or C$_{18}$ reverse phase column/ and
analysed by MS, IR and CD spectroscopy.

RESULTS
In vivo experiments

80 % of the applied dose was excreted into urine during
the first 24 hours; 7,5 % of the dose appeared in bile during
6 hours. There was no significant accumulation of RGH 4615 in
the rat. Distribution of radioactivity was nearly homogeneous
in different tissues.

The structure of identified metabolites of RGH 4615 is
shown in Figure 1.

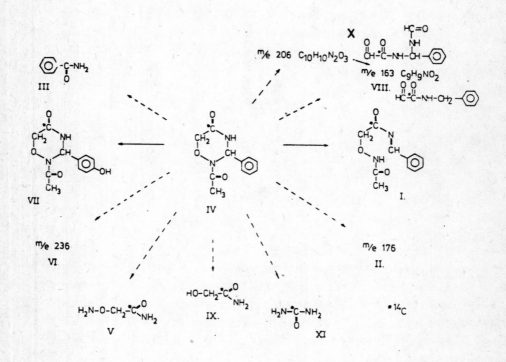

Figure 1. Identified metabolites from urine and liver
          perfursion fluid

Amounts of isolated metabolites are shown in Table 1.

Table 1. Identified metabolites of RGH 4615. Values represent
% of excreted radioactivity in biological samples.

Metabolites	I	II	III	IV	V	VI	VII	VIII	IX	X	XI	Total
Urine	1.6	3.6	C.1	32	0.3	0.9	2.4	0.8	0.9	26	1.5	70
Brain 30'				0.5								
120'				0.4								
Bile				26.6								
Liver-perfusion-fluid	16.1	2.2		53.4		0.2	0.7			1.2		73.8

Unreacted RGH 4615 in the urine was found to be partially
resolved with the /+/ enantiomer content being increased up to
30 % of optical purity.

One of the metabolic routes is aromatic hydroxylation
(metabolite VII). Para-position for the OH group was proved
by IR spectra. The appearance of urea(XI) among metabolites
was surprising. Besides HPLC, TLC and IR its identity was also
proved by urease enzymatic studies when $^{14}CO_2$ could be detected. Other metabolites were isolated as aminooxy-aceticacid-
amide (V) glycolic acidamide (X) and a ring-opened product (I).

Appearance of radio-inactive benzamide (III) among the
metabolites of RGH 4615 must be the result of oxidation at
position 2, subsequent ring opening to diacyl-amine and hydrolysis of the latter compound (Figure 2).

In vitro experiments

In vitro transformation of RGH 4615, as well as of N-benzyl-
amides and benzylidene-bis-amides was investigated in brain
and liver homogenates. We found that benzamide was a common
metabolite of all compounds studied (Fig.2) including
N-benzylamides of chloro-propionic acid (Beclamid), glycolic
acid, acetamidoxy-acetic acid and diphenyl-acetic acid.

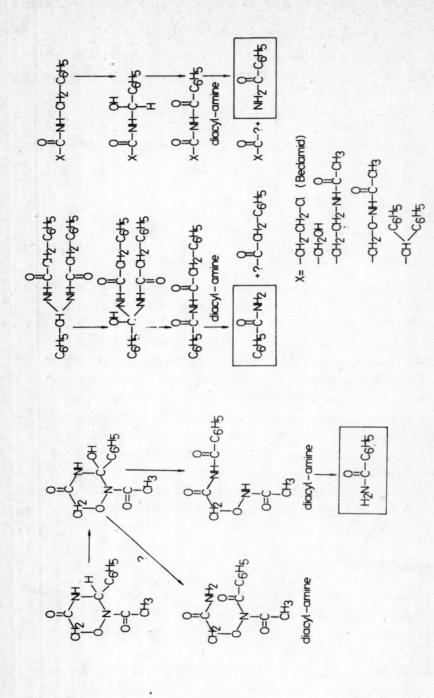

Figure 2. *In vitro* biotransformation of RGH 4615 and some open chain model compounds.

Since benzamide could be formed by hydrolysis of diacyl-amines, it may be assumed that the anticonvulsive effect of RGH 4615 and of other N-benzyl-amides is related to the presence of diacyl-amines in brain. These diacyl-amines show hydrolytic and acylating reactivity resembling antiepileptic drugs of diacyl-amine structure, such as hydantoines, oxazolidinediones, acyl-ureas, etc.

This work was supported by the Chemical Works of Gedeon Richter Ltd. Budapest.

REFERENCES

1/   European Patent 55484A1., L. Kisfaludy, L. Ürögdi,
     A. Patthy, L. Dancsi, J. Szillbereky, E. Moravcsik,
     H. Tüdős, L. Ötvös, Zs. Tegyey, É. Pálosi, Á. Sarkadi,
     L. Szporny.
     European Patent 55490A1. L. Kisfaludy, L. Ürögdi,
     A. Patthy, E. Moravcsik, H. Tüdős, L. Ötvös, Zs. Tegyey,
     É. Pálosi, A. Sarkadi, L. Szporny.

Bio-Organic Heterocycles
van der Plas H.C., Ötvös L., and Simonyi M. eds

BIOTRANSFORMATION OF 5-ETHYL- AND 5-ISOPROPYL-2'-DEOXYURIDINES IN MICE AND RATS

SZINAI I., BIHARI M.,* UJSZÁSZY K., VERES Zs.,
SZABOLCS A., HOLLY S., GÁCS-BAITZ E. and ÖTVÖS L.

Central Research Institute for Chemistry, The Hungarian
Academy of Sciences, Budapest, Pf. 17, H-1525, Hungary

*Chemical Works of Gedeon Richter Ltd., Budapest, Hungary

5- Ethyl- and 5-isopropyl-2'-deoxiuridines (e^5dUrd, ip^5dUrd) have been shown to be potent inhibitors of Herpes simplex virus replication. Both compounds are pharmacologically active in nucleoside form. While e^5dUrd is the active component of AEDURID used widespreadly in topical treatment in dermatology and ophtalmology ip^5dUrd is recently under clinical trial. The combination of e^5dUrd with 5-fluorouracil has been shown to have cytostatic effect on human colorectal xenografts (1). According to earlier studies, e^5dUrd was a good substrate of pyrimidine nucleoside phosphorylases in contrast to ip^5dUrd. 5-Ethyl- and 5-isopropyluracil, the products of these *in vitro* enzyme reactions were not substrates of dihydro-uracil dehydrogenase [2,3]. We studied the metabolism of e^5dUrd and ip^5dUrd in mice and rats.

METHODS

$2\text{-}^{14}C\text{-}e^5$dUrd and $2\text{-}^{14}C\text{-}ip^5$dUrd were administered i.p. in single doses of 200 and 210 mg/kg to male CFLP mice and male Wistar rats, respectively. Urine and faeces were collected from 0 to 72 hours. Urine collected during 24 hours was purified on Extrelut column and by TLC (silica gel). Metabolites were eluted from TLC plates by Camag eluchrom instrument. The structure of radioactive metabolites was determined by UV, MS, IR, NMR and CD spectroscopy

RESULTS AND DISCUSSION

The excretion of radioactivity is shown in Figure 1. Unchanged e^5dUrd, ip^5dUrd and their metabolites were excreted quickly into urine (85-95 %) from both species during a day. No radioactivity appeared in expired CO_2 from e^5dUrd and ip^5dUrd in constrast to thymidine (Figure 1).

Figure 1.: Excretion of radioactivity

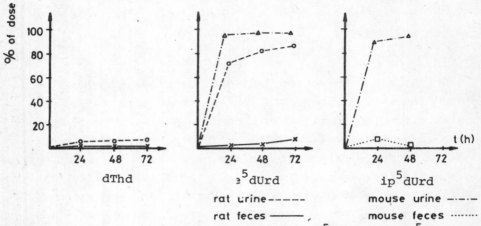

dThd e^5dUrd ip^5dUrd

rat urine - - - - -   mouse urine - · - · -
rat feces ————— ,   mouse feces ·········

The main routes of biotransformation of e^5dUrd and ip^5dUrd are A./ The cleavage of N-glycosidic bond by pyrimidine nucleoside phosphorylases, and B./ hydroxylation of the alkyl side-chain. Transformation of the two nucleoside analogues was qualitatively similar with different rates along route A./ (Figure 2). We isolated e^5dUrd (52,4 %) and ip^5dUrd (80 %) in unchanged form as well as 5-ethyluracil (25,4 %) and 5-isoprpyluracil (1-2 %) from urine of mice and rats. Two other hydroxylated metabolites 5-(1-hydroxyethyl)-uracil (17,4 %) and 5-[(1-methyl-2-hydroxy)-ethyl]-2'-deoxyuridine (12-15 %) were also identified. Hydroxylation of ethyluracil occured at α-position of the alkyl side-chain while β-hydroxyderivative was formed from ip^5dUrd.

The hydroxylation of the 5-ethyl group of e^5dUrd and/or ethyluracil yielded 5-(1-hydroxyethyl)-uracil with the same chirality in both mice and rats according to CD spectra.(Fig.3.) The determination of absolute configuration is envisaged.

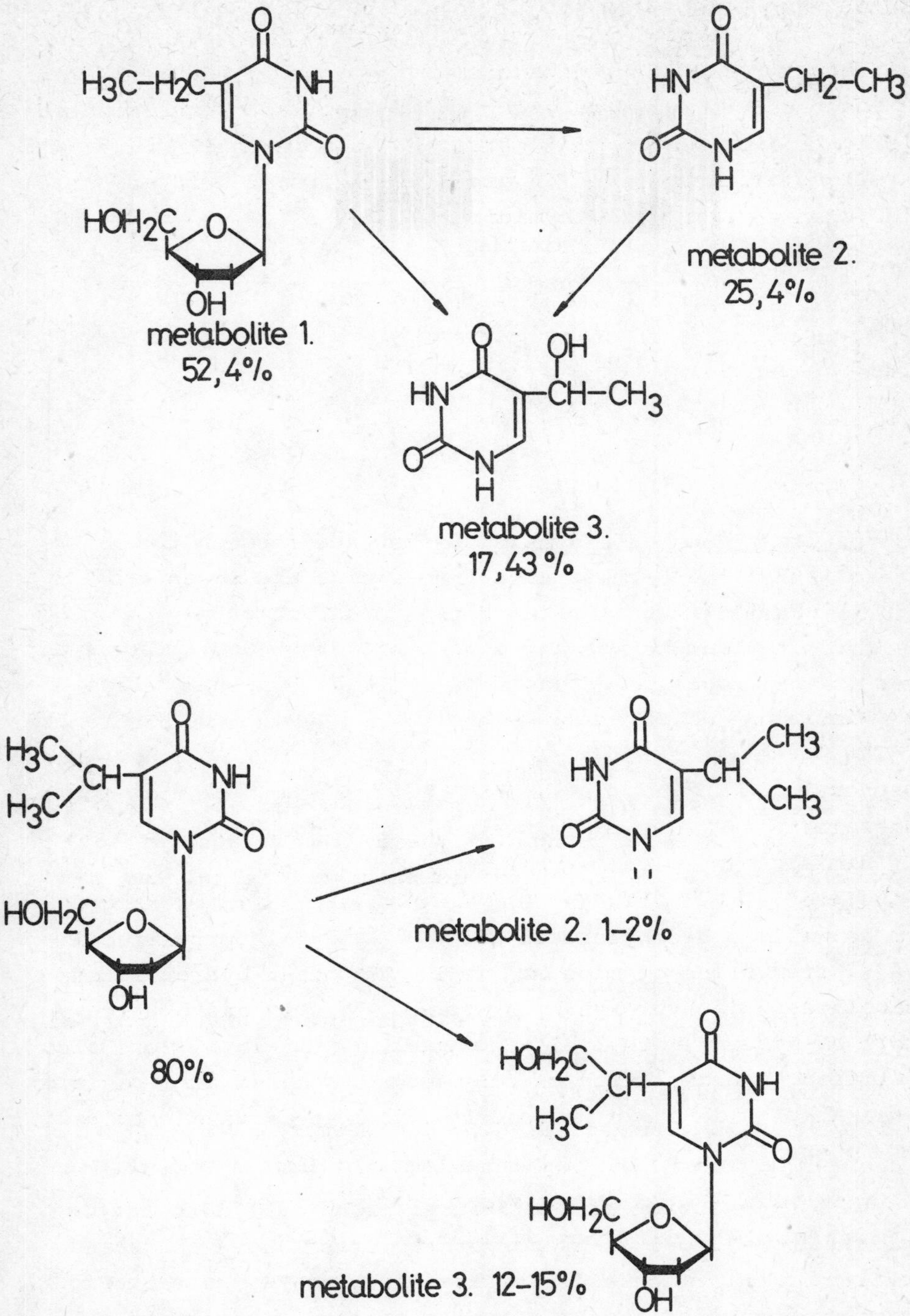

Figure 2.: Metabolic pathways of e⁵dUrd and ip⁵dUrd

Figure 3. CD spectra of 5-(1-hydroxyethyl)-uracil identified
from urine of mice and rats

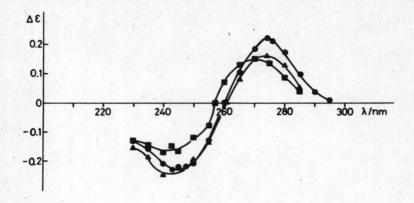

Our results showed an essential difference between the
catabolism of the 5-substituted nucleoside analogues and
natural pyrimidine nucleosides. This observation may help
designing new antiherpes and citostatic drugs and understanding
their pharmacological action.

REFERENCES

1. Kopper, L., Magyarosi, E., Jeney, A., Lapis, K., Szabolcs A.
   and Ötvös, L. Potentiation of the 5-fluorouracil effect
   with 5-ethyl-2'-deoxyuridine in human colorectal tumor
   xenografts
   Oncology, in press.
2. Newmark, P., Stephens, J.D. and Barrett, H.W.
   Substrate specificity of the dihydro-uracil dehydrogenase
   and uridine phosphorylase of rat liver
   Biochim. Biophys, Acta, 62 414-416 (1962)
3. Barrett, H.W., Munavalli, S.N. and Newmark, P.
   Synthetic pyrimidines as inhibitors of uracil and thymine
   degradation by rat liver supernatant
   Biochim. Biophys, Acta, 91 199-204 (1964)

Bio-Organic Heterocycles
van der Plas H.C., Ötvös L., and Simonyi M. eds

THE EFFECT OF 5-ALKYL-PYRIMIDINES AND 5-ALKYL-PYRIMIDINE NUCLEOSIDES ON THE CATABOLISM OF 5-FLUOROURACIL *IN VIVO*

VERES Zs., SZABOLCS A., SZINAI I., KOVÁCS T., DÉNES G. and ÖTVÖS L.

Central Research Institute for Chemistry, The Hungarian Academy of Sciences, Budapest, Pf. 17, H-1525, Hungary

5-Fluorouracil (FU) has been used as anticancer agent for many years. Owing to the catabolic pathway of this compound, CO_2 appears in the expired air [1]. The rapid catabolism is a factor limiting the use of FU in the treatment of cancer [2]. Different means suggested for enhancing the effect of FU include the prolonged release of FU by Ftorafur which acts as a depot form of FU [3]; the coadministration of FU with thymidine [4] and the inhibition of dihydro-uracil dehydrogenase by 5-cyanouracil [5].
We have studied the effect of 5-substituted pyrimidines and 5-substituted pyrimidine nucleosides on the catabolism of FU *in vivo.*

METHODS

[2-^{14}C] FU alone, or together with either 5-alkyl-pyrimidine nucleosides or 5-alkyl-pyrimidines was administered i.p. to male WISTAR rats. The animals were placed immidiately into metabolism chambers (SYMAX) for collection of urine and expired $^{14}CO_2$. $^{14}CO_2$ was trapped in 1-amino-3-methoxy-propane and then counted in a scintillation cocktail containing toluene. Pyrimidine nucleoside phosphorylases were prepared from mouse (male, CFLP) liver and rat (male, CFY) intestinal mucosa. Tissues were homogenised in 3 volumes of 0.02 M potassium phosphate buffer (pH 8.0) containing 10 mM 2-mercaptoethanol and 1 mM EDTA. The homogenates were centrifuged at 4° for 1.5 hr at 100,000 g and the cytosols were treated with ammonium

sulfate. The precipitate obtained between 35-65 % saturation
was resuspended in potassium phosphate buffer containing
2-mercaptoethanol and EDTA, as described before, and dialysed
against the same buffer.
The phosphorolysis of natural pyrimidine nucleosides and
pyrimidine nucleoside analogues was carried out according to
the method of Yamada [6].

RESULTS

The distribution of radioactivity formed from $[2-^{14}C]FU$
between urine and CO_2 was dose dependent. On decreasing the
dose of FU, more counts were recovered as expired CO_2 and less
were found in the urine, while the total recovery was relatively
constant (in percent of the injected dose, Table 1).

Table 1. Recovery of radioactivity from rats after administration of $[2-^{14}C]FU$.

| | Dose of $[2-^{14}C]FU$ (mmole/kg) | | | | |
	1.54	1.15	0.77	0.28	0.03
$^{14}CO_2$ expired within 8 hours (% of dose administered)	47	48	61	74	78.5
Radioactivity excreted with urine within 24 hours (% of dose administered)	31	43	23	12	9

The catabolism of $[2-^{14}C]FU$ into expired $^{14}CO_2$ was reduced by
thymidine (dThd), thymine (Thy), uridine (Urd), uracil (Ura),
5-ethyl-2'-deoxyuridine (e^{5}dUrd), 5-ethyluridine (e^{5}Urd),
5-ethyluracil (e^{5}Ura) and 5-(1-dl-hydroxyethyl)uracil (he^{5}Ura)
(Table 2). The N-glycosidic bond in e^{5}dUrd and e^{5}Urd is
cleaved by pyrimidine nucleoside phosphorylases giving e^{5}Ura
(Fig.1). Arabinofuranosyl-5-ethyluracil (ara-e^{5}Urd) and the
α-anomer of e^{5}dUrd being no or weak substrates of pyrimidine
nucleoside phosphorylases do not influence the catabolism of

FU. Hence, base formation seems important for this effect.
FU is degraded in rat liver by a series of reactions, the first
of which being a reduction of the 5,6-double bond by dihydro-
uracil dehydrogenase [1]. Though e^5Ura and he^5Ura are not
substrates of the dehydrogenase [7,8] they likely bind to the
enzyme and decrease the degradation of FU. 5-Hexyluracil
(hex^5Ura) and 5-isopropyluracil (ip^5Ura) do not reduce the
amount of $^{14}CO_2$ formed from $[2-^{14}C]$FU.

Table 2. Effect of pyrimidine bases and pyrimidine nucleosides
         on the catabolism of FU.

	$^{14}CO_2$ expired within 8 hours	Radioactivity excreted with urine within 24 hours
	(% of dose* administered)	
Control	78.5	9
dThd	14	71
Thy	4	66
Urd	27.5	51
Ura	44.5	34.5
e^5dUrd	41.5	37
αe^5dUrd	76	20.5
e^5Urd	46	34
ara-e^5Urd	76	12
e^5Ura	31	44
he^5Ura	11.5	70
ip^5Ura	77.5	6
hex^5Ura	72	12

* Dose of $[2-^{14}C]$FU was 0.03 mmole/kg body weight, that
of he^5Ura was 2.86 mmole/kg and dose of all other compounds was 5.71 mmole/kg.

Inhibiton of the degradation of FU resulted in increasing
toxicity of this compound. The LD_{50} value of FU alone was
30 mg/kg when it was administered to male CFLP mice for
7 days (once a day) while the LD_{50} was 3 mg/kg when FU was
administered together with e^5dUrd or e^5Ura (Fig.2).

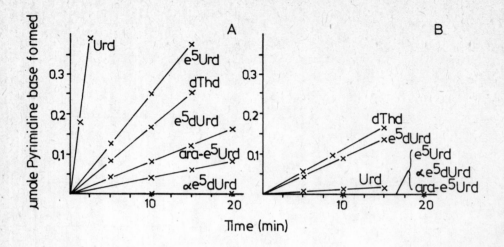

Fig.1. Phosphorolysis of natural and 5-alkyl-pyrimidine nucleosides in the presence of uridine (A) and thymidine (B)
phosphorylase. The substrate concentration was 3.33 mM.

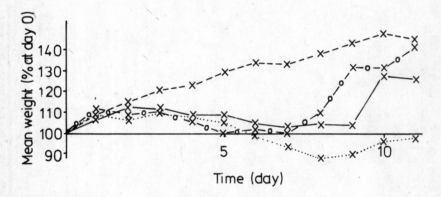

Fig.2. Weight change for groups of mice treated with FU alone
or in combination with e^5dUrd or e^5Ura. Compounds were
administered i.p. for 7 days (once a day). Key: x---x
control; x——x 30 mg/kg FU; x-o-x 3 mg/kg FU + 250 mg/kg
e^5dUrd; x...x 3 mg/kg FU + 137 mg/kg e^5Ura.

Our results suggest that the biological half life of FU can be prolonged significantly by blocking its catabolism.

REFERENCES

1. Chaudhuri,N.K., Mukherjee,K.L. and Heidelberger,C. Studies on Fluorinated Pyrimidines.VII. The Degradative Pathway. Biochem.Pharmacol., 1, 328-341 (1958)
2. Heidelberger,C. Fluorinated Pyrimidines. Progr.Nucleic Acid Res.Mol.Biol., 4, 1-50 (1965)
3. Garibjanian,B.T., Johnson,R.K., Kline,I., Vadlamudi,S., Gang,M., Venditti,J.M. and Goldin,A.Comparison of 5-Fluorouracil and Ftorafur. II. Therapeutic Response and Development of Resistance in Murine Tumors. Cancer Treat. Rep., 60, 1347-1361 (1976)
4. Spiegelman,S., Nayak,R., Sawyer,R., Stolfi,R. and Martin,D. Potentiation of the Anti-Tumor Activity of 5FU by Thymidine and its Correlation with the Formation of /5FU/RNA. Cancer, 45, 1129-1134 (1980)
5. Gentry,G.A., Morse,P.A.Jr. and Dorsett,M.T. *In Vivo* Inhibition of Pyrimidine Catabolism by 5-Cyanouracil. Cancer Res., 31, 909-912 (1971)
6. Yamada,E.W.Pyrimidine Nucleoside Phosphorylases of Rat Liver. J.Biol.Chem., 243, 1649-1655 (1968)
7. Newmark,P., Stephens,J.D. and Barrett,H.W.Substrate Specificity of the Dihydro-Uracil Dehydrogenase and Uridine Phosphorylase of Rat Liver. Biochim.Biophys.Acta, 62, 414-416 (1962)
8. Barrett,H.W., Munavalli,S.N. and Newmark,P.Synthetic Pyrimidines as Inhibitors of Uracil and Thymine Degradation by Rat-Liver Supernatant. Biochim.Biophys.Acta, 91, 199-204 (1964)

STUDIES ON MITOMYCIN ANALOGS; A NOVEL APPROACH TO THE PYRROLO [1,2-a] INDOLE SKELETON

VERBOOM W. and REINHOUDT D.N.

Laboratory of Organic Chemistry, Twente University
of Technology, P.O. Box 217, 7500 AE Enschede,
The Netherlands

The mitomycins form an interesting class of antitumor anti-biotics. In spite of its relatively high toxicity mitomycin C (1) is currently employed clinically in the treatment of several cancers. Therefore work is in progress by many groups to synthesize less toxic analogs. One of the key problems in the synthesis of such compounds is the construction of the pyrrolo-[1,2-a]indole skeleton 2. [1]

Until recently it was generally accepted that in apolar solvents enamines react with electron-deficient acetylenes like dimethyl acetylenedicarboxylate (DMAD) to give (unstable) *cis*-fused 3-(dialkylamino)cyclobutenes which could ring open in a disrotatory mode to *cis,cis*-1,3-alkadienes. Recently we have proven that this ring opening proceeds in a symmetry allowed conrotatory fashion with formation of *cis,trans*-1,3-alkadienes which in some cases isomerize to *cis,cis*-1,3-alkadienes by means of a [1,5] hydrogen shift. [2] However, in methanol these 3-(1-pyrrolidinyl)cyclobutenes underwent a thermal rearrangement to pyrrolizine derivatives. Moreover, pyrrolizines were also formed directly by reaction of (1-pyrrolidinyl)enamines with DMAD in a protic polar solvent like methanol. [3] For instance, reaction of enamine 3 with DMAD in methanol afforded

E = COOMe

3   4   5

the *cis*-pyrrolizine 4 in high yield. This product was also obtained by dissolving in methanol of the corresponding cyclobutene, prepared from 3 and DMAD in an apolar solvent. We observed that in both cases the reaction proceeded via the Michael adducts 5. This result led us to investigate if in general 1-(1-pyrrolidinyl)-1,3-butadienes could be converted into pyrrolizines. We found that the 1-(1-pyrrolidinyl)-1,3-butadienes 6a,b cyclized to the pyrrolizines 7a,b, respectively [4], by heating in 1-butanol or in acetonitrile in the presence of zinc chloride. For this particular study we also prepared 8, in which one of

6a, R = OPh   7a, R = OPh   8   9a (α-E)
b, R = Ph   b, R = Ph   b (β-E)

the double bonds of the 1-(1-pyrrolidinyl)-1,3-butadiene moiety constitutes part of a heteroaromatic system, by reaction of 2-(1-pyrrolidinyl)benzo[b]thiophene with DMAD in methanol. Both (E)-8 and (Z)-8, which were not interconverted under the reaction conditions, underwent a thermal rearrangement viz. in refluxing toluene to exclusively the *trans*-thieno[3,2-b]pyrrolizine 9b and in refluxing 1-butanol to a 2:1 isomer mixture of the *cis*- and *trans*-thieno[3,2-b]pyrrolizines 9a and 9b, respectively. In this particular case we were able to study the mechanism of the pyrrolizine formation in detail. The rate of reaction generally increases with increasing polarity of the solvent and is dependent on the configuration of the 1-(1-pyrrolidinyl)-1,3-butadiene ($k_{(E)-\underline{8}}/k_{(Z)-\underline{8}} = 5.2$). The formation of pyrrolizines can be explained by two consecutive reactions. The first comprises a concerted antarafacial [1,6] hydrogen shift that generates a 1,5-dipolar species, e.g. $\underline{10a}$, that may undergo stereomutation to $\underline{10b}$; in the second reaction a disrotatory electrocyclization of the 6π-electron system takes place to give

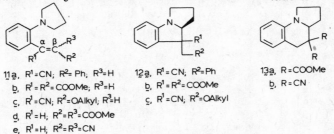

$$\underline{8} \xrightarrow{[1,6]-H}$$

10a   10*   10b

Disrot.   Disrot

9b   9a

Scheme 1

the corresponding pyrrolizines. Stereoselective reactions of the deuterium labelled *E/Z*-isomers of $\underline{8}$ proved that the cyclization takes place from a helical conformation. [4] In relation to the mitomycins we considered 2-vinyl-*N*,*N*-dialkylanilines $\underline{11}$, a potentially interesting class of starting materials for the synthesis of analogs. In these compounds one of the double bonds of the 1-(1-pyrrolidinyl)-1,3-butadiene moiety *constitutes part of an aromatic ring*. We found that compounds $\underline{11a\text{-}c}$ with an elec-

$\underline{11a}$, R^1=CN; R^2= Ph; R^3=H
    $\underline{b}$, R^1=R^2= COOMe; R^3=H
    $\underline{c}$, R^1=CN; R^2=OAlkyl; R^3=H
    $\underline{d}$, R^1=H; R^2=R^3=COOMe
    $\underline{e}$, R^1=H; R^2=R^3=CN

$\underline{12a}$, R^1=CN; R^2=Ph
    $\underline{b}$, R^1=R^2=COOMe
    $\underline{c}$, R^1=CN; R^2=OAlkyl

$\underline{13a}$, R=COOMe
    $\underline{b}$, R=CN

347

tron-withdrawing group at the α-position cyclized to a *cis,trans* mixture of the corresponding pyrrolo[1,2-*a*]indoles 12a-c under various reaction conditions depending on the nature of the substituents. In the case of 11c even refluxing in mesitylene for several days was necessary to perform the reaction. [5] Starting from 11d,e, with two electron-withdrawing groups at the β-position, another type of cyclization occurred namely formation of the pyrrolo[1,2-*a*]quinolines 13a,b. In these cases pyrrolo[1,2-*a*]indole formation is not possible because of the lack of stabilization of the negative end of the 1,5-dipole required for this type of cyclization [6]. Preparation of substituted pyrrolo-[1,2-*a*]indoles and further modifications of these compounds for the synthesis of new mitomycin analogs is currently under investigation.

REFERENCES

[1] T. Kametani and K. Takahashi, Synthesis of pyrrolo[1,2-*a*]indoles and related systems, Heterocycles 9, 293-351 (1978).

[2] D.N. Reinhoudt, W. Verboom, G.W. Visser, W.P. Trompenaars, S. Harkema and G.J. van Hummel, Reactivity of cis-fused 3-(dialkylamino)cyclobutenes in polar and apolar solvents. Synthesis, X-ray structures and reactions of *cis,cis*- and *cis,trans*-1,3-cycloalkadienes, J. Am. Chem. Soc. 106, 1341-1350 (1984).

[3] W. Verboom, G.W. Visser, W.P. Trompenaars, D.N. Reinhoudt, S. Harkema and G.J. van Hummel, Synthesis of pyrrolizines by intramolecular capture of 1,4-dipolar intermediates in reactions of enamines with dimethyl acetylenedicarboxylate, Tetrahedron 37, 3525-3533 (1981).

[4] D.N. Reinhoudt, G.W. Visser, W. Verboom, P.H. Benders and M.L.M. Pennings, In situ generation of 1,5-dipoles by concerted [1,6] hydrogen transfer. Stereoselective thermal rearrangement of 1-(1-pyrrolidinyl)-1,3-butadienes to pyrrolizines, J. Am. Chem. Soc. 105, 4775-4781 (1983).

[5] W.C. Dijksman, W. Verboom, D.N. Reinhoudt, C.G. Hale, S. Harkema and G.J. van Hummel, Novel applications of the "*t*-amino effect" in heterocyclic chemistry ; Synthesis of 1-alkylindoles, Tetrahedron Lett. 25, 2025-2028 (1984).

[6] W. Verboom, D.N. Reinhoudt, R. Visser and S. Harkema, "*tert*-Amino effect" in heterocyclic synthesis. Formation of *N*-heterocycles by ring-closure reactions of substituted 2-vinyl-*N,N*-dialkylanilines, J. Org. Chem. 49, 269-276 (1984).

Bio-Organic Heterocycles
van der Plas H.C., Ötvös L., and Simonyi M. eds

STERIC STRUCTURE-SUBSTRATE PROPERTIES OF (E)-5-(2-BROMOVINYL)-2'-DEOXYURIDINE-5'-TRIPHOSPHATE ANALOGUES IN DNA POLYMERIZATION REACTION

ÖTVÖS L., KOVÁCS T., SÁGI J., SZEMZŐ A. and SZABOLCS A.

Central Research Institute for Chemistry, The Hungarian Academy of Sciences, Budapest, Pf. 17, H-1525, Hungary

Summary

The selective antiherpes virus (HSV-1) action of (E)-5-(2-bromovinyl)-2'-deoxyuridine (E-bv^{5}dU) was recently correlated with the incorporation of E-bv^{5}dU into HSV-1 DNA [1].

To elucidate the role of steric structure of the side-chain of E-bv^{5}dU in DNA polymerase-catalysed reaction we have synthesized some structural analogues of the 5'-triphosphate of E-bv^{5}dU. A positive correlation was found between incorporation of analoques of E-bv^{5}dU into DNA in E. coli DNA polymerase I-catalysed polymerization reaction and their antiviral activity in cell culture published earlier [2,3].

Introduction

(E)-5-(2-bromovinyl)-2'-deoxyuridine is one of the most potent and selective antiviral agents in both cell culture and experimental animals [2]. First step in selective inhibition of virus DNA replication is assumed to be the phosphorylation of this nucleoside by virus-induced thymidine kinase in virus-infected cells [4]. The antiviral activity well correlated with the incorporation of 5'-triphosphate of E-bv^{5}dU into HSV-1 DNA catalyzed by HSV-1 induced DNA polymerase [1]. To investigate the role of molecular structure of 5-substituent in E-bv^5-dU we have synthesized the following structural analogues of 5'-triphosphate of E-bv^{5}dU (Fig.1.) and studied the rate and extent of their incorporation into DNA in E. coli DNA polymerase I-catalysed polymerization reaction.

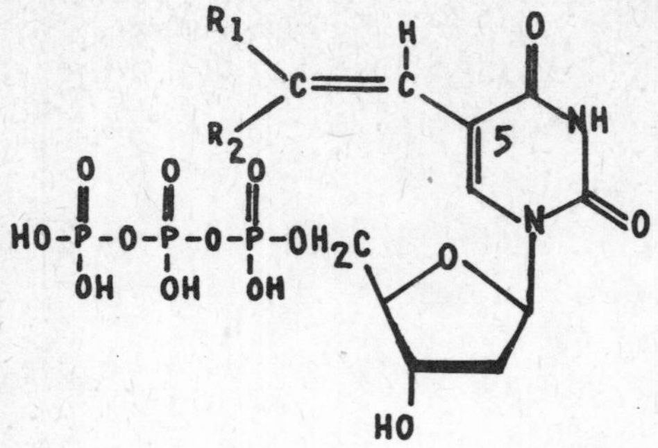

R_1	R_2	Abbr.
H	H	v^5dU
Br	H	$E\text{-}bv^5dU$
H	Br	$Z\text{-}bv^5dU$
Br	Br	b_2v^5dU

Fig.1.: 5-Substituted 2'-deoxyuridine-5'-triphosphates

The use of the E. coli polymerase I was based on the similarity in substrate properties of DNA polymerase enzymes from different sources for 5-substituted dUTPs [5,6,7]. E.g., mean K_m value for dTTP of HSV-1 DNA polymerase is about 14 times lower and mean K_i value for $E\text{-}bv^5dUTP$ of HSV-1 polymerase is about 22 times lower than those of the human α-polymerase. Thus, both dTTP and $E\text{-}bv^5dUTP$ have higher binding to the HSV-1 DNA polymerase, than to the α-polymerase, but K_m/K_i ratios (relative binding) are similar [5,6]. Rate of incorporation by five different enzymes of $E\text{-}bv^5dU$ is also similar [6]. Substrate specificity for 5-alkyl-dUTPs of mammalian DNA polymerases and the E. coli DNA polymerase I enzyme was also comparable [6,7].
Based on these data, we assume that the relative incorporation rate of dUTP analogues (a comparison with dTTP) determined in E. coli DNA polymerase I-catalysed polymerization reaction may be extrapolated to HSV-I DNA polymerase system.

Materials and Methods

Modified nucleosides were prepared by Wittig-reaction, the corresponding nucleotides were synthesized by the procedure described by Ludwig [8].

Escherichia coli MRE 600 DNA polymerase I Klenow fragment enzyme (6900 units/mg), dTTP and poly[d(A-T)] were purchased from Boehringer-Mannheim GmbH. [³H]dATP (17 Ci/mmol) was from New England Nuclear.

Enzymatic polymerization of deoxyuridines was followed by the [^{3}H]dAMP incorporation into acid-insoluble polymer product in the presence of poly[d(A-T)] template-primer.

Reaction mixture: 50 mM phosphate buffer (pH 7.4), 5 mM MgCl$_2$, 0.1 mM [^{3}H]dATP, 0.25 mM dNTP, 0.05 mM(P) template-primer and 0,5 mg enzyme/50 µl of mixture.

The polymerization reaction was carried out in the same way as described by Sági et al [9].

Results and Discussion

Table 1. and Fig.2. show the time-course of incorporation of E-bv^{5}dU analogues into acid-insoluble polymer product cataly-sed by E. coli DNA polymerase I Klenow fragment enzyme. In comparison with dTTP, at saturating substrate concentrations the extent of incorporation of v^{5}dUTP and E-bv^{5}dUTP were 100% and 96%, respectively. However, in cases of Z-bv^{5}dUTP and b$_2$v^{5}dUTP decreased rate of [^{3}H]dAMP incorporation was found.

Table 1.: Time-course of incorporation of E-bv^{5}dU analogues into poly[d(A-T)]

Substrates: [^{3}H]dATP'+	Time of incubation				
	30 min nmol	60 min nmol	120 min nmol	300 min nmol	24 h nmol
dTTP	1.35	1.85	2.00	2.09	2.40
v^{5}dUTP	1.32	1.78	2.00	2.00	2.40
E-bv^{5}dUTP	0.84	1.20	1.55	1.84	2.20
Z-bv^{5}dUTP	0.30	0.42	0.46	0.50	0.62
b$_2$v^{5}dUTP	0.27	0.41	0.50	0.58	0.84

The kinetic data observed emphasize the importance of structure and configuration of substituted 5-vinyl side chain in polymerization reaction. Based on the data on the extent of incorporation, orientation of side-chain in cases of v^{5}dU and E-bv^{5}dU - as determined by X-ray crystallography [10,11]-

does not hinder formation of phosphodiester bond.

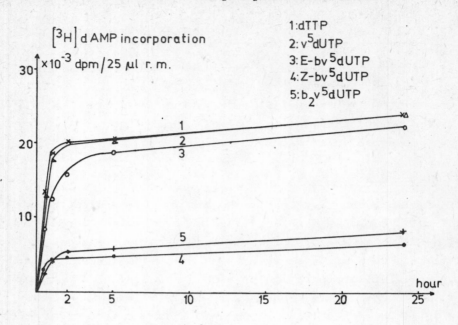

Fig.2.: Incorporation of [³H]dAMP into poly[d(A-T)] in the presence of E-bv⁵dU-analogues

On the contrary, data on initial rate and final incorporation of Z-bv⁵dU and b₂v⁵dU show that Br atom in Z-position inhibits polymerization reaction.

Influence of structure and configuration of the 5-substituent is in agreement with the concept of "orientational steric substituent effect" [12].

Incorporation of E-bv⁵dU analogues into a synthetic DNA in E. coli DNA polymerase I-catalysed polymerization reaction can be well correlated with their published antiviral activity as expressed by minimal inhibitor concentration (Table 2.). Thus, incorporation of E-bv⁵dU analogues into DNA may have a decisive role in the antiviral capacity. For the application of a compound as a drug, however, selectivity also has to be taken into account; v⁵dU is not selective since normal cell thymidine kinases can phosphorylate it [13].

Table 2.: Relative rate and extent of incorporation of E-bv^{5}dU analogues and their ·minimal inhibition concentrations

Substrates: [^{3}H]dATP +	Incorporation of analogs		Minimal inhibition concentration [2,3] MIC[ug/mL]
	rate at 30 min, %	extent at 24 h, %	
dTTP	100	100	
v^{5}dUTP	96	100	0.018
E-bv^{5}dUTP	62	92	0.005–0.008
Z-bv^{5}dUTP	22	25	0.1
b$_2$v^{5}dUTP	20	35	0.1

References

1. Mancini, W.R., DeClercq, E. and Prusoff, W.H.: The Relationship between ·Incorporation of /E/-5-/2-Bromovinyl/-2'-deoxyuridine into Herpes Simplex Virus Type 1 DNA with Virus Infectivity and DNA Integrity.
   J.Biol.Chem. 258 /2/ 792-795 (1983)
2. DeClercq, E.: Antiviral Activity of 5-Substituted Pyrimidine Nucleoside Analogues
   Pure and Appl.Chem. 55 /4/ 623-636 (1983)
3. Goodchild, J., Porter, R.A., Raper, R.H., Sim, I.S., Upton, R.M., Viney J., and Wadsworth H.J.; Structural Requirements of Olefinic 5-Substituted Deoxyuridines for Antiherpes Activity.
   J.Med.Chem. 26 1252-1257 (1983)
4. Cheng, Y.-C., Dutschman, G., DeClercq, E., Jones A.S., Ra im, S.G.Verhelst, G., and Walker, R.T.: Differential Affinities of 5-/2-Halogenovinyl/-2'-Deoxyuridines for Deoxythymidine Kinases of Various Origins.
   Mol.Pharm. 20 230-233 (1981)

5. Allaudeen, H.S., Kozarich, J.W., Bertino, J.R. and DeClercq, E.: On the Mechanism of Selective Inhibition of Herpes Virus Replication by /E/-5-/2-bromovinyl/-2'-deoxyuridine. Proc.Natl.Acad.Sci. USA 78 5 2698-2702 (1981)

6. Ruth, J. L. and Cheng, Yung-Chi.: Nucleoside Analogues with Clinical Potential in Antivirus Chemotherapy. The effect of Several Thymidine and 2'-Deoxycytidine Analogue 5'-triphosphates on Purified Human (α,β) and Herpes Simplex Virus (Types 1,2) DNA Polymerases Mol.Pharm. 20 415-422 (1981)

7. Sági, J., Nowak, R., Zmudzka, B., Szemző, A. and Ötvös L.: A Study of Substrate Specificity of Mammalian and Bacterial DNA Polymerases with 5-Alkyl-2'-deoxyuridine-5'-triphosphates Biochim. et Biophys Acta 606 196-201 (1980)

8. Ludwig, J.: A New Route to Nucleoside 5'-triphosphates. Acta Biochim. et Biophys. Acad.Sci.Hung. 16 /3-4/ 131-133 (1981)

9. Sági, J., Szabolcs, A., Szemző, A., and Ötvös, L.: /E/-5-/2-bromovinyl/-2'-deoxyuridine-5'-triphosphate as a DNA Polymerase Substrate. Nucl.Acids Res. 9 /24/ 6985-6994 (1981)

10. Hamor, T.A., O'Leary, M.K. and Walker, R.T.: Antiviral Nucleic Acid Derivatives. II. Crystal Structure of 5-Vinyl-2'-deoxyuridine. Acta Cryst. B34, 1627-1630 (1978)

11. Párkányi, L., Kálmán, A., Kovács, T., Szabolcs, A. and Ötvös, L.: Crystal and Molecular Structure of /E/-5-/2-bromovinyl/-2'-deoxyuridine. Nucl.Acids Res. 11 /22/ 7999-8005 (1983)

12. Ötvös, L., Elekes, I., Kraicsovits, F., and Moravcsik, E.: Explanation of Steric Substituent Effects in Reaction of Biopolymers in Steric Effects in Biomolecules Ed: G. Náray-Szabó, Akadémiai Kiadó, Budapest, pp.305-325 (1982)

13. Langen, P. and Bärwolff, D.: On the Mode of Action of 5-Vinyl-2'-deoxyuridine. Biochem.Pharm. 24 1907-1910 (1975)

SUBSTRATE SPECIFICITY OF E. COLI RNA POLYMERASE FOR 5-ALKYL-UTPs

SÁGI J., SZEMZŐ A., SZABOLCS A. and ÖTVÖS L.

Central Research Institute for Chemistry, The Hungarian Academy of Sciences, Budapest, Pf. 17, H-1525, Hungary

Enzymatic synthesis of ribopolynucleotides containing one or more either natural or modified nucleotides has been carried out so far almost exclusively with polynucleotide phosphorylase enzyme from the corresponding nucleoside diphosphates. Preparation of some modified polynucleotides, e.g. poly(5-ethyluridine) could only be realized, however, under special conditions [1].

We were interested in preparing poly(5-alkyluridine) copolymers of defined sequence. To avoid difficulties, we planned to use another enzyme, the DNA template-directed RNA polymerase of Escherichia coli. This report discusses a structure - activity study of 5-alkyl-UTPs in the RNA polymerase-catalyzed polymerization reaction.

MATERIALS AND METHODS

5-Alkyluridines and the corresponding 5'-triphosphates (r^5UTP) were synthesized by Szemző et al. [2]. Poly[d(A-pr^{5}U)] was prepared by E. coli DNA polymerase I enzyme as described earlier [3]. E. coli MRE 600 RNA polymerase (700 units/mg), UTP, and poly[d(A-T)] were purchased from Boehringer (Mannheim), [^{3}H]ATP (1 TBq/mmol) was from Amersham. Activation of poly[d(A-T)] was carried with pancreatic DNase [4].

The reaction mixture [4] in 82 µl contained 300 µM [^{3}H]ATP (38 dpm/pmol); 300-300 µM of 5-alkyl-UTP, CTP and GTP; 200 µM(P) of synthetic DNA or 250 µg/ml calf thymus DNA template. Reactions were started by addition of 16 µg RNA polymerase. Samples (10 or 50 µl) taken were precipitated onto

GF/C filters by acid, then washed, dried and counted. In inhibition experiments 100 μM UTP was used.

RESULTS AND DISCUSSION

I. Initial incorporation rate

Substrate specificity of E. coli RNA polymerase was characterized by determining the initial rate and extent of incorporation of UTP-analogues into acid-insoluble product.

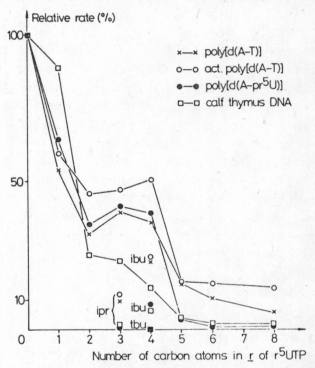

Fig.1: Initial rate of incorporation of UTP-analogues into RNA product

With the exception of 5-t-butyl-UTP, each 5-alkyl-UTP examined was a substrate of the E. coli RNA polymerase (Fig.1 and Table 1). Incorporation rate of the analogues, however, was lower than that of the UTP (n=o) and depended on both substitution and, to some extent, the structure of DNA template applied. The fall in relative rate was sharp when the

Table 1.: Initial rates of transcription with 5-alkyl-UTPs

TEMPLATES

SUBSTRATES: [^{3}H]ATP +	poly[d(A-T)]		poly[d(A-T)] activated		poly[d(A-pr^{5}U)]		calf thymus DNA	
	a	b	a	b	a	b	a	b
-	0.03	-	0.02	-	0.03	-	0.05	-
UTP	1.83	100	0.57	100	2.80	100	1.52	100
5-methyl-UTP	1.00	54	0.35	60	1.81	64	1.35	89
5-ethyl-UTP	0.61	32	0.27	46	1.00	35	0.36	24
5-n-propyl-UTP	0.74	39	0.28	47	1.18	41	0.33	22
5-n-butyl-UTP	0.67	36	0.30	51	1.11	39	0.20	13
5-n-pentyl-UTP	0.30	15	0.11	16	0.12	3	0.05	3
5-n-hexyl-UTP	0.22	11	0.11	16	0.05	1	0.03	2
5-n-octyl-UTP	0.14	6	0.10	14	0.06	1	0.03	2
5-isopropyl-UTP	0.20	10	0.09	12	0.05	1	0.02	1
5-isobutyl-UTP *	0.43	22	0.16	25	0.27	9	0.09	6
5-t-butyl-UTP	0.03	0	0.02	0	0.03	0	0.01	0

a = Transcription rate: nmol of [^{3}H]AMP incorporation/10 min.
b = Relative rate (%). * Isobutyl = 2-methyl-propyl.

5-hydrogen atom of uracil in UTP was replaced by methyl or ethyl group, and UTPs substituted with long alkyl chain (pentyl to octyl) were only very weak substrates. Reduced rates may come from the increasing steric inhibition ("orientational and compressional steric substituent effect" [5]) by the 5-substituent on the formation of phosphodiester linkage. The hydrophobicity of the 5-alkyl substituent may also affect the rate of transcription.

Isoalkyl analogues of UTP were weaker substrates than the corresponding n-alkyl r^{5}UTPs. This reflects the steric effect

of the substituent on transcription rate. 5-t-butyl-UTP was
no substrate of E. coli RNA polymerase enzyme regardless of
the DNA template applied.

Structure of the DNA template had also an effect on the
transcription rate. Activation of poly[d(A-T)] by pancreatic
DNase decreased absolute rates, while the relative rates
increased with all substrate analogues. Using 5-propyluracil-
containing template analogue of poly [d(A-T)], the poly-
[d(A-pr^{5}U)] which is less thermostable [6] and more active [4]
than poly[d(A-T)], "strongly" substituted derivatives of UTP
(long-chain and isoalkyl) were much weaker or no substrates of
the enzyme. Transcription rate decreased most drastically in
the presence of calf thymus DNA. Effect of template structure
on rate reflects presumably the differences in conformation of
DNA templates.

II. Time-course of analogue incorporation

5-Methyl-UTP could completely replace UTP in the transcription reaction after 4 hours of incubation (Fig.2). Extent
of incorporation of 5-n-propyl- and 5-ethyl-UTPs were about
two third of that of UTP. Copolymerization of ATP with 5-n-
pentyl-, hexyl- and octyl-UTPs, however, reached a low
plateau value after 1 hour, their incorporation being limited.
Finally, no reaction was observed with 5-t-butyl-UTP even
after 22 hours of incubation. Limitation in the extent of
incorporation of UTP-analogues by RNA polymerase may be the
consequence of an unusual steric structure of the newly formed
ribopolynucleotide product. This modified RNA cannot be
propagated further by RNA polymerase.

III. Inhibition of transcription

Effects of the weak substrate analogue 5-n-octyl-UTP and
the non-substrate 5-t-butyl-UTP on the UTP and [^{3}H]ATP copolymerization on poly[d(A-T)] template, and UTP, GTP, CTP and
[^{3}H]-ATP copolymerization on calf thymus DNA template catalyzed
by E. coli RNA polymerase enzyme were also studied (Fig.3).
5-n-Octyl-UTP proved to be a strong inhibitor of transcription.

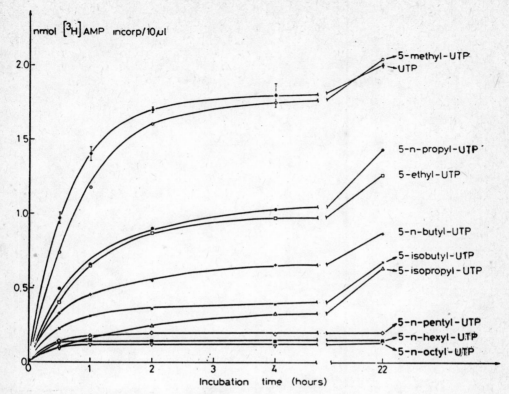

Fig.2.: Incorporation of UTP-analogs into RNA product as a
function of time on poly [d(A-T)] template

In the presence of calf thymus DNA, 80% inhibition was observed
at 0.1 mM concentration and 90% at 0.3 mM. The small degree
of inhibition exerted by 5-t-butyl-UTP proves that it is no
substrate of the RNA polymerase.

In conclusion, RNA polymerase of E. coli is able to
copolymerize a wide range of 5-alkyl-UTPs with ATP on different
templates. Rate and extent of incorporation of UTP-analogues
are, however, reduced compared to the natural substrate UTP.
The decrease depends on carbon chain length and bulkiness of
the substituent. Long-chain (pentyl to octyl) and branched-chain
(isopropyl and t-butyl) analogues were weak or no substrates.
Structure of the DNA template applied has also an effect on the
rate of transcription. 5-n-Octyl-UTP proved to be a strong
inhibitor of RNA synthesis on calf thymus DNA.

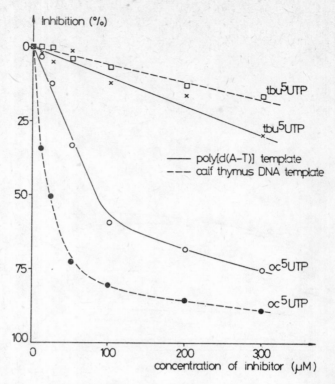

Fig.3.: Inhibition of transcription by 5-n-octyl- (oc^{5}UTP) and 5-t-butyl-UTPs (tbu^{5}UTP).

REFERENCES

1. Swierkowski,M. and Shugar, D. Poly-5-Ethyluridylic Acid, a Polyuridýlic Acid Analogue, J.Mol.Biol., <u>47</u>, 57-67 (1970).
2. Szemző, A., Szabolcs, A., Sági, J. and Ötvös,L⁻. Preparation of 5-Alkyluridines and Their 5'-Mono- and -Triphosphates. J.Carbohydrates,Nucleosides,Nucleotides, <u>7</u>, 365-379 (1980).
3. Sági,J., Szabolcs,A., Szemző,A⁻ and Ötvös,L. Modified Poly-nucleotides,I. Investigation of the Enzymatic Polymerization of 5-Alkyl-dUTPs. Nucleic Acids Res., <u>4</u>, 2767-2777 (1977).
4. Sági,J. and Ötvös,L. Modified Polynucleotides,IV. Template Activity of 5-Alkyluracil-Containing Poly[d(A-r^{5}U)] Copoly-mers for DNA and RNA Polymerases. Nucleic Acids Res., <u>7</u>, 1593-1601 (1979).

5. Ötvös,L., Sági,J., Szemző,A. and Szabolcs,A. Stereochemical Investigation of the Enzyme Catalyzed DNA Synthesis. 11th IUPAC Symp. Papers, Vol.1, Bioorg.Chem.pp. 91-94 (1978).

6. Sági,J., Brahms,S., Brahms,J. and Ötvös,L. Effect of 5-Alkyl Substitution of Uracil on the Thermal Stability of Poly[d(A-r^{5}U)] Copolymers. Nucleic Acids Res. $\underline{6}$, 2839-2848 (1979).

THE SYNTHESES AND PROPERTIES OF TRICYCLIC AZINES AS POTENTIAL INTERFERON INDUCERS AND ANTIVIRAL AGENTS

SZULC Z., MŁOCHOWSKI J., FIKUS M.,* INGLOT A.D. and SZULCZ B.**

Institute of Organic and Physical Chemistry, Technical University of Wroclaw, 50-370 Wroclaw, Poland
  *Institute of Biochemistry and Biophysics, Polish Academy of Sciences, 02-532 Warsaw, Poland
  **L. Hirszfeld Institute of Immunology and Experimental Therapy, Polish Academy of Sciences, 53-114 Wroclaw, Poland

Tilorone i and its analogs as well as CMA 2 are the most active low molecular weight interferon (IFN) inducers [1,4]. The relationship between the molecular structure and activity of some IFN inducers have been discussed by several authors [5,8]. It has been suggested that the capacity of a compound to induce IFN may depend on its ability to intercalate with DNA [6]. We reported synthesis of novel tilorone analogs having 1,8-diazafluorenone moiety instead of fluorenone [7]. Among them, the compound 3, isoelectronic with tilorone, was investigated and it was found to be weak IFN inducer in vitro and in vivo but it has antiviral activity in vitro comparable to that of tilorone. In this work we elaborated the synthesis of a novel, strong intercalating agent, Vivakorfen 4 and its analogs 8-14 being also derivatives of 4,7-phenanthroline with various side chains. The starting di-N-oxide 5 was prepared according to our procedure [3], then converted into dicarboxylic acid 7 and aminoester 8. Deoxygenation of 5 with phosphorus oxychloride afforded 6 and substitution of chlorine atoms with nucleophiles such as amines or aminoalcohols gave desired derivatives 9-14. The ability of 4 to strong intercalation with DNA and

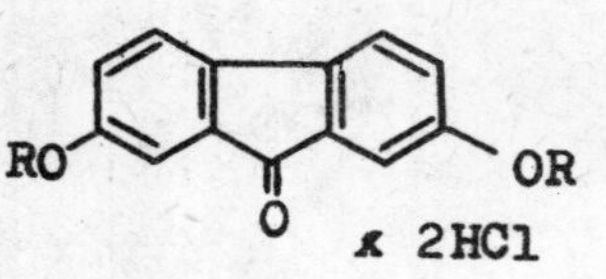

1, R=-CH$_2$CH$_2$NEt$_2$ x 2HCl

CH$_2$COOH

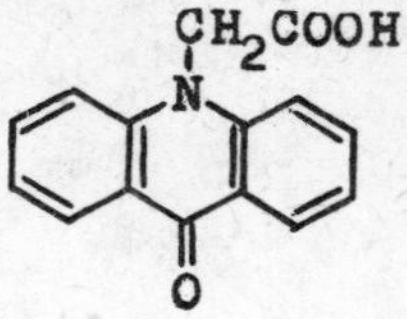

2

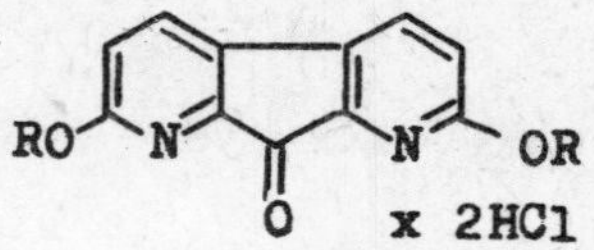

3, R=-CH$_2$CH$_2$NEt$_2$ x 2HCl

OPh

PhO   **9**, 45%

NH$_3$, PhOH
160-165°C

N-H, △

X

X=   N-

10, 72%

5

POCl$_3$, △

Cl

Cl   **6**, 30%

Et$_2$NH
(Ph)$_2$O
150-160°C

Cl   **11**, 15%

X

X

+

X

12, X=Et$_2$N-, 35%

KCN
PhCOCl
NaOH, △

COOH

HOOC   **7**

1. SOCl$_2$, △
2. Et$_2$N(CH$_2$)$_2$OH

COOR

ROOC   x 2HCl

8, R=Et$_2$N(CH$_2$)$_2$-, 25%

Et$_2$N(CH$_2$)$_n$OH
n=2,3 or 5

DMSO, NaH

OR

RO   x 2HCl

4, R = Et$_2$N(CH$_2$)$_2$-, 57%

13, R = Et$_2$N(CH$_2$)$_3$-, 52%

14, R = Et$_2$N(CH$_2$)$_5$-, 46%

364

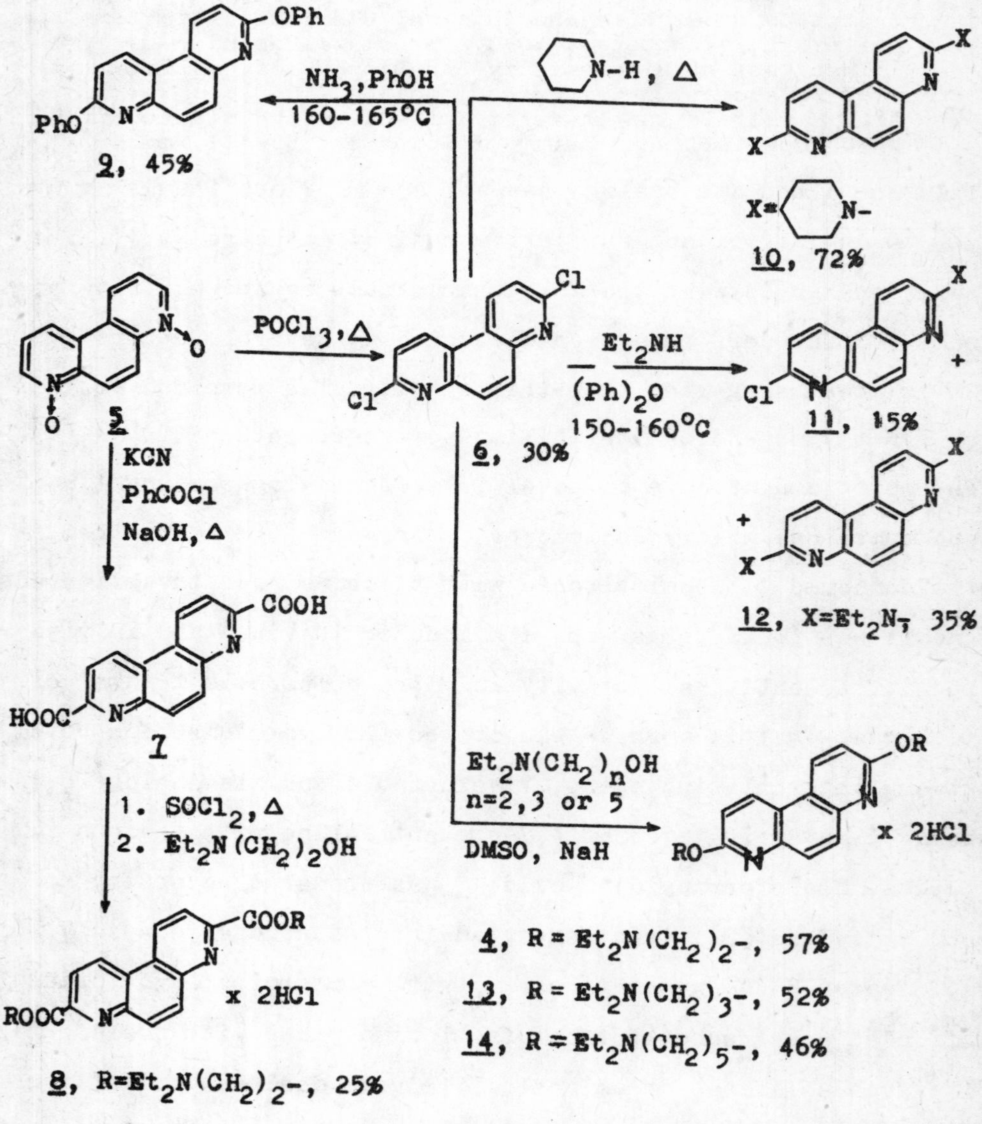

its usefulness as a fluorescent probe was recently revealed[2].
Now, we have attempted to define the relationship between molecular structure of the compounds, their ability to bind with
DNA, to inhibit DNA polymerase, to induce IFN and antiviral
activity. The results obtained are shown in the Table.

Table. Biological and biochemical properties
of selected compounds[a]

Compound	Toxicity in vitro for L cells (μg/ml)	IFN induction in the mouse in vivo	Antiviral[b] activity (μg/ml)	T_m (°C) of calf thymus DNA	% Inhibition of E.coli polymerase
1	14	strong	7	84	85
2	1100	very strong	250	72	not done
3	68	weak (inconsistent)	31	80	63
4	34	inactive	15	$86(86,5)^c$	68
8	1100	inactive[d]	none	not done	not done
13	34	inactive[d]	none	$82,5^c$	57
14	17	inactive[d]	none	82^c	68
Control	–	–	–	$72(66)^c$	0

[a]The conditions of biological and biochemical experiments were
the same   as reported previously [7].
[b]Minimal concentration inhibiting by 50% the cytopathogenic
effect of vesicular stomatitis virus(VSV) in the mouse L cells
cultured in vitro.
[c]T_m-the mid-point of thermal transition, was determined in
5mM tris-HCl buffer, pH=7.
[d]In vitro in culture of bone marrow macrophages.

The compounds 1,3,4,13 and 14 interact with DNA directly and
they inhibit the DNA polymerase. However, it contrast to tilorone compounds 3,4,8,13 and 14 do not act as IFN inducers.
Two of them (3,4) have antiviral activity against VSV comparable to that of tilorone but others (8,13,14) are inactive.
On the other hand, CMA is a very strong IFN inducer but it

does not intercalate with DNA. Thus, direct interaction of the compound with DNA is not required for IFN induction.

References

[1] Chandra P. and Wright G.J., Tilorone Hydrochloride: The Drug Profile, Top. Curr. Chem., 1977, 72, 125-127.

[2] Fikus M., Gołaś T., Szulc Z. and Młochowski J., Vivakorfen 3,8-bis[2-(dietylamino)etoxy]-4,7-fenantroliny chlorowodorek - nowa sonda fluorescencyjna w badaniach DNA, XIX Zjazd Polskiego Towarzystwa Biochemicznego, Szczecin, Poland, 26-28, IX, 1983.

[3] Kloc K. and Młochowski J., Investigation on reactivity of phenanthrolines. IV. Rocz.Chem., 1975, 49, 1621-1627.

[4] Kramer M.J., Taylor J.L. and Grossberg S.E., Induction of Interferon in mice by 10-carboxymethyl-9-acridanone, Methods Enzymol., 1981, 78 (Interferons, Pt.A) 284-287.

[5] Mayer G.D., Structural and biological relationship of low-molecular-weight interferon inducers, Pharmacol Ther., 1980, 8(1), 173-192.

[6] Sturm J., Schreiber L. and Daune M., Binding of ligands to a one-dimesional heterogenous lattice. Intercalation of tilorone with DNA, Biopolymers 1981, 20(4), 765-785.

[7] Szulc Z., Młochowski J., Fikus M. and Inglot D.A., Synthesis of potential interferon inducers and DNA intercalators. Part.I, Heterocycles 1984, 22(1), 73-78.

[8] Torrence P.T. and De Clerq E., Interferon Inducers: general survey and classification, Methods Enzymol., 1981, 78 (Interferons, Pt.A) 291-299.

METABOLISM OF UXEPAM® (RGH-3331) IN RATS

TEGYEY Zs., VERECZKEY L.,* TAMÁS J., RÖHRICHT J.,*
KISFALUDY L.* and ÖTVÖS L.

Central Research Institute for Chemistry, The Hungarian
Academy of Sciences, Budapest, Pf. 17, H-1525, Hungary
*Chemical Works of G. Richter Ltd., Budapest, Hungary

A valuable pharmacological property of (-)-dihydro-di-
azepam (RGH-3330) was reported to be the much greater loss in
muscle relaxant activity than in the tranquilizing effect, as
compared to diazepam (Kisfaludy et al. 1973; Pálosi et al.
1973). The secondary amino group in position 4 of RGH-3330
allowed the preparation of several N-acyl derivatives
(Röhricht et al. 1974). Among these compounds ($\pm$)7-chloro-
1,3,4,5 - tetrahydro-1-methyl-4-carbamoyl-5-phenyl-2H-1,4-
benzodiazepin - 2-one (UXEPAM^R; RGH-3331; I) seemed to be a
very good anxiolitic agent with slight muscle relaxant effect.
This fact justified the study of *in vivo* metabolism of the
drug.

For metabolic studies 2-^{14}C-Uxepam (Tegyey et al. 1979)
was administered in a single oral dose of 20 mg/kg
(0,37 MBq/animal) to RG-Wistar rats of both sexes. The excretion of radioactivity in urine and faeces is summarized
in Table 1. 64 % of the total radioactivity is excreted during
48 hr and this amount increases only by 5 % up to 96 hr.
Because of the small radioactivity values excreted in the
last 24 hr we did not continue the collection over 96 hr.
The bile duct was cannulated under urethane narcosis
(1 g/kg) and bile was collected at 2 hr intervals for 8 hr.
The average biliary excretion of radioactivity was only 7,5 %
during this period (Table 2).

Metabolites were isolated from urine and bile. Conjugated metabolites were hydrolysed by β-glucuronidase, then

Table 1. Radioactivity excreted in urine and faeces after
oral administration of ^{14}C-I (number of animals: 4)

time of collection (hr)	radioactivity % of dose	
	urine	faeces
0 - 24	18,06	1,27
24 - 48	7,79	37,25
48 - 72	0,17	3,54
72 - 96	0,02	0,62
0 - 96	26,04	42,68

Table 2. Radioactivity excreted in bile after oral
administration of ^{14}C-I

time of collection (hr)	radioactivity % of dose	number of animals
0 - 2	1,66 ± 1,09	7
2 - 4	2,52 ± 1,03	7
4 - 6	3,77 ± 1,81	3
6 - 8	1,98 ± 0,49	5
0 - 8	7,50 ± 3,95	6

extracted together with the unconjugated metabolites. Chloroform was used for extraction at pH 7 followed by n-butanol
at pH 2. About 60 % of excreted radioactivity could be
extracted with chloroform and only 2 % with n-butanol.
Separation and purification of metabolites was accomplished
by column and thin layer chromatography (Ötvös et al. 1978).

Table 3 contains the principal mass-spectrometric data
of metabolites and their amount in urine and bile. For
compounds, I, V and VI, structural assignments were established
by comparing their mass spectra with those of authentic
samples.

In order to detect unchanged compound I excreted, dried

faeces was extracted by chloroform in a Soxhlet apparatus and aliquot of the chloroform solution was chromatographed on TLC in two dimensions with reference substance. Radiometry showed I to be present in 8,15 % of the radioactivity excreted in faeces.

Table 3. <u>Metabolites of UXEPAM[R]</u> (I)

compound	% of urinary ^{14}C (48 hr)	% of biliary ^{14}C (8 hr)	mass spectrometric data m/z (I %)
I	11,07	3.19	331(11)i, 329(32)M, 287(11), 286(23), 270(50), 269(70), 258(10), 257(22), 243(18), 241(24), 228(100)
II	21,54	3,62	317(20)i, 315(50)M, 270(20), 268(20), 257(40), 256(60), 255(100), 229(50), 227(45), 216(50), 214(70), 197(20), 195(50)
III	-	7,81	333(7)i, 331(20)M, 272(40), 271(100), 243(30), 205 (30)
IV	0,60	-	313(0,6)M, 282(50), $C_{15}H_{12}N_3O_3$ 281(30), 267(2), 253(100), 239(4), 255(11)
V	0,50	5,67	286(35)i, 284(90)M, 283(80), 256(100), 255(30)
VI	-	1,92	302(14)i, 300(39)M, 271(100), 269(10), 257(30), 256(20), 255(17)
VII	4,30	4,36	152(80)M $C_7H_8N_2O_2$, 126(10), 109(100), 81(25), 80(35)
VIII	-	28.80	112(100)M

Figure 1 summarizes the probable metabolic pathway of UXEPAM[R] (I) in rats. A small amount of the drug is excreted intact. Analogously to metabolism of other benzodiazepine derivatives (Schwartz et al. 1965; de Silva and Puglisi 1970), N^1-demethylation plays an important part in the biotransformation of I. The II ⟶ III transformation is a later step

in the metabolism of the drug. We have not been able to detect
dihydro-diazepam as a metabolite, but hydrolysis of the
carbamoyl moiety followed by oxidation gives diazepam (V). V
was expected to be a main metabolite of I, however, its amount
is markedly smaller than that of II and III. Structure of two
metabolites isolated from the n-butanol extract (VII, M: 152,
$C_7H_8N_2O_2$ and VIII, M: 112) could not be determined.

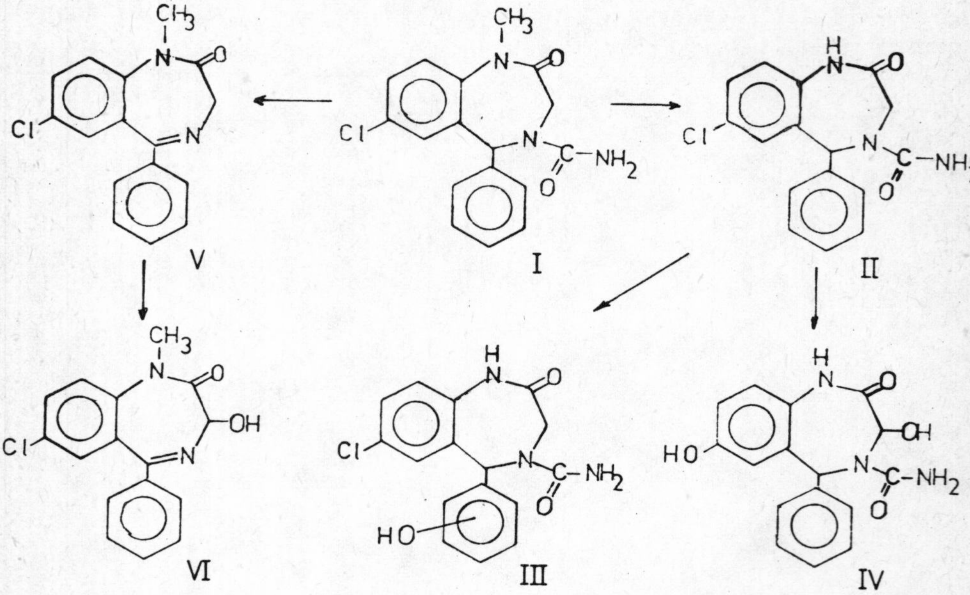

Figure 1. Probable metabolic pathway of UXEPAM[R] (I)

REFERENCES

Kisfaludy,L., Röhricht,J., Ürögdi,L., Pálosi,É. and Szporny,L.
      (1973). Hungarian Patent 160769
Ötvös,L., Tegyey,Zs., Vereczkey,L., Ledniczky,M., Tamás,J.,
      Pálosi,É. and Szporny,L. (1978). Metabolism of Levorotary
      4,5-Dihydro-diazepam in the Rat. Drug.Metab.Disp., 6,
      213-217
Pálosi,É., Ürögdi,L., Szporny,L. and Kisfaludy,L. (1973).
      1,3,4,5-Tetrahidro-1,4-benzodiazepin-2-on Optikailag
      Aktiv Módosulatainak Farmakológiai Vizsgálata
      Acta.Pharm.Hung., 43, 218-223

Röchricht,J., Kisfaludy,L., Ürögdi,L., Pálosi,É., Szeberényi,L.
    and Szprony,L. (1974). Hungarian Patent 171033

Schwartz,M.A., Koechlin,B.A., Postma,E., Palmer,S. and Krol,G.
    (1965). Metabolism of Diazepam in Rat, Dog and Man
    J.Pharm.Exp.Therap., $\underline{149}$, 423-435

de Silva,J.A.F. and Puglisi,C.V. (1970). Determination of
    Medazepam (Nobrium), Diazepam (Valium) and Their Major
    Biotransformation Products in Blood and Urine by
    Electron Capture Gas-Liquid Chromatography
    Anal.Chem., $\underline{42}$, 1725-1736

Tegyey,Zs., Maksay,G. and Ötvös,L. (1979). Synthesis of
    2-^{14}C-labelled-3H-1,4-Benzodiazepines
    J.Lab.Comp., $\underline{16}$, 377-385

MECHANISM OF BIOCHEMICAL TRANSFORMATION OF PHENAZEPAM

GOLOVENKO N.Ya.

Physico-Chemical Institute, Academy of Sciences of the Ukrainian SSR, Odessa, USSR

Phenazepam (I), a derivative of the 1,4-benzodiazepine series possesses high tranquilizing and marked anti-convulsive activities as well as hypnosedative properties; it also potentiates the effect of anesthetics and narcotic analgetics. Using relatively large doses of phenazepam results in insignificant degree of myorelaxation and ataxia (Bogatsky et al., 1980). By means of the combined radio-chromatography and mass-spectrometry methods, we have found (Golovenko et al., 1982.) that ^{14}C-I undergoes intensive metabolism in the rat and mouse yielding 3-hydroxy-derivative (II) and aromatic hydroxylation products (III-VI):

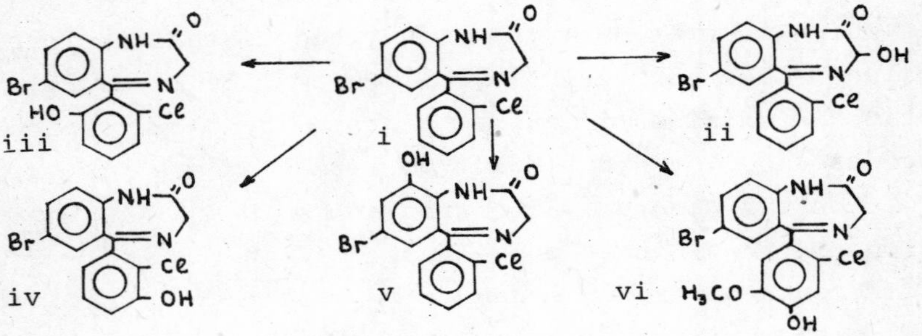

During 24 hours 11.2% of the phenolic metabolites and 0.42% of metabolite II are excreted from the rat. For the mouse, these values are 13.3% and 12.8%, respectively (Golovenko et al., 1979). Thus, for rat organism the oxidation of the aromatic ring is preferred. Similar results were obtained in the experiments in vitro using isolated hepato-

25 Plas

cytes (Golovenko et al., 1983).

In order to study biochemical mechanisms of I oxidation by cytochrome P-450 dependent enzymes, we have used microsomes of rat and mouse liver. Monooxygenases of mouse liver oxidize the heterocycle with higher rate to form II, while those of rat liver oxidize the aromatic ring to form III (Golovenko et al., 1980). We suggested that this phenomenon was due to the difference in the content of certain cyctochrome P-450 isoforms in the liver of the two species. Intact microsomes of mouse liver (1.1±0.3 nmol/mg protein) and of rat liver (0.64±0.05 nmol/mg protein) are non-equivalent. Phenobarbital and 3-methylcholanthrene administration to rats resulted in hemoprotein induction (1.38±0.08 and 0.91 nmol/mg protein, respectively). In this case the rate of metabolite formation increased by 5 to 6 times. Meanwhile, both inductors proved uneffective towards metabolite III.

Thus, the increase of cytochrome P-450 content in rat hepatocytes results in a shift of direction for I hydroxylation. Both this fact and the difference in the kinetic parameters of II and III formation testify the fact that both processes are catalyzed by at least two enzymes.

The rate of I oxidation by rat and mouse microsomes depends on the presence of NADPH and NADH in the incubation medium. Here a NADPH-dependent electron-transport system prevails which is proved by the high yield of reaction products while using NADPH. Consequently, phenazepam like other xenobiotics is oxidized in animal organism by NADPH-dependent monooxygenases (Golovenko, 1981). No rate increase for II and III formation was observed at equimolar NADPH and NADH concentrations in the incubation medium.

Experiments on oxygen consumption from the incubation medium have shown that molecular oxygen takes part in I oxidation which agrees with the general regularities of xenobiotics oxidation:

$$I + NADPH + H^+ + O_2 \longrightarrow II/III + NADP^+ + H_2O$$

The nature of hemoprotein interaction with I and II corresponds to the second type of binding. The spectral changes of the hemoprotein-benzodiazepine complex reveal that these depend on the substrate concentration. Enzyme saturation is reached at concentration 300 μmol (I), and 200 μmol (II). In general, for I and II low K_s values are obtained which are characteristic of the majority of benzodiazepines (Golovenko, Meteshkin, 1978) and differ from K_m values. This, perhaps, proves that I and II are bound by two centers of hemoprotein: heme and apoenzyme. The interaction with the former is associated with spectral changes of the hemoprotein, while the latter provides the formation of enzyme-substrate complex and the oxidation of I. The differences in typical characteristics (K_s and ΔD_{max}) of the complexes as well as the shift of spectral maxima and minima caused by I and II are due to their electron structure. Analysis on the charges of different atoms shows that the distribution of electron density in the heterocycle of II differs from that of I.

REFERENCES

Bogatsky A.V., Andronati S.A., Golovenko N.Ya. (1980) Tran-
quilizers (1,4-benzodiazepines and related structures).
Naukova dumka, Kiev, 280 p.

Golovenko N.Ya. (1981) Mechanisms of the Reactions of Xenobiotics Metabolism in Biologic Membranes. Naukova dumka,
Kiev, 220 p.

Golovenko N.Ya., Meteshkin Yu.V. (1978) Interaction of Albino Rat Liver Microsomal Cytochrome P-450 with 1,4-benzodia-
zepines. Doklady AN UkrSSR, Ser B, N 2, p.154-157.

Golovenko N.Ya., Meteshkin Yu.V., Kuznetsova S.V (1983)
Oxidation of Phenazepam in Isolated Rat Hepatocytes. Voprosy
Meditsinskoi Khimii, 29, p.49-53.

Golovenko N.Ya., Meteshkin Yu.V., Yakubovskaya L.N. (1980)
Catalytic Properties of Monooxygenases Oxidizing ^{14}C-phena-
zepam. Voprosy Meditsinskoi Khimii, 26, p.637-640.

Golovenko N.Ya., Zinkovsky V.G. (1982) Comparative Metabolism and Pharmacokinetics of Phenazepam. In: Phenazepam, ed.
A.V.Bogatsky, Naukova dumka, Kiev, p.32-86.

Golovenko N.Ya., Zinkovsky V.G., Bogatsky A.V., Andronati
S.A., Seredinin S.B., Yakubovskaya L.N. (1979) Comparative
Study of Excretion of Phenazepam Metabolites in Single and
Multiple Administration to Albino Rats. Khimiko-Farmatsev-
ticheskii Zhurnal, 13, p.154-157.

Bio-Organic Heterocycles
van der Plas H.C., Ötvös L., and Simonyi M. eds

ENZYME-CATALYZED SYNTHESIS OF HETEROCYCLIC COMPOUNDS

ANDRONATI S.A., and DAVIDENKO T.I.

Physico-Chemical Institute, Academy of Sciences
of the Ukrainian SSR, Odessa, USSR

To use the actual advantages of enzyme systems in organic synthesis, we considered the following reactions: hydroxylation, reduction, N-demethylation, hydrolysis, and acetylation of substituted 1,4-benzodiazepine-2-ones, 1,5-benzodiazocines, fluorenones, as well as 5,6, and 7-membered oxygen containing heterocycles (T.I.Davidenko, E.V.Sevastyanova, 1980; T.I.Davidenko, N.P.Sereda, 1980).

To work out biotechnologic techniques of the synthesis of pharmacologically active derivatives of 1,4-benzodiazepine-2-ones we have studied the microbiological nitroreduction and hydroxylation. Usually the reduction was realized by hydrogen in the presence of catalysts - Adams platinum, palladium or Raney nickel (S.A.Andronati et al., 1982). We have shown that microbiological transformation of nitrosubstituted 1,2-dihydro-3 H -1,4-benzodiazepine-2-ones by cells of E.coli, BKMB-471, 835, 870 results in the formation of products featuring the structure of aminoderivatives of 1,2-dihydro-3H-1,4-benzodiazepine-2-ones.

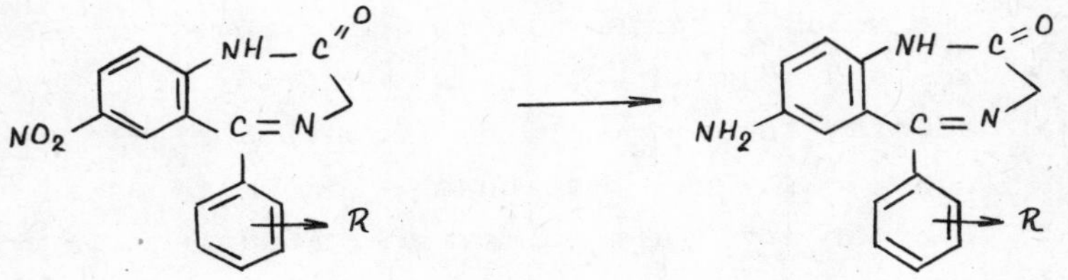

The yields of the corresponding aminoderivatives are 70-80%. No tetrahydroderivatives are formed. The composition of the compounds was proved by IR, NMR, UV-spectroscopy and mass spectrometry. Optimal yields of aminoderivatives were obtained using either pH 7.0-8.0 buffer solution or nutrient medium. The composition of the nutrient medium (g/100 ml of water): 0.6 g of peptone, 0.6 g of Na_2HPO_4, 0.3 g of KH_2PO_4, 0.1 g of NaCl, 0.001 g of $CaCl_2$, 0.4 g of glucose. Concentration of nitroderivatives of 1,2-dihydro-3H-1,4-benzodiazepine-2-ones is 0.3-3 mg/100 ml of nutrient medium.

Since nitroreductase is an intracellular enzyme difficult to isolate while the substrates and the final products are not high-molecular compounds, it was of interest to study the nitroreduction by E.coli cells immobilized into polyacrylamide gel (PAAG). Nitroreductase activity of immobilized cells is 80% of the corresponding native cells.

PAAG was shown to possess considerable inhibiting effect. both at 37°C (incubation period 12 hrs) and at 20°C (incubation period 6 hrs), while 50,9% and 77.4% of 7-amino-1,2-dihydro-3H-1,4-benzodiazepine-2-one is formed, respectively. A decrease of nitroreductase activity of the cells is observed at immobilization into PAAG, which is larger at 37°C with 12 hrs incubation period and in case of smaller quantity of cells (0,2 g). Maximal quantity of the final product is formed using 5% acrylamide. It is possible to immobilize 1,5 g of E.coli cells into 5 g of PAAG.

Study of the effect of temperature shows that both for free and immobilized cells, the best yields of amino derivatives of 1,4-benzodiazepine-2-ones are observed at 20°C. Dependence on the pH of E.coli cells immobilized into PAAG does not differ from that of the free ones. Nitroreductase activity of the cells is preserved for 15 transformations.

The regio- and stereospecific introduction of hydroxyl group is of special interest for the formation of 3-oxy-1,4-benzodiazepine-2-ones. The existing chemical methods are either multistaged or difficult to perform. Studying microbiological transformation of 7-bromo-5-(o,m,p-chloro)phenyl- and 3-methyl-7-bromo-5-(o,m,p-chloro)phenyl-1,2-dihydro-3H-1,4-

benzodiazepine-2-ones by Actinomyces shows the possibility of formation of the corresponding 3-hydroxyderivatives.

IR, UV and mass spectra as well as the data of microanalysis proved the formation of the hydroxy product. Thus, in IR spectra the stretching vibrations were observed of the free and associated hydroxyl groups within the region of 3480-3600 cm^{-1}, while the bands of the free and associated N-H at 3390 and 3180 cm^{-1}; the intensive band of the stretching vibrations of carbonyl group is observed at 1692 cm^{-1}, the less intensive band of the -C=N- bond at 1610 cm^{-1}. The UV spectrum of the compound is characterized by the presence of the absorption band with λ_{max} 230-231 nm. Mass spectra proved the molecular masses of the compounds. The substances are optically active. Thus, for 3-Me-7-bromo-5-(o-chlorophenyl)-1,2-dihydro-3H-1,4-benzodiazepine-3-ol-2-one $[\alpha]_D^{20}$ = +103°.

Act. roseochromogenes, BKMA-612, Str.viridis, BKMA-607, and Act.lavendulae, BKMA-608 possess the greatest transforming activity. The yields of 3-hydroxyderivatives of 1,4-benzodiazepine-2-ones are 30-40%. The composition of the nutrient medium (g/100 ml water): maize extract - 1.0, glucose - 0.5, NaCl - 0.5, $(NH_4)_2SO_4$ - 0.3, $CaCO_3$ - 0.5, starch - 1.5, $FeSO_4$ - 0.3 at pH 6.8-7.0. Immobilization of Actinomyces into PAAG in this case gives no satisfactory results. Immobilization into polyvinyl alcohol in the presence of a co-substrate reduces the activity of immobilized cells to 30% of the activity of native cells. The best results are obtained at Actinomyces immobilization into carrageenan gel (hydroxylase activity for Str.viridis - 65%, Act.lavendulae - 35%, Act.roseochromogenes - 29% of that of native cells). At storing (4°C) during 49 days for Act. roseochromogenes 100% preservation of the initial activity was observed.

REFERENCES

Andronati S.A., Avrutsky G.Ya., Bogatsky A.V. (1982) Phenazepam, Naukova dumka, Kiev, 288 p.

Davidenko T.I., Sereda N.P. (1980) Microbiologic synthesis of 2-(acylamino-P-propionylamide)-benzophenones. Khimiko-Farmatsevticheskii Zhurnal, N 8, p.63-66.

Davidenko T.I., Sevastyanova E.V. (1980) Enzymatic hydroxylation of fluorenones. Khimiko-Farmatsevticheskii Zhurnal, N 7, p.66-68.

POLYMERIC DERIVATIVES OF QUINIDINE

AZORI M., PATÓ J. CSÁKVÁRI É., FEHÉRVÁRI F. and
TÜDŐS F.

Central Research Institute for Chemistry,
The Hungarian Academy of Sciences,
Budapest, Pf. 17, H-1525, Hungary

Quinidine (Q) is a naturally occuring alkaloid widely used in
the management of chronic cardiac arrhythmias. Its main disad-
vantage is the short half-life associated with metabolic elimination [1]. In order to improve the pharmacological properties
of Q we have synthetized its polymeric derivatives where the
drug is linked to a polymeric carrier by means of covalent bond
cleavable in biological environment.

Applying the same carrier, we aimed to examine the influence
of the mode of coupling on the physicochemical properties of the
polimer-prodrug, and the rate of release of the drug.

The carrier used is a non-toxic, water soluble copolymer with
an average molecular weight of 20.000 [2]. It was prepared by
radical copolymerization of N-vinylpyrrolidone with maleic anhydride [3]. Covalent binding of Q by direct acylation of its
OH group with the polyanhydride, as well as, the coupling of Q
via 6-aminocaproic acid (Aca) spacer were described earlier in
[3] and [4], respectively.

For studying the enzymatic cleavage of Q a specific amino
acid for chymotrypsin, phenylalanine was built in between the
spacer and Q. The synthesis was carried out according to the reaction scheme.

$$\text{Boc-Phe-OH} + \text{Q-OH} \xrightarrow{\text{DCC/DMP}} \text{Boc-Phe-OQ} \xrightarrow{\text{HCl/MeOH}} \text{H-Phe-OQ}$$

$$\xrightarrow[\text{EEDQ}]{\text{Trt-Aca-OH}} \text{Trt-Aca-Phe-OQ} \xrightarrow{\text{H}^+} \text{H-Aca-Phe-OQ} \xrightarrow{\text{coupling}}$$

METHODS

2.66 g t.-butyloxycarbonyl phenylalanine (BocPheOH), 3.24 g
quinidine (Q-OH), and 120 mg 4-dimethylaminopyridine (DMP) were
dissolved in 30 cm^3 dry dichloromethane then 2.2 g dicyclohexyl-
carbodiimide (DCC) in 5 cm^3 dichloromethane was added.

After stirring overnight the solution was filtered, extracted by NaOH and dried over MgSO$_4$. After evaporation of the solvent an oily product was obtained which was purified on a silica
gel column (eluent: chloroform/acetone 1:1). The resulting Boc-
phenylalanine quinidine ester has R$_f$ = 0.78 (in chloroform/acet-
one 1:1). IR(KBr): 1740 (CO ester), 1705 (CO urethane), 1510,
1500 cm^{-1} (C-C Ar.),

Butyloxycarbonyl-group was removed by 10% hydrochloric acid
in methanol at room temperature. The solvent was evaporated, the
residue was dissolved in water, then pH of the solution was adjusted to 10 with NaOH. The solute was extracted by ether and
dried over MgSO$_4$. After evaporation of ether, the product,
phenylalanine quinidine ester (Phe-OQ) was chromatographycally
homogeneous. R$_f$ = 0.48, IR 1740 cm^{-1} (CO ester), 3350 cm^{-1} (NH$_2$).

3.3 g PheOQ and 2.6 g Trt-Aca were dissolved in 30 cm^3 dry
dichloromethane and 1.97 g N-ethoxycarbonyl-2-ethoxydihidro-
quinoline (EEDQ) was added with stirring. After 6 hours the solution was extracted by HCl, NaOH, and H$_2$O successively. The organic phase was dried, the solvent was evaporated. The residue,
Trt.AcaPheQ dissolved in 20 cm^3 acetic acid (70 %) and boiled
for 5 min. then cooled and filtered. The product, aminohexanoyl-
phenylalanine quinidine ester crystallized immediatelly by adding NaOH to the filtrate. After recrystallization from methanol
m.p.: 73 - 75 °C; IR(KBr): 1680 (CO amide), 1770 (CO ester),
3200 (CONH), 3450 cm^{-1} (NH$_2$). Yield: 2.1 g (49 %).

The coupling reaction (Fig.1) was carried out according to
the method described in [2] in dimethylformamide solution, at
40 °C, for 4 hrs. The products were precipitated by ether and
reprecipitated by DMF/ether for purification. The unreacted anhydride groups were hydrolized by aqueous NaOH. Drug content was
determined _via_ characteristic UV absorption of Q at λ_{max} = 345 nm
(ε = 4880 mol$^{-1} \cdot$l$\cdot$cm^{-1}).

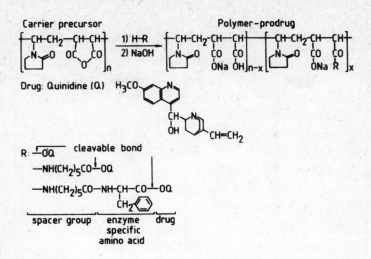

Fig. 1. Reaction scheme of coupling.

Hydrolysis measurements of polymeric derivatives of Q shown in Fig.1 were carried out at 37 $^\circ$C, pH 8,4 (adjusted by NaOH). Initial concentrations of polymer-prodrug samples were: 5 - 8 mg/cm^3. Release of Q was followed by GPC method using Sephadex G-50 with UV detection (see Fig.2).

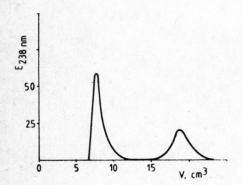

Fig.2. Illustrative GPC curve of a polymer bound quinidine. Sephadex G-50, eluent 4-ethyl-morpholine/HCl, pH = 7,2.

Fig.3. Hydrolysis - time pro-file. Sample: directly bound quinidine. 37 $^\circ$C, pH = 8,4; c = 5 mg/cm^3.

Thin layer electrophoresis (densitometrically detected) and potentiometric titration in pH-stat have also been checked as hydrolysis following methods. Reproducibility was poor in the former case, while polyelectrolyte behaviour of the carrier caused troubles in the latter one [5].

Hydrolysis of directly bound Q is shown in Fig.3 from which the first order kinetic constant. $k_1 = 2.8 \cdot 10^{-6}$ s^{-1} was obtained. Half-life data of the derivatives studied are summarized in Tab. 1. The last $t_{1/2}$ value in Tab.1 refers to chymotrypsin cataly- zed hydrolysis of the sample. Enzyme and substrate concentrations were: $[E] = 1$ mg/cm^3, $[S] = 15$ mg/cm^3.

Table 1. Hydrolysis data of polymeric derivatives of quinidine obtained at 37 ^{O}C, pH = 8.4

Mode of coupling	Drug content mg/g	$t_{1/2}$ hours
Direct	100	69
via Aca	118	36 – 42
– Aca-Phe	63	20
– Aca-Phe	63	< 2*

*Chymotrypsin catalysed hydrolysis

REFERENCES

1. Ueda, C.T., Ballard, B.E. and Rowland,M. Concentration-time Effects on Quinidine Disposition Kinetics in. Rhesus Monkeys. J. Pharmacol. Exp. Ther., 200, 459-468 (1977)
2. Csákvári, É., Azori, M. and Tüdős, F. Physico-chemical Studies of Polymeric Carriers 1.-2., Polym. Bull., 5, 413-416; ibid. 5, 673-677 (1981)
3. Pató, J., Azori, M. and Tüdős, F. Polymeric Prodrugs, 1. Makromol. Chem. Rapid Commun., 3, 643-647 (1982)
4. Pató, J., Azori, M. and Tüdős, F. Polymeric Prodrugs, 2. Makromol. Chem. Rapid Commun., 4, 25-28 (1983)
5. Csákvári, É., Azori, M. and Tüdős, F. Physico-Chemical Studies of Polymeric Carriers 3., Polym. Bull. (1984) in press.

Bio-Organic Heterocycles
van der Plas H.C., Ötvös L., and Simonyi M. eds

8,16-DIHETEROSTEROIDS AS IMMUNOMODULATORS

KUZMITSKIY B.B., LAKHVICH F.A., KHRIPACH V.A.,
ZHURAVKOV Yu.L. and AKHREM A.A.

Institute of Bioorganic Chemistry, Byelorussian SSR
Academy of Sciences, 220600, Minsk, Zhodinskaya, 5/2,
USSR

The study of immunomodulating activity of heterosteroids is of considerable interest both in view of finding new biologically active compounds and for determining the structure-activity relationships.

We have investigated the immunostimulating and immuno-depressive properties of heterocyclic steroids of the 8,16-diaza- and 8-aza-16-oxagonane series, whose synthesis and various kinds of biological action have been previously described [1].

I $\quad$ $R^1=R^2=R^5=H$; $R^3R^4=O$; $X=NH$

II $\quad$ $R^1=R^2=H$; $R^3R^4=O$; $R^5=Me$; $X=NH$

III $\quad$ $R^1=R^5=H$; $R^2=OMe$; $R^3R^4=O$; $X=NH$

IV $\quad$ $R^1=R^2=R^3=R^4=R^5=H$; $X=NH$

V $\quad$ $R^1=R^2=R^3=R^4=R^5=H$; $X=NH$; 13,14-dihydro-

VI $R^1 = R^2 = H$; $R^3 R^4 = O$; $R^5 = Me$; $X = O$

VII $R^1 = R^2 = R^3 = R^4 = H$; $R^5 = Me$, $X = O$

VIII $R^1 = R^2 = OMe$; $R^3 R^4 = O$; $R^5 = Me$; $X = O$

IX $R^1 = R^2 = OMe$; $R^3 = R^4 = H$; $R^5 = Me$; $X = O$

X $R^1 = R^2 = R^3 = H$; $R^4 = OH$; $R^5 = Me$; $X = O$

XI $R^1 = R^2 = OMe$; $R^3 = H$; $R^4 = OH$; $R^5 = Me$; $X = O$

The present investigation of 8,16-diheterosteroids covers their effects on: i) the primary humoral immune response in CBA mice in vivo, ii) PWM-induced immunoglobulin synthesis in human peripheral blood lymphocyte (PBL) cultures, iii) the proliferative response of PBL to PHA and allogenic cells in one-way mixed lymphocyte cultures (MLC), and iv) the natural cytotoxicity and cell-mediated lympholysis (CML) of PBL in vitro.

8,16-Diazasteroids I-III have been found to increase the humoral immune response both in vivo and in vitro (Table 1). Also, they stimulate the PHA- and allogenic cells-induced lymphocyte proliferation, and both natural and cell-mediated cytotoxicity of PBL (Table 2).

The influence of the methyl group at C_{15} and methoxy group at C_3 has been shown: compounds II and III reveal higher activity than 8,16-diazasteroids without these groups (I).The absence of carbonyl function in position 12 of compound IV leads to its depressive effect on the immune response. The reduction of the double bond in compound V increases the observed inhibitory effect on the humoral and cell-mediated immune response both in vivo and in vitro.

As different from 8,16-diazasteroids, 8-aza-16-oxasteroids show one-way effect on immune response. All the compounds of the series studied here inhibit both humoral and cell-mediated immunity in vivo and in vitro, the activity of natural killers remaining unchanged (Tables 1 and 2).

Table 1. Effect of 8,16-diheterosteroids on humoral immune response

Compound	Prim. imm. resp., AFC/10° nucl.cells, 0,1 LD_{50}, % Eff.	P (n=6)	PWM-ind. Ig-synthesis in PBL, % Eff.		
			10^{-6}M	10^{-5}M	10^{-4}M
I	+60.7 ± 9,43	<0,001	+17,7	+54,0	+40,4
II	+97,7 ± 22,28	<0,01	+52,2	+75,9	+50,1
III	+54,3 ± 10,12	<0,01	+67,4	+93,0	+52,5
IV	-33,2 ± 6.41	<0,001	-26,2	-43,3	-74,0
V	-70,6 ± 0,91	<0,001	-35,3	-60,2	-83,4
VI	-21,5 ± 5,95	<0,01	-47,6	-53,1	-69,0
VII	-39,5 ± 1,21	<0,001	-56,7	-63,3	-72,7
VIII	-25,0 ± 1,69	<0,001	-54,8	-62,7	-72,9
IX	-28,8 ± 2,70	<0,001	-62,4	-68,1	-77,0
X	-19,5 ± 3,01	<0,001	-52,0	-64,3	-73,5
XI	-46,0 ± 1,76	<0,001	-63,3	-73,1	-78,1

It could be mentioned that the introduction of methoxy groups into ring A (VIII, IX) does not essentially affect the immunodepressive activity of 8-aza-16-oxasteroids, and the presence of the carbonyl function in position 12 decreases this activity.

A partial reduction of the 12-carbonyl function results in an increase of immunodepressive activity of 8-aza-16-oxasteroids only in the presence of 2,3-dimethoxy groups (XI), and not otherwise (X).

Both stimulation and inhibition of immune response by 8,16-diheterosteroids are found to be dose-, concentration- and T-cell-dependent.

Table 2. Effect of 8,16-diheterosteroids (10^{-5} M) on cell-mediated immunity

Compound	PHA-ind. DNA-synth. in PBL % Eff.	MLC % Eff.	CML % Eff.	NK-activity % Eff.
I	+73,7	+60,7	+49,8	+26,1
II	+86,7	+96,3	+92,5	+90,9
III	+79,5	+80,9	+76,8	-
IV	-97,1	-67,7	-62,0	-69,5
V	-96,7	-79,8	-77,1	-76,5
VI	-63,1	-43,4	-42,0	- 5,2
VII	-72,3	-69,2	-64,5	-4,7
VIII	-63,7	-	-	-
IX	-79,5	-	-	-
X	-43,8	-60,5	-54,6	- 4,8
XI	-74,0	-78,8	-80,2	- 6,6

Several compounds of the investigated 8,16-diheterosteroids show good prospects for clinical use in immunotherapy.

REFERENCES

1. Akhrem A.A., Lakhvich F.A., Lis L.G., Kuzmitskiy B.B. Total synthesis, structure and properties of 8-azasteroids, a new class of biologically active compounds. Vesti Akad. Navuk BSSR, ser. khim. navuk, 1982, N.6, s. 81-90.

Bio-Organic Heterocycles
van der Plas H.C., Ötvös L., and Simonyi M. eds

SOME METABOLITES IN CEPHALOSPORIN BIOSYNTHESIS

WELWARD L., MICHÁLKOVÁ E. and DUDEK M.

Drug Research Institute, Modra, Czechoslovakia

SUMMARY

DL-Methionine is the source of sulphur in cephalosporin
C biosynthesis by the strain Cephalosporium acremonium. A
fraction of DL-methionine is oxidatively deaminated to 2-keto-
4-methylthiobutyric acid by the action of D-amino acid oxidase
from Cephalosporium acremonium. In submerged cultivation, the
reduced form of this compound, 2-hydroxy-4-methylthio-
butyric acid accumulates extracellularly. Its concentration
increases with the time of cultivation. The concentration
level of 2-hydroxy-4-methylthiobutyric acid in filtrates of
the cultivation broth is stabilized from 100 to 110 hours
of cultivation. In this cultivation phase the biosynthetic
production rate of cephalosporin C is at its maximum and a
rapid fall of DL-methionine concentration occurs.

INTRODUCTION

Cephalosporin C is produced by fermentation process using
different mutant strains of Cephalosporium acremonium
(Acremonium chrysogenum). The antibiotic can be isolated from
the culture broth in the form of its zinc salt complex. The
pattern of antibiotic synthesis by Cephalosporium acremonium
is typical to the Cephalosporium type. All biosynthetic
cephalosporins feature the D-α-aminoadipyl side chain. Some
mutants of this strain are dependent on organic sulphur source
and can effectively utilize only DL-methionine for the

synthesis of the antibiotic. The possible pathway suggested for the biosynthesis [1-3] is, as follows:

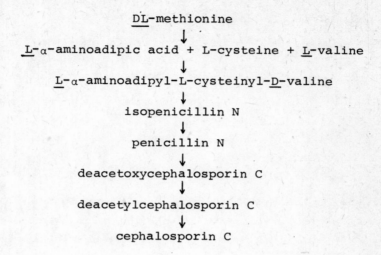

DL-methionine

↓

L-α-aminoadipic acid + L-cysteine + L-valine

↓

L-α-aminoadipyl-L-cysteinyl-D-valine

↓

isopenicillin N

↓

penicillin N

↓

deacetoxycephalosporin C

↓

deacetylcephalosporin C

↓

cephalosporin C

During the growth cycle of Cephalosporium acremonium in submerged culture, D-amino acid oxidase is produced. By action of this enzyme D-methionine undergoes oxidative deamination yielding 2-oxo-4-methylthiobutyric acid, from which the corresponding 2-hydroxy acid is formed. This acid is accumulated in the filtrate of the broth.

$$CH_3-S-CH_2-CH_2-\underset{\underset{NH_2}{|}}{CH}-COOH$$

↓

$$CH_3-S-CH_2-CH_2-\underset{\underset{O}{\|}}{C}-COOH$$

↓

$$CH_3-S-CH_2-CH_2-\underset{\underset{OH}{|}}{CH}-COOH$$

2-Hydroxy-4-methylthiobutyric acid forms insoluble zinc salt and can be coprecipitated with cephalosporin C in the isolation process.

METHODS

In our laboratories we determined the following metabolites in the culture broth of Acremonium chrysogenum during the
cultivation process:
cephalosporin C (major product of the biosynthesis),
deacetylcephalosporin C (minor component),
DL-methionine, 2-hydroxy-4-methylthiobutyric acid.

HPLC: cephalosporin compounds (Lichrosorb RP-18, 10 μm,
       250 x 4.6 mm, 3.5 % methanol in 0.01 M KH_2PO_4);
       DL-methionine (transformed to its dansylderivative:
       Lichrosorb RP-18, 10 μm, 250 x 4.6 mm, 60% methanol
       in 0.01 M KH_2PO_4).

Potentiometric redox titration: 2-hydroxy-4-methylthiobutyric
acid by indirect bichromatometric method.

RESULTS

The concentration level of 2-hydroxy-4-methylthiobutyric
acid increases during the first 100 hours of cultivation after
which it remains constant. This cultivation phase is
characterized by the highest biosynthetic production rate of
cephalosporins. Simultaneously, the concentration of
DL-methionine decreases rapidly (Fig.1). It was determined
that 35% DL-methionine from the cultivation medium is
transformed into the ineffective 2-hydroxy derivative by
D-amino acid oxidase from the strain Acremonium chrysogenum
during its growth.

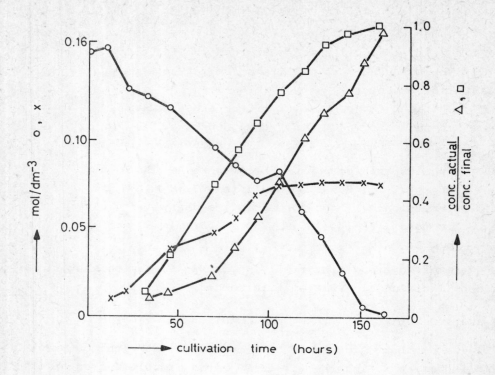

Fig.1. Concentration of characteristic compounds in the
course of cultivation.

       –△–   deacetylcephalosporin C
       –□–   cephalosporin C
      –x–   2-hydroxy-4-methylthiobutyric acid
      –o–   DL-methionine

REFERENCES

1. Flynn, E.H. (Ed.), Cephalosporins and Penicillins.
    Chemistry and Biology. Academic Press, New York,
    London, 1972.
2. Corman, M. and Hubert, F., Ann. Rep. Ferment, Proc. 1,
    327-346 (1977).
3. Abraham, E.P., J.Antib. XXX Suppl. 1977, 1-22.

BIOMIMETIC SYNTHESIS OF ACRIDONE ALKALOIDS[ӿ]

RÓZSA Zs., SZENDREI K. and LEWIS J. R.[*]

Department of Pharmacognosy, Medical University,
Szeged, Hungary

*Department of Chemistry, University of Aberdeen,
 Aberdeen, Scotland

The acridone alkaloids are found solely in the Rutaceae plant family and

in recent years the successful callus tissue culturing of Ruta graveolens

developed by Kozovkina et al (1979) has now enabled Baument et al (1982)

to undertake biosynthetic studies using labelled metabolites.

Anthranilic acid, its N-methyl derivative and acetate are incorporated

into rutacridone (1) in a manner supporting the proposal by Adams et al

(1976, 1977), based on a biomimetic conversion of 2-methylamino-2,4,6-

trimethoxybenzophenone (2) to 1,3-dimethoxy-10-methylacridone (3) and by

the isolation of (2) from natural sources (Casey and Malhotra, 1975), that

aminobenzophenones are precursors to acridone alkaloids.

(1)   (2)   (3)

[ӿ] A joint research project sponsored by the British Council under its
academic links scheme with Eastern Europe.

Figure 1 depicts the main acridone alkaloids reported by Szendrei et al
(1976) to be produced by callus tissue culturing of <u>Ruta graveolens</u> and
we report on the synthesis and cyclisation of aminobenzophenones with
relevant substitution patterns to these alkaloids.

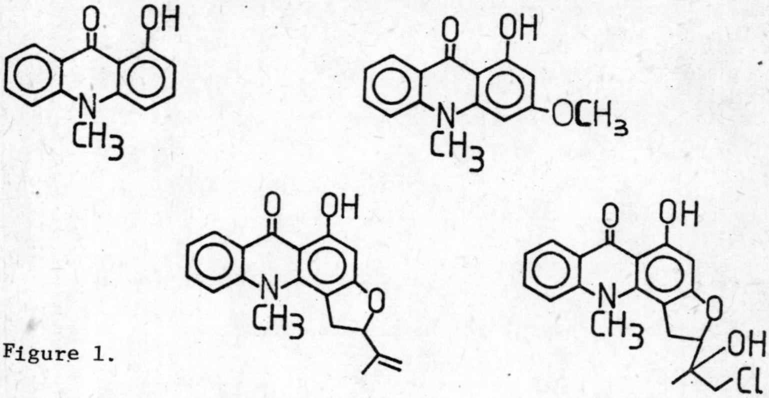

Figure 1.

Friedel-Crafts condensation of 2-nitrobenzoic acid with 3,5-dimethoxy-
phenol in trifluoroacetic anhydride at 15° gave both the unsymmetrical
and symmetrical nitrobenzophenones (4; $R=NO_2$) and (5; $R=NO_2$) in 20%
and 35% yield respectively.

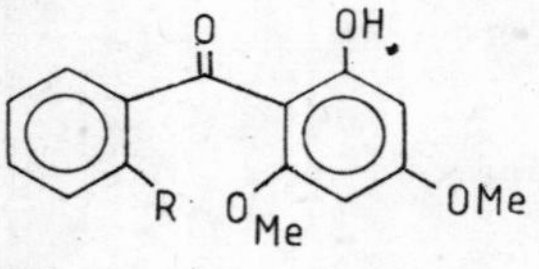

(4)

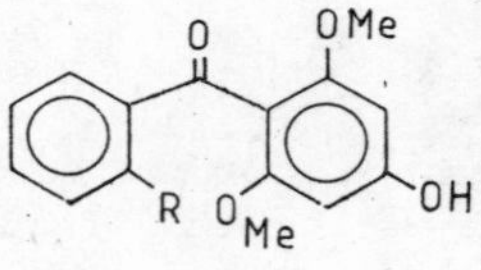

(5)

<u>Reduction of</u> (4; $R=NO_2$). Treatment of (4; $R=NO_2$) with Zn dust in
ethanol at 15° gave not only the required amine (4; $R=NH_2$) but also
small quantities (∿1%) of 1-hydroxy-3-methoxyacridan-9-one.

<u>Methylation-cyclisation of</u> (4; $R=NH_2$). The amine (4; $R=NH_2$) on treat-
ment with methyl iodide in acetone at 15° gave four products - the
dimethylamine (4; $R=NMe_2$), the mono methylamine (4; $R=NHMe$), the
trimethoxy-N-dimethylamine (6) and the trimethoxy-N-methylamine (2).

394

(6)   (2)   (7)

If the methylation was performed at 56°, acridones (7; R=H or Me) were also produced (yield - 1.5% and 3.6% respectively). Cyclisation of the methylamine (4; R=NHMe) could also be achieved (yield 60%) by heating in ethanol containing a trace of acetic acid. Contrary to our findings with other aminobenzophenones (Adams et al), cyclisation did not easily occur using base (NaH-DMSO).

Reduction of (5; R=NO$_2$). The symmetrical nitrobenzophenone on Zn/EtOH reduction gave two products, yields varying with the mode of addition of the Zn dust. The expected product, the amine (5; R=NH$_2$), was accompanied by a yellow compound with m.p. 272-6°(decomp.). Spectral data for this compound suggest its structure to be (8).

(8)

λ_{max}(MeOH) 242, 292, 364 nm

ν_{max}(KBr) 1727, 1648 cm^{-1}

δ(d$_5$DMSO)   7.8-6.9 m(Ar-H),

3.62s(CH=) 3.51s (OMe), 9.9 s(OH).

Methylation of amine (5; R=NH$_2$). Treatment of the symmetrical amine with methyl iodide in acetone at 15° gave the N-dimethyl-(5; R=NMe$_2$) and the N-methylamine (5; R=NHMe).

Cyclisation of amine (5; R=NHMe). The amine on heating at 100° with NaH-DMSO gave the acridone (7; R=Me) in low yield (∿1%).

References.

Adams, J.H., Gupta, P., Khan, M.S., Lewis, J.R. and Watt, R.A. (1976)
J. Chem. Soc. Perkin 1, 2089.

Adams, J.H., Gupta, P., Khan, M.S., and Lewis, J.R. (1977) J. Chem. Soc.
Perkin 1, 1, 2173.

Baument, A., Kozovkina, I.N., Krauss, G., Hicke, M., and Grüger, D.
(1982) Cell Plant Reports, 1, 168.

Casey, A.C. and Malhotra, A., (1975) Tetrahedron Letters, 401.

Kozovkina, I.N., Chemysheva, T.P., and Al'termann, I.E., (1979)
Fiziologiya Rastenii, 26, 492.

Szendrei, K., Rózsa, Z., Reisch, J., Kozovkina, I.N. and Minker, E.,
(1976) Herba Hungaria, 15, 23.

SYNTHESIS OF ION SELECTIVE CROWN ETHERS AND THEIR EFFECT ON THE PERMEABILITY OF SOME LIPOSOMES

ÁGAI B., BITTER I., TŐKE L., HELL Z., SZŐGYI M.* and CSERHÁTI T.**

Dept. of Org.Chem.Techn., Techn. Univ. Budapest, Hungary

*Inst. of Biophys., Semmelweis Med. Univ. Budapest, Hungary

**Plant Prot. Inst. of Hung.Acad.Sci., Budapest, Hungary

One of the most valuable feature of many of the ionophores is their ability to complex selectively with various cations. This ability has provided a basis for some ion-selective electrodes as well as preferential transport of certain cations across membranes. Ionophores can be classified on the basis of structural features, which in turn determine the mechanism by which they facilitate ion conductance as /a/ neutral, /b/ charged, and /c/ channel-forming ionophores [3].

Pressman /1964/ found that certain antibiotics could induce the selective movement of K^+ into rat liver mitochondria [4]. These antibiotics - which are neutral at physiological pH - could also increase the permeability of synthetic lipid bilayers to K^+. Among depsipeptide antibiotics valinomycin exhibits the highest selectivity to K^+ although others /Enniatin A and B, macrotetralide nactins/ may also act as discriminatory cation carriers by forming complexes with the alkali metal cations.

The difficulties of obtaining macrocyclic antibiotics from fungal sources or of synthesizing polypeptides, caused a requirement for more accesible probes for the process of ion transfer. Pedersen /1967/ reported the synthesis of a group of macrocyclic polyethers [5], now termed crown ethers, which seemed to be excellent ligands for complexing alkali ions. A detailed study of the complexation properties of these polyether ligands led to the discovery of three types of complexes: 1:1 /doughnut shaped/, 2:1 /sandwich/ and 3:2 /club-sand-wich/ crown-cation stoichiometry depending on the relation-ship of the ionic radii of the metals and the number of crown oxygen atoms [2]. Recently we have reported on the synthesis of a great number of modified crown ethers [1]. We aimed at preparing new derivatives of enhenced potassium selectivity /Eq.1./.

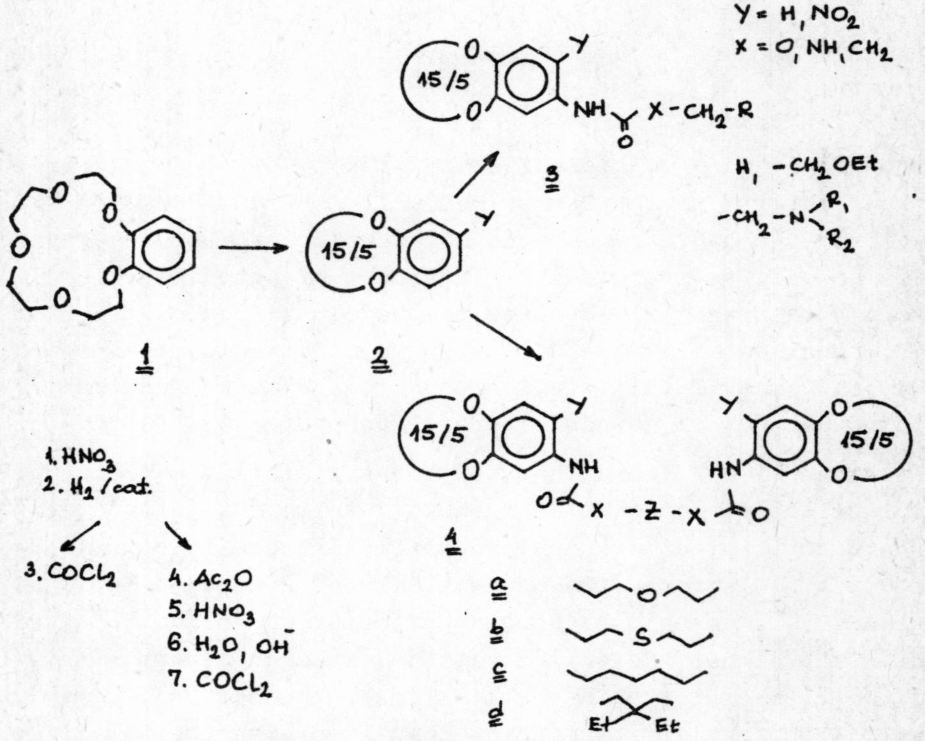

Eq.1. Synthesis of modified crown ethers

The selectivity constants of our new ligands were measured in PVC liquid-membrane electrode. We have observed that bis-crown ethers show much greater selectivity to K^+ than mono-crowns. /Fig.1./ Benzo-15-crown-5 can form only 2:1 /sandwich/complex with K^+ which is stronger if the two crown units are linked with chain. Since we have found as selective ligands to K^+ as valinomycin we wanted to investigate how they influence the permeability of model membranes. ^{42}K-efflux from dipalmitoyl-phosphatidyl-cholin /DPPC/ liposomes has been measured modify-

ing the membrane permeability with crown ethers and valinomycin as well.

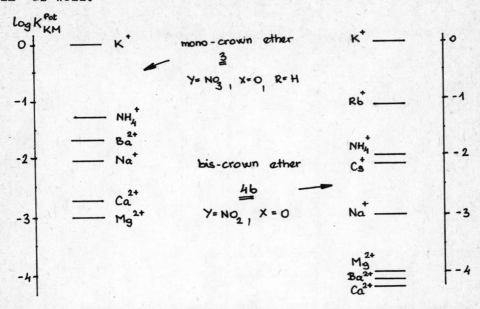

Fig.1. Cation selectivity in liquid membrane electrode.

METHOD

To measure the permeability, liposomes were formed from DPPC
in 0,16 m KCl solution containing tracer amounts of ^{42}KCl by
sonication. The crown ethers were added to the liposomes in
1:200 and 1:100 molar ratio, respectively. The efflux rate
was measured by 15 min. periods. ^{42}K-content was determined by
scintillation counter [6].

RESULTS AND DISCUSSION

The permeability time-constants are listed in Table 1. /crown
ethers:DPPC molar ratio 1:100/. The measurements with ^{42}K,
^{86}Rb and ^{24}Na indicates that the most significant effect was
found in case of potassium. Compounds 4 exhibit similar membrane disturbing effect as valinomycin /Fig.2./. Preliminary
DSC measurements providing data of the main transition temperature of DPPC-water dispersion are in accordance with the
penetration data.

Table 1. Permeability time-constants
of liposomes modified with crown ethers

| Compound $\underline{4}$ | | p_t /x10^{-4}/s/ | | |
| X = O | ^{42}K | ^{86}Rb | ^{24}Na |
Y	Z			
H	b	12,5	8,2	
NO_2	d	11,9	4,4	1,4
H	c	7,1	5,5	
NO_2	b	5,9	4,4	1,0
control		0,3	0,3	0,5

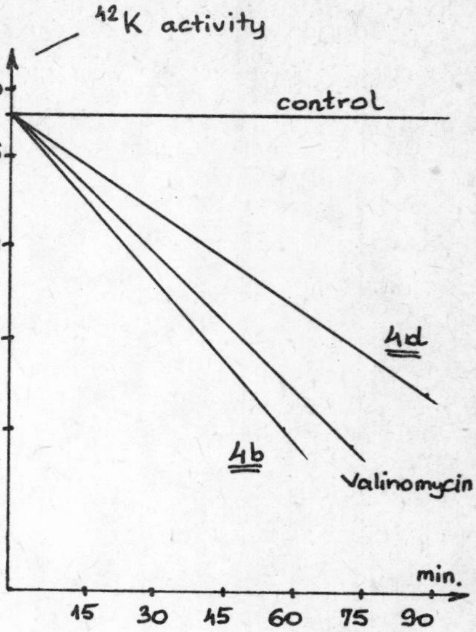

Fig.2. The ^{42}K-activity of liposomes
plotted as a function of time

On the basis of our results we conclude that crown ethers may
substitute ionophore antibiotics in different membrane transfer processes.

REFERENCES
[1] a/ Agai B., Bitter I., Tőke L., Csongor É.: Acta Chim.Acad.
        Sci. Hung. 1982, 110 /1/ 25-28.
    b/ Agai B., Bitter I., Tőke L., Csongor E.: Acta Chim.
        Acad.Sci. Hung. 1982, 110 /1/ 29-33
[2] Frensdorff H.K.: J.Am.Chem.Soc. 93,600 /1971/
[3] Izatt R.M., Christensen J.J.: Progress in Macrocyclic
    Chemistry, Vol.I. 219-254, Chapter V. John Wiley and Sons,
    New York-Chichester-Brisbane-Toronto, 1979.
[4] Moore C., Pressman B.C.: Biochem.Biophys.Res.Comm.1964,15
    562
[5] Pedersen C.J.: J.Am.Chem.Soc. 89,7017 /1967/
[6] Szőgyi M., Tölgyesi F., Cserháti T.: Phys.Chem.Transmemb-
    rane Ion Motions 1983,29-35.

Bio-Organic Heterocycles
van der Plas H.C., Ötvös L., and Simonyi M. eds

α-CHYMOTRYPSIN CATALYZED HYDROLYSIS OF β-PYRIMIDINYL PROPIONIC ACID ESTERS

TELEGDI J., TÜDŐS H., KRAICSOVITS F. and ÖTVÖS L.

Central Research Institute for Chemistry, The Hungarian Academy of Sciences, Budapest, Pf. 17, H-1525, Hungary

SUMMARY

A series of β-pyrimidinyl propionic acid esters was synthesized. Kinetic measurements of enzymatic hydrolysis proved that these compounds are substrates of α-chymotrypsin, but compared with β-phenyl-α-acetamino-propionic acid ester (NAPEE), their rate of hydrolysis is lower by orders of magnitude.

INTRODUCTION

α-Chymotrypsin catalyzes, in general, the hydrolysis of esters. Specific substrates for this enzyme are esters of β-aryl propionic acid derivatives. While α-chymotrypsin hydrolyzes esters of aralkyl acids having L configuration, the D enantiomers do not react with the enzyme but are specific inhibitors (Zerner at al. 1964).

We report here kinetic results on α-chymotrypsin-catalyzed reactions of esters containing β-pyrimidinyl ring in the acyl moiety (Table 1).

MATERIALS AND METHODS

β-Pyrimidinyl propionic acid esters were synthesized by condensation of β-bromo-derivatives of acid esters with bis-O—trimethylsilyl-uracil at 110-120 °C in argon atmosphere. Racemic β-pyrimidinyl-α-amino acids were prepared by the method of Doel et al. (1969) from uracil-1-yl-acetaldehyde with

Strecker's method. The N-acyl compounds of the β-pyrimidinyl-
α-amino acids were obtained by the Schotten-Bauman procedure
in alkalic solution or by acetic acid anhydride in acetic acid.
The esterification was carried out in absolute alcohol with
dry hydrogen chloride gas at room temperature.

All synthesized compounds gave sharp melting point
(Table 1). Purity was checked by analysis as well as thin
layer chromatography. Structures were confirmed by IR and NMR
spectrometry.

<u>Table 1.:</u> β-Pyrimidinyl propionic acid esters

No.	R_1	R_2	R_3	MP/°C/
1	H	H	C_2H_5	96–97
2	H	CH_3	CH_3	82–83
3	H	$NH_2 \cdot HCl$	C_2H_5	209–210
4	H	$NH-CO-CH_3$	C_2H_5	212–213
5	C_2H_5	$NH_2 \cdot HCl$	C_2H_5	196–197
6	$n-C_3H_7$	$NH_2 \cdot HCl$	C_2H_5	210–211
7	$n-C_3H_7$	$NH-CO-CH_3$	C_2H_5	198–199
8	$i-C_3H_7$	$NH_2 \cdot HCl$	C_2H_5	181–183
9	$i-C_3H_7$	$NH-CO-CH_3$	C_2H_5	180–182
10	$n-C_6H_{13}$	$NH_2 \cdot HCl$	C_2H_5	189–190
11	$n-C_6H_{13}$	$NH-CO-CH_3$	C_2H_5	196–197

Salt-free, chrystalline α-chymotrypsin was purchased from Koch-Light Laboratories, Ltd. N-Acetyl-L-phenylalanine ethyl ester (NAPEE) was the same as described by Bernhard (1955). Kinetic measurements were accomplished with pH-stat (Radiometer Copenhagen).

RESULTS AND DISCUSSION

I. Comparison of the hydrolysis of NAPEE and of β-pyrimidinyl propionic acid esters

The primary aim of our investigation was to elucidate whether or not the compounds containing β-pyrimidinyl group are substrates of α-chymotrypsin.

The Lineweaver-Burk plots (Fig.1) shows that α-acetamino-β-pyrimidinyl propionic acid ester (4) is hydrolyzed by the enzyme. As Fig.1 shows the hydrolysis rates of NAPEE and α-acetamino-β-pyrimidinyl propionic acid ester differ by an order of magnitude.

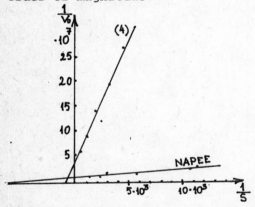

Fig.1. Lineweaver-Burk plots for compounds NAPEE and (4).

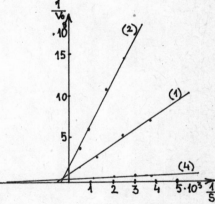

Fig.2. Lineweaver-Burk plots of compounds (1), (2) and (4)

The enzymatic hydrolysis of (4) was found to stop at 50% conversion in agreement with the expectation on stereo-specificity, i.e. it appears that (4) of L configuration is only subject to hydrolysis.

In the case of β-pyrimidinyl compounds containing free amino group, enzymatic hydrolysis was not observed.

The effect of α-substitution in β-pyrimidinyl propionic acid esters is shown by Fig.2. Further substantial decrease in hydrolysis rate was found for compounds (1) and (2).

As seen on Fig.1 and Fig.2, β-pyrimidinyl propionic acid ester derivatives are very poor substrates of α-chymotrypsin. We wanted to know whether the slow hydrolytic rate is due to kinetic control or to unfavourable binding properties.

II. Binding properties of β-pyrimidinyl compounds

In enzyme inhibition measurements NAPEE served as good substrate. The Dixon plot (Dixon 1953) displays the dependence of hydrolysis rate on the concentration of inhibitors at a constant concentration of the good substrate (Fig.3,4).

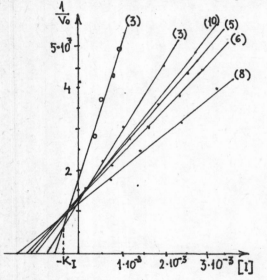
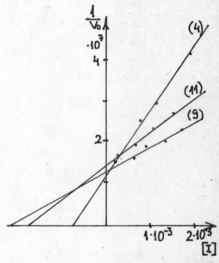

Fig.3. Dixon plots of derivatives containing free NH_2 groups. The concentration of NAPEE is 2.5×10^{-2}M (•) or 1.7×10^{-2}M (o).

Fig.4. Dixon plots of α-acyl-amino derivatives. The concentration of NAPEE is 2.5×10^{-2}M.

The following conclusions may be drawn:
   a/ α-chymotrypsin binds all β-pyrimidinyl propionic acid esters having different alkyl groups at position 5 (Fig. 3,4);

b/ the affinity to the enzyme of these derivatives containing free α-amino group is of comparable magnitude (Fig.3);

c/ compound (3) is found to be a competitive inhibitor (Fig.3);

d/ differences in enzyme-inhibitor binding are greater for α-acetylamino derivatives.

On the basis of the above results it may be concluded that the slow rate of hydrolysis of β-pyrimidinyl propionic acid ester derivatives by α-chymotrypsin implies kinetic control.

REFERENCES

Bernhard, S.A. (1955); New method for the determination of the amidase activity of trypsin: kinetics of the hydrolysis of benzoyl-L-arginin amide.
Biochem. J. 59, 506-509

Dixon, M. (1953); The determination of enzyme inhibitor constants. Biochem. J. 55, 170-171

Doel, M.T., Jones, A.S. and Taylor, N. (1969); An approach to the synthesis of peptide analogues of oligonucleotides. Tetrahedron Lett. 2285-2288

Zerner, B. and Bender M.L. (1964); The kinetic consequences of the acyl-enzyme mechanism for the reactions of specific substrates with chymotrypsin. J. Am. Chem. Soc. 86, 3669-3679

PREPARATION AND ANTIBACTERIAL ACTIVITY OF N4-SUBSTITUTED 5-AZACYTIDINES

PÍSKALA A., ČIHÁK A., HANNA N.B.* and ŠKUTCHAN J.

Institute of Organic Chemistry and Biochemistry,
Czechoslovak Academy of Sciences,
Prague, Czechoslovakia

INTRODUCTION

5-Azacytidine [1] is currently used in clinical treatment of acute leukemia [2] and its derivative 2'-deoxy-5-azacytidine [3] has also been shown to be an effective antileukemic agent in children [4]. As a part of our continuing interest in 5-azacytosine nucleosides we set out to prepare the N4-substituted derivatives of 5-azacytidine.

RESULTS AND DISCUSSION

A series of N4-substituted 5-azacytidines Ia-g was prepared by a modification of the isocyanate procedure used earlier for the preparation of 5-azacytidine [1].

The synthesis of nucleosides Ia-g started with isocyanate II which is well available by reaction of 2,3,5-tri-O-benzoyl-D-ribosyl chloride with silver cyanate. Addition of O-methyl-isourea to the isocyanate II afforded isobiuret III which on cyclocondensation with triethyl orthoformate gave methoxytriazinone IV. Methanolysis of the latter compound led to the

* Present address: National Research Centre, Dokki, Cairo,
Egypt.

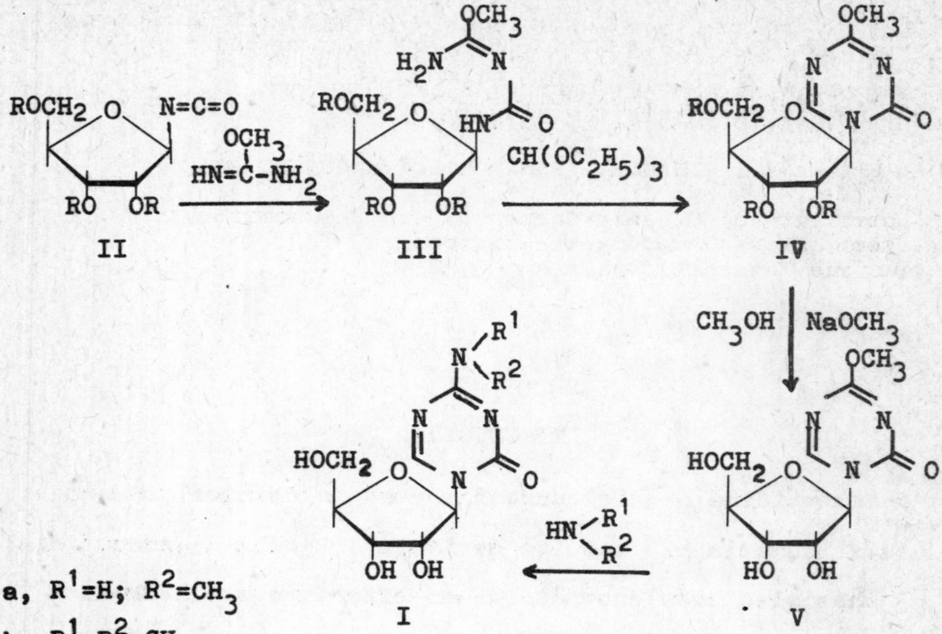

a, R^1=H; R^2=CH$_3$

b, R^1=R^2=CH$_3$

c, R^1=H; R^2=iso.C$_3$H$_7$

d, R^1=H; R^2=n-C$_4$H$_9$

e, R^1=R^2=n-C$_4$H$_9$

f, R^1=H; R^2=C$_6$H$_5$CH$_2$

g, R^1=H; R^2=

R = C$_6$H$_5$CO

free nucleoside V which was reacted with different amines in
methanol to give nucleosides Ia-g. In cases using n-propyl-,
n-butyl-, sec-butyl, benzyl and furfurylamine a part of the
primary products decomposed to the respective amidinoureas
which formed molecular compounds VId-h with the non-decomposed

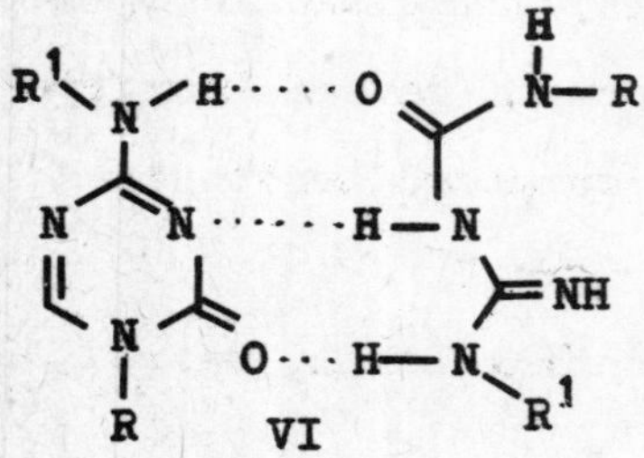

VIa, R=R^1=CH$_3$

b, R=β-D-ribopyranosyl; R^1=H

c, R=β-D-glucopyranosyl; R^1=H

d, R=β-D-ribofuranosyl; R^1=n-C$_3$H$_7$

408

VIe, R=β-D-ribofuranosyl; R^1=n-C_4H_9

f, R=β-D-ribofuranosyl; R^1=sec-C_4H_9

g, R=β-D-ribofuranosyl; R^1=$C_6H_5CH_2$

h, R=β-D-ribofuranosyl; R^1=furfuryl

primary products. The molecular compounds VIe,g were also prepared by cocrystallization of the nucleosides Id,f with the corresponding amidinoureas obtained by hydrolysis of nucleosides Id,f with ammonia in water. Analogous molecular compounds were also prepared with 1-methyl-5-azacytosine (VIa), 1-β-D-ribopyranosyl-5-azacytosine (VIb) and 1-β-D-glucopyranosyl-5-azacytosine (VIc). The formation of the molecular compounds VIa-h is presumably based on strong hydrogen bonds between 5-azacytosine and amidinourea. The nature of these molecular complexes resembles the base pair cytosine-guanine in the double stranded DNA. The components of the molecular compounds were separated by column chromatography on silica gel or more conveniently by use of Amberlite IRC 50 $[H^+]$ ion exchange resin.

When tert-butylamine was used for aminolysis of methoxy derivative V the displacement proceeded much slower due to steric hindrance even at higher temperature and mainly a fission of the triazine ring occurred with the formation of 4-methyl-1-β-D-ribofuranosylisobiuret (VII). On reaction of the isobiuret VII with N,N-dimethylformamide dimethyl acetal in methanol, N^4,N^4-dimethyl-5-azacytidine (Ib) was formed. The intermediate V could not be isolated. An analogous cyclocondensation of the blocked isobiuret III led to 2´,3´,5´-tri-O-benzoyl - N^4, N^4-dimethyl-5-azacytidine (VIII) which was methanolysed to the free nucleoside Ib.

The N^4-substituted 5-azacytidines Ia–g and the molecular compounds VIe–h were tested for their antibacterial activity using a culture of <u>E</u>. <u>coli</u> <u>B</u> growing on a mineral medium with glucose. None of these compounds was as active as 5-azacytidine. N^4-methyl-5-azacytidine (Ia) exhibited the highest degree of inhibition (100% at concentration 0.1 mg/ml).

REFERENCES

[1] Pískala A. and Šorm F. (1964) Nucleic acids components and their analogues. LI. Synthesis of 1-glycosyl derivatives of 5-azauracil and 5-azacytosine. Collect.Czech. Chem.Commun. <u>29</u>, 2060-2076.

[2] Von Hoff D.D., Slavík M. and Muggia F. (1976) 5-Azacytidine. A new anticancer drug with effectiveness in acute myelogenous leukemia. Ann.Intern.Med. <u>85</u>, 237-245.

[3] Pískala A. and Šorm F. (1978) Anomeric 4-amino-1-(2-deoxy-D-erythro-pentofuranosyl)-s-triazin-2(1H)-ones (2-deoxy-5-azacytidine and its α-D anomer) in <u>Nucleic Acid Chemistry</u> (Townsend L.B. and Tipson R.S., Eds), Part 1, pp. 443-449, Wiley-Interscience, New York.

[4] Rivard G.E., Momparler R.L., Demers J., Benoit P., Raymond R., Lin K.T. and Momparler L.F. (1981) Phase I study on 5-aza-2-deoxycytidine in children with acute leukemia. Leukemia Res. <u>5</u>, 453-462.

AUTHOR INDEX

Numbers in parentheses refer to pages of discussion.

SUBJECT INDEX

Numbers in parentheses refer to pages of discussion.

414

418